Tsau Young Lin, Setsuo Ohsuga, Churn-Jung Liau, Xiaohua Hu (Eds.)

Foundations and Novel Approaches in Data Mining

Studies in Computational Intelligence, Volume 9

Editor-in-chief
Prof. Janusz Kacprzyk
Systems Research Institute
Polish Academy of Sciences
ul. Newelska 6
01-447 Warsaw
Poland
E-mail: kacprzyk@ibspan.waw.pl

Further volumes of this series can be found on our homepage: springeronline.com

Vol. 1. Tetsuya Hoya
Artificial Mind System – Kernel Memory Approach, 2005
ISBN 3-540-26072-2

Vol. 2. Saman K. Halgamuge, Lipo Wang (Eds.)
Computational Intelligence for Modelling and Prediction, 2005
ISBN 3-540-26071-4

Vol. 3. Bożena Kostek
Perception-Based Data Processing in Acoustics, 2005
ISBN 3-540-25729-2

Vol. 4. Saman K. Halgamuge, Lipo Wang (Eds.)
Classification and Clustering for Knowledge Discovery, 2005
ISBN 3-540-26073-0

Vol. 5. Da Ruan, Guoqing Chen, Etienne E. Kerre, Geert Wets (Eds.)
Intelligent Data Mining, 2005
ISBN 3-540-26256-3

Vol. 6. Tsau Young Lin, Setsuo Ohsuga, Churn-Jung Liau, Xiaohua Hu, Shusaku Tsumoto (Eds.)
Foundations of Data Mining and Knowledge Discovery, 2005
ISBN 3-540-26257-1

Vol. 7. Bruno Apolloni, Ashish Ghosh, Ferda Alpaslan, Lakhmi C. Jain, Srikanta Patnaik (Eds.)
Machine Learning and Robot Perception, 2005
ISBN 3-540-26549-X

Vol. 8. Srikanta Patnaik, Lakhmi C. Jain, Spyros G. Tzafestas, Germano Resconi, Amit Konar (Eds.)
Innovations in Robot Mobility and Control, 2005
ISBN 3-540-26892-8

Vol. 9. Tsau Young Lin, Setsuo Ohsuga, Churn-Jung Liau, Xiaohua Hu (Eds.)
Foundations and Novel Approaches in Data Mining, 2005
ISBN 3-540-28315-3

Tsau Young Lin
Setsuo Ohsuga
Churn-Jung Liau
Xiaohua Hu
(Eds.)

Foundations and Novel Approaches in Data Mining

Professor Tsau Young Lin
Department of Computer Science
San Jose State University
San Jose, CA 95192
E-mail: tylin@cs.sjsu.edu

Dr. Churn-Jung Liau
Institute of Information Science
Academia Sinica, Taipei 115, Taiwan
E-mail: liaucj@iis.sinica.edu.tw

Professor Setsuo Ohsuga
Honorary Professor
The University of Tokyo, Japan
E-mail: ohsuga@fd.catv.ne.jp

Professor Xiaohua Hu
College of Information Science and
Technology, Drexel University
Philadelphia, PA 19104, USA
E-mail: thu@cis.drexel.edu

ISBN 978-3-642-06650-4
e-ISBN 978-3-540-31229-1

ISSN print edition: 1860-949X
ISSN electronic edition: 1860-9503

Springer is a part of Springer Science+Business Media
springeronline.com

Softcover reprint of the hardcover 1st edition 2006

Preface

This volume is a collection of expanded versions of selected papers originally presented at the second workshop on Foundations and New Directions of Data Mining (2003), and represents the state-of-the-art for much of the current research in data mining. The annual workshop, which started in 2002, is held in conjunction with the IEEE International Conference on Data Mining (ICDM). The goal is to enable individuals interested in the foundational aspects of data mining to exchange ideas with each other, as well as with more application-oriented researchers. Following the success of the previous edition, we have combined some of the best papers presented at the second workshop in this book. Each paper has been carefully peer-reviewed again to ensure journal quality. The following is a brief summary of this volume's contents.

The six papers in Part I present theoretical foundations of data mining. The paper *Commonsense Causal Modeling in the Data Mining Context* by L. Mazlack explores the commonsense representation of causality in large data sets. The author discusses the relationship between data mining and causal reasoning and addresses the fundamental issue of recognizing causality from data by data mining techniques. In the paper *Definability of Association Rules in Predicate Calculus* by J. Rauch, the possibility of expressing association rules by means of classical predicate calculus is investigated. The author proves a criterion of classical definability of association rules. In the paper *A Measurement-Theoretic Foundation of Rule Interestingness Evaluation*, Y. Yao, Y. Chen, and X. Yang propose a framework for evaluating the interestingness (or usefulness) of discovered rules that takes user preferences or judgements into consideration. In their framework, measurement theory is used to establish a solid foundation for rule evaluation, fundamental issues are discussed based on the user preference of rules, and conditions on a user preference relation are given so that one can obtain a quantitative measure that reflects the user-preferred ordering of rules. The paper *Statistical Independence as Linear Dependence in a Contingency Table* by S. Tsumoto examines contingency tables from the viewpoint of granular computing. It finds that the degree of independence, i.e., rank, plays a very important role in extracting a probabilistic model from a given contingency table. In the paper *Foundations of Classification* by J.T. Yao, Y. Yao, and Y. Zhao, a granular computing model is suggested for learning two basic issues: concept formation and concept relationship identification. A classification rule induction method is proposed to search for a suitable covering of a given universe, instead of a suitable partition. The paper *Data Mining as Generalization: A Formal Model* by E. Menasalvas and A. Wasilewska presents a model that formalizes data mining as the process of information generalization. It is shown that only three generalization operators, namely, classification operator, clustering operator, and association operator are needed to express all Data Mining algorithms for classification, clustering, and association, respectively.

The nine papers in Part II are devoted to novel approaches to data mining. The paper *SVM-OD: SVM Method to Detect Outliers* by J. Wang et al. proposes a new SVM method to detect outliers, SVM-OD, which can avoid the parameter that caused difficulty in previous ν-SVM methods based on statistical learning theory (SLT). Theoretical analysis based on SLT as well as experiments verify the effectiveness of the proposed method. The paper *Extracting Rules from Incomplete Decision Systems: System ERID* by A. Dardzinska and Z.W. Ras presents a new bottom-up strategy for extracting rules from partially incomplete information systems. System is partially incomplete if a set of weighted attribute values can be used as a value of any of its attributes. Generation of rules in ERID is guided by two threshold values (minimum support, minimum confidence). The algorithm was tested on a publicly available data-set "Adult" using fixed cross-validation, stratified cross-validation, and bootstrap. The paper *Mining for Patterns Based on Contingency Tables by KL-Miner – First Experience* by J. Rauch, M. Šimůnek, and V. Lín presents a new data mining procedure called *KL-Miner*. The procedure mines for various patterns based on evaluation of two–dimensional contingency tables, including patterns of a statistical or an information theoretic nature. The paper *Knowledge Discovery in Fuzzy Databases Using Attribute-Oriented Induction* by R.A. Angryk and F.E. Petry analyzes an attribute-oriented data induction technique for discovery of generalized knowledge from large data repositories. The authors propose three ways in which the attribute-oriented induction methodology can be successfully implemented in the environment of fuzzy databases. The paper *Rough Set Strategies to Data with Missing Attribute Values* by J.W. Grzymala-Busse deals with incompletely specified decision tables in which some attribute values are missing. The tables are described by their characteristic relations, and it is shown how to compute characteristic relations using the idea of a block of attribute-value pairs used in some rule induction algorithms, such as LEM2. The paper *Privacy-Preserving Collaborative Data Mining* by J. Zhan, L. Chang and S. Matwin presents a secure framework that allows multiple parties to conduct privacy-preserving association rule mining. In the framework, multiple parties, each of which has a private data set, can jointly conduct association rule mining without disclosing their private data to other parties. The paper *Impact of Purity Measures on*

Knowledge Extraction in Decision Trees by M. Lenič, P. Povalej, and P. Kokol studies purity measures used to identify relevant knowledge in data. The paper presents a novel approach for combining purity measures and thereby alters background knowledge of the extraction method. The paper *Multidimensional On-line Mining* by C.Y. Wang, T.P. Hong, and S.S. Tseng extends incremental mining to online decision support under multidimensional context considerations. A multidimensional pattern relation is proposed that structurally and systematically retains additional context information, and an algorithm based on the relation is developed to correctly and efficiently fulfill diverse on-line mining requests. The paper *Quotient Space Based Cluster Analysis* by L. Zhang and B. Zhang investigates clustering under the concept of granular computing. From the granular computing point of view, several categories of clustering methods can be represented by a hierarchical structure in quotient spaces. From the hierarchical structures, several new characteristics of clustering are obtained. This provides another method for further investigation of clustering.

The five papers in Part III deal with issues related to practical applications of data mining. The paper *Research Issues in Web Structural Delta Mining* by Q. Zhao, S.S. Bhowmick, and S. Madria is concerned with the application of data mining to the extraction of useful, interesting, and novel web structures and knowledge based on their historical, dynamic, and temporal properties. The authors propose a novel class of web structure mining called web structural delta mining. The mined object is a sequence of historical changes of web structures. Three major issues of web structural delta mining are proposed, and potential applications of such mining are presented. The paper *Workflow Reduction for Reachable-path Rediscovery in Workflow Mining* by K.H. Kim and C.A. Ellis presents an application of data mining to workflow design and analysis for redesigning and re-engineering workflows and business processes. The authors define a workflow reduction mechanism that formally and automatically reduces an original workflow process to a minimal-workflow model. The model is used with the decision tree induction technique to mine and discover a reachable-path of workcases from workflow logs. The paper *A Principal Component-based Anomaly Detection Scheme* by M.L. Shyu et al. presents a novel anomaly detection scheme that uses a robust principal component classifier (PCC) to handle computer network security problems. Using this scheme, an intrusion predictive model is constructed from the major and minor principal components of the normal instances, where the difference of an anomaly from the normal instance is the distance in the principal component space. The experimental results demonstrated that the proposed PCC method is superior to the k-nearest neighbor (KNN) method, the density-based local outliers (LOF) approach, and the outlier detection algorithm based on the Canberra metric. The paper *Making Better Sense of the Demographic Data Value in the Data Mining Procedure* by K.M. Shelfer and X. Hu is concerned with issues caused by the application of personal demographic data mining to the anti-terrorism war. The authors show that existing data values rarely represent an individual's multi-dimensional existence in a form that can be mined. An abductive approach to data mining is used to improve data input. Working from the "decision-in," the authors identify and address challenges associated with demographic data collection and suggest ways to improve the quality of the data available for data mining. The paper *An Effective Approach for Mining Time-Series Gene Expression Profile* by V.S.M. Tseng and Y.L. Chen presents a bio-informatics application of data mining. The authors propose an effective approach for mining time-series data and apply it to time-series gene expression profile analysis. The proposed method utilizes a dynamic programming technique and correlation coefficient measure to find the best alignment between the time-series expressions under the allowed number of noises. It is shown that the method effectively resolves the problems of scale transformation, offset transformation, time delay and noise.

We would like to thank the referees for reviewing the papers and providing valuable comments and suggestions to the authors. We are also grateful to all the contributors for their excellent works. We hope that this book will be valuable and fruitful for data mining researchers, no matter whether they would like to discover the fundamental principles behind data mining, or apply the theories to practical application problems.

San Jose, Tokyo, Taipei, and Philadelphia
April, 2005

T.Y. Lin
S. Ohsuga
C.J. Liau
X. Hu

References

1. T.Y. Lin and C.J. Liau(2002) Special Issue on the Foundation of Data Mining, *Communications of Institute of Information and Computing Machinery*, Vol. 5, No. 2, Taipei, Taiwan.

Contents

Part I Theoretical Foundations

Part II Novel Approaches

Part III Novel Applications

Part I

Theoretical Foundations

Commonsense Causal Modeling in the Data Mining Context

Lawrence J. Mazlack

Applied Artificial Intelligence Laboratory
University of Cincinnati
Cincinnati, OH 45221-0030
mazlack@uc.edu

Abstract. Commonsense causal reasoning is important to human reasoning. Causality itself as well as human understanding of causality is imprecise, sometimes necessarily so. Causal reasoning plays an essential role in commonsense human decision-making. A difficulty is striking a good balance between precise formalism and commonsense imprecise reality. Today, data mining holds the promise of extracting unsuspected information from very large databases. The most common methods build rules. In many ways, the interest in rules is that they offer the promise (or illusion) of causal, or at least, predictive relationships. However, the most common rule form (association rules) only calculates a joint occurrence frequency; they do not express a causal relationship. Without understanding the underlying causality in rules, a naïve use of association rules can lead to undesirable actions. This paper explores the commonsense representation of causality in large data sets.

1. Introduction

Commonsense causal reasoning occupies a central position in human reasoning. It plays an essential role in human decision-making. Considerable effort has been spent examining causation. Philosophers, mathematicians, computer scientists, cognitive scientists, psychologists, and others have formally explored questions of causation beginning at least three thousand years ago with the Greeks.

Whether causality can be recognized at all has long been a theoretical speculation of scientists and philosophers. At the same time, in our daily lives, we operate on the commonsense belief that causality exists.

Causal relationships exist in the commonsense world. If an automobile fails to stop at a red light and there is an accident, it can be said that the failure to stop was the accident's cause. However, conversely, failing to stop at a red light is not a certain cause of a fatal accident; sometimes no accident of any kind occurs. So, it can be said that knowledge of some causal effects is imprecise. Perhaps, complete knowledge of all possible factors might lead to a crisp description of whether a causal effect will occur. However, in our commonsense world, it is unlikely that all possible factors can be known. What is needed is a method to model imprecise causal models.

Another way to think of causal relationships is counterfactually. For example, if a driver dies in an accident, it might be said that had the accident *not* occurred; they would still be alive.

Our common sense understanding of the world tells us that we have to deal with imprecision, uncertainty and imperfect knowledge. This is also the case of our scientific knowledge of the world. Clearly, we need an algorithmic way of handling imprecision if we are to computationally handle causality. Models are needed to algorithmically consider causes. These models may be symbolic or graphic. A difficulty is striking a good balance between precise formalism and commonsense imprecise reality

1.1 Data mining, introduction

Data mining is an advanced tool for managing large masses of data. It analyzes data previously collected. It is *secondary* analysis. Secondary analysis precludes the possibility of experimentally varying the data to identify causal relationships.

There are several different data mining products. The most common are *conditional rules* or *association rules*. Conditional rules are most often drawn from induced trees while association rules are most often learned from tabular data. Of these, the most common data mining product is association rules; for example:

- *Conditional rule:*
  IF Age < 20
  THEN Income < $10,000
  with {belief = 0.8}
- *Association rule:*
  Customers who buy beer and sausage
  also tend to buy mustard
  with {confidence = 0.8}
  in {support = 0.15}

At first glance, these structures seem to imply a causal or cause-effect relationship. That is: *A customer's purchase of both sausage and beer* *causes* *the customer to also buy mustard.* In fact, when typically developed, association rules do not *necessarily* describe causality. Also, the strength of causal dependency may be very different from a respective association value. All that can be said is that associations describe the strength of joint co-occurrences. Sometimes, the relationship might be causal; for example, if someone eats salty peanuts and then drinks beer, there is probably a causal relationship. On the other hand, if a crowing rooster probably does not cause the sun to rise.

1.2 Naïve association rules can lead to bad decisions

One of the reasons why association rules are used is to aid in making retail decisions. However, simple association rules may lead to errors.

For example, it is common for a store to put one item on sale and then to raise the price of another item whose purchase is assumed to be associated. This may work if the items are truly associated; but it is problematic if association rules are blindly followed [Silverstein, 1998].

Example: ßAt a particular store, a customer buys:

- *hamburger* without hot dogs *33%* of the time
- *hot dogs* without hamburger *33%* of the time
- both *hamburger* and *hot dogs 33%* of the time
- *sauerkraut* only if *hot dogs* are also purchased[1]

This would produce the transaction matrix:

	hamburger	hot dog	sauerkraut
t_1	1	1	1
t_2	1	0	0
t_3	0	1	1

This would lead to the associations:

- (hamburger, hot dog) = 0.5
- (hamburger, sauerkraut) = 0.5
- (hot dog, sauerkraut) = 1.0

If the merchant:

- Reduced price of hamburger (as a sale item)
- Raised price of sauerkraut to compensate (as the rule *hamburger* ⇒ *sauerkraut* has a high confidence.
- The offset pricing compensation would not work as the sales of sauerkraut would not increase with the sales of hamburger. Most likely,

[1] *Sauerkraut* is a form of pickled cabbage. It is often eaten with cooked sausage of various kinds. It is rarely eaten with hamburger.

the sales of hot dogs (and consequently, sauerkraut) would likely decrease as buyers would substitute hamburger for hot dogs.

1.3 False causality

Complicating causal recognition are the many cases of false causal recognition. For example, a coach may win a game when wearing a particular pair of socks, then always wear the same socks to games. More interesting, is the occasional false causality between music and motion. For example, Lillian Schwartz developed a series of computer generated images, sequenced them, and attached a sound track (usually Mozart). While there were some connections between one image and the next, the music was not scored to the images. However, on viewing them, the music appeared to be connected. All of the connections were observer supplied.

An example of non-computer illusionary causality is the choreography of Merce Cunningham. To him, his work is non-representational and without intellectual meaning[2]. He often worked with John Cage, a randomist composer. Cunningham would rehearse his dancers, Cage would create the music; only at the time of the performance would music and motion come together. However, the audience usually conceived of a causal connection between music and motion and saw structure in both.

1.4 Recognizing causality basics

A common approach to recognizing causal relationships is by manipulating variables by experimentation. How to accomplish causal discouvery in purely observational data is not solved. (Observational data is the most likely to be available for data mining analysis.) Algorithms for discouvery in observational data often use correlation and probabilistic independence. If two variables are statistically independent, it can be asserted that they are not causally related. The reverse is not necessarily true.

Real world events are often affected by a large number of potential factors. For example, with plant growth, many factors such as temperature, chemicals in the soil, types of creatures present, etc., can all affect plant growth. What is unknown is what causal factors will or will not be present in the data; and, how many of the underlying causal relationships can be discouvered among observational data.

---

2 "Dancing form me is movement in time and space. Its possibilities are bound only by our imaginations and our two legs. As far back as I can remember, I've always had an appetite for movement. I don't see why it has to represent something. It seems to me it is what it is ... its a necessity ... it goes on. Many people do it. You don't have to have a reason to do it. You can just do it." --- http://www.merce.org:80/dancers

Some define cause-effect relationships as: When α occurs, β always occurs. This is inconsistent with our commonsense understanding of causality. A simple environment example: When a hammer hits a bottle, the bottle *usually* breaks. A more complex environment example: When a plant receives water, it *usually* grows.

An important part of data mining is understanding, whether there is a relationship between data items. Sometimes, data items may occur in pairs but may not have a deterministic relationship; for example, a grocery store shopper may buy both bread and milk at the same time. Most of the time, the milk purchase is not caused by the bread purchase; nor is the bread purchase caused by the milk purchase.

Alternatively, if someone buys strawberries, this may causally affect the purchase of whipped cream. *Some* people who buy strawberries want whipped cream with them; of these, the desire for the whipped cream varies. So, we have a conditional primary effect (whipped cream purchase) modified by a secondary effect (desire). How to represent all of this is open.

A largely unexplored aspect of mined rules is how to determine when one event causes another. Given that α and β are variables and there appears to be a statistical covariability between α and β, is this covariability a causal relation? More generally, when is any pair relationship causal? Differentiation between covariability and causality is difficult.

Some problems with discouvering causality include:

- Adequately defining a causal relation
- Representing possible causal relations
- Computing causal strengths
- Missing attributes that have a causal effect
- Distinguishing between association and causal values
- Inferring causes and effects from the representation.

Beyond data mining, causality is a fundamentally interesting area for workers in intelligent machine based systems. It is an area where interest waxes and wanes, in part because of definitional and complexity difficulties. The decline in computational interest in cognitive science also plays a part. Activities in both philosophy and psychology [Glymour, 1995, 1996] overlap and illuminate computationally focused work. Often, the work in psychology is more interested in how people *perceive* causality as opposed to whether causality actually exists. Work in psychology and linguistics [Lakoff, 1990] [Mazlack, 1987] show that categories are often linked to causal descriptions. For the most part, work in intelligent computer systems has been relatively uninterested in grounding based on human perceptions of categories and causality. This paper is concerned with developing commonsense representations that are compatible in several domains.

2. Causality

Centuries ago, in their quest to unravel the future, mystics aspired to decipher the cries of birds, the patterns of the stars and the garbled utterances of oracles. Kings and generals would offer precious rewards for the information soothsayers furnished. Today, though predictive methods are different from those of the ancient world, the knowledge that dependency recognition attempts to provide is highly valued. From weather reports to stock market prediction, and from medical prognoses to social forecasting, superior insights about the shape of things to come are prized [Halpern, 2000].

Democritus, the Greek philosopher, once said: "Everything existing in the universe is the fruit of chance and necessity." This seems self-evident. Both randomness and causation are in the world. Democritus used a poppy example. Whether the poppy seed lands on fertile soil or on a barren rock is chance. If it takes root, however, it will grow into a poppy, not a geranium or a Siberian Husky [Lederman, 1993].

Beyond computational complexity and holistic knowledge issues, there appear to be inherent limits on whether causality can be determined. Among them are:

- *Quantum Physics:* In particular, Heisenberg's uncertainty principle
- *Observer Interference:* Knowledge of the world might never be complete because we, as observers, are integral parts of what we observe
- *Gödel's Theorem:* Which showed in any logical formulation of arithmetic that there would always be statements whose validity was indeterminate. This strongly suggests that there will always be inherently unpredictable aspects of the future.
- *Turing Halting Problem:* Turning (as well as Church) showed that any problem solvable by a step-by-step procedure could be solved using a Turing machine. However, there are many routines where you cannot ascertain if the program will take a finite, or an infinite number of steps. Thus, there is a curtain between what can and cannot be known mathematically.
- *Chaos Theory:* Chaotic systems appear to be deterministic; but are computationally irreducible. If nature is chaotic at its core, it might be fully deterministic, yet wholly unpredictable [Halpern 2000, page 139].
- *Space-Time:* The malleability of Einstein's space time that has the effect that what is "now" and "later" is local to a particular observer; another observer may have contradictory views.
- *Arithmetic Indeterminism:* Arithmetic itself has random aspects that introduce uncertainty as to whether equations may be solvable. Chatin [1987,

1990] discovered that Diophantine equations may or may not have solutions, depending on the parameters chosen to form them. Whether a parameter leads to a solvable equation appears to be random. (Diophantine equations represent well-defined problems, emblematic of simple arithmetic procedures.)

Given determinism's potential uncertainty and imprecision, we might throw up out hands in despair. It may well be that a precise and complete knowledge of causal events is uncertain. On the other hand, we have a commonsense belief that causal effects exist in the real world. If we can develop models tolerant of imprecision, it would be useful. Perhaps, the tools that can be found in soft computing may be useful.

2.1 Nature of causal relationships

The term *causality* is used here in the every day, informal sense. There are several strict definitions that are not wholly compatible with each other. The formal definition used in this paper is that if one thing (event) occurs because of another thing (event), we say that there is a dependent or causal relationship.

Fig. 1. Diagram indicating that β is causally dependent on α.

Some questions about causal relationships that would be desirable to answer are:

- To what degree does α cause β? Is the value for β sensitive to a small change in the value of α?
- Does the relationship always hold in time and in every situation? If it does not hold, can the particular situation when it does hold be discouvered?
- How should we describe the relationship between items that are causally related: probability, possibility? Can we say that there is a causal strength between two items; causal strength representing the degree of causal influence that items have over each other?

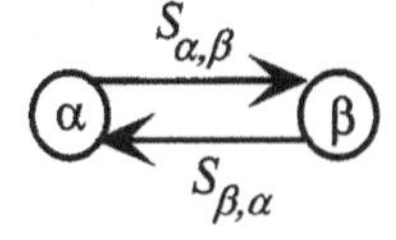

Fig. 2. Mutual dependency.

- Is it possible that there might be mutual dependencies; i.e., $\alpha \rightarrow \beta$ as well as $\beta \rightarrow a$? Is it possible that they do so with different strengths? They can be described as shown in *Fig. 2.* where $S_{i,j}$ represents the strength of the causal relationship from i to j . Often, it would seem that the strengths would be best represented by an approximate belief function. There would appear to be two variations:

- *Different causal strengths for the same activity, occurring at the same time:*

  For example, α could be *short men* and β could be *tall women.* If $S_{\alpha,\beta}$ meant the strength of desire for a social meeting that was caused in *short men* by the sight of *tall women,* it might be that $S_{\alpha,\beta} > S_{\beta,\alpha}$.

  On the other hand, some would argue that causality should be completely asymmetric and if it appears that items have mutual influences it is because there is another cause that causes both. A problem with this idea is that it can lead to eventual regression to a first cause; whether this is true or not, it is not useful for commonsense representation.

- *Different causal strengths for symmetric activities, occurring at different times:*

  It would seem that if there were causal relationships in market basket data, there would often be imbalanced dependencies. For example, if a customer first buys strawberries, there may be a reasonably good chance that she will then buy whipped cream. Conversely, if she first buys whipped cream, the subsequent purchase of strawberries may be less likely. This situation could also be represented by *Fig 2.* However, the issue of time sequence would be poorly represented. A graph representation could be used that implies a time relationship. Nodes in a sequence closer to a root could be considered to be earlier in time than those more distant from the root. Redundant nodes would have to be inserted to capture every alternate sequence. For example, one set of nodes for when strawberries are bought before whipped cream and another set when whipped cream is bought before strawberries. However, this representation is less elegant and not satisfactory when a time differential is not a necessary part of causality. It also introduces multiple nodes for the same object (e.g., strawberries, whipped cream); which at a minimum introduces housekeeping difficulties.

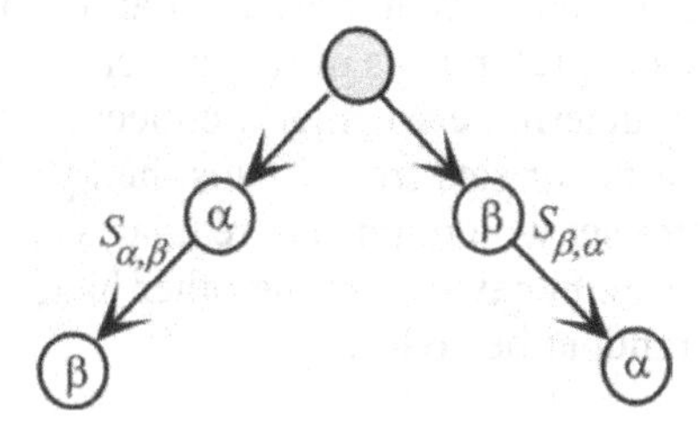

Fig. 3. Alternative time sequences for two symmetric causal event sequences where representing differing event times necessary for representing causality. Nodes closer to the root occur before nodes more distant from the root. Causal strengths may be different depending on sequence.

It is potentially interesting to discouver the absence of a causal relationship; for example, discouvering the lack of a causal relationship in drug treatment's of disease. If some potential cause can be eliminated, then attention can become more focused on other potentials.

Prediction is not the same as causality. Recognizing whether a causal relationship existed in the past is not the same as predicting that in the future one thing will occur because of another thing. For example, knowing that α was a causal (or deterministic) factor for β is different from saying whenever there is α, β will deterministically occur (or even probalistically occur to a degree λ). There may be other necessary factors.

Causal necessity is not the same thing as causal sufficiency; for example, in order for event δ to occur, events α,β,φ need to occur. We can say that α, by itself, is necessary, but not sufficient.

Part of the difficulty of recognizing causality comes from identifying relevant data. Some data might be redundant; some irrelevant; some are more important than others. Data can have a high dimensionality with only a relatively few utilitarian dimensions; i.e., data may have a higher dimensionality than necessary to fully describe a situation. In a large collection of data, complexity may be unknown. Dimensionality reduction is an important issue in learning from data.

A causal discovery method cannot transcend the prejudices of analysts. Often, the choice of what data points to include and which to leave out, which type of curve to fit (linear, exponential, periodic, etc.), what time increments to use (years, decades, centuries, etc.) and other model aspects depend on the instincts and preferences of researchers.

It may be possible to determine whether a collection of data is random or deterministic using attractor sets from Chaos theory [Packard, 1980]. A low dimensional attractor set would indicate regular, periodic behavior and would indicate determinate behavior. On the other hand, high dimensional results would indicate random behavior.

2.2 Types of causality

There are at least three ways that things may be said to be related:

- *Coincidental:* Two things describe the same object and have no determinative relationship between them.
- *Functional:* There is a generative relationship.
- *Causal:* One thing causes another thing to happen. There are at least four types of causality:
    - *Chaining:* In this case, there is a temporal chain of events, $A_1,A_2,\ldots, A_n$, which terminates on A_n. To what degree, if any, does A_i (i=1,..., n-1) cause A_n? A special case of this is a backup mechanism or a *preempted alternative.* Suppose there is a chain of casual dependence, A_1 causing A_2; suppose that if A_1 does not occur, A_2 still occurs, now caused by the alternative cause B_1 (which only occurs if A_1 does not).
    - *Conjunction (Confluence):* In this case, there is a confluence of events, $A_1,\ldots, A_n$, and a resultant event, B. To what degree, if any, did or does A_i cause B? A special case of this is *redundant causation.*

Say that either A_1 or A_2 can cause B; and, both A_1 and A_2 occur simultaneously. What can be said to have caused B?

• *Network:* A network of events.

• *Preventive:* One thing prevents another; e.g., *She prevented the catastrophe.*

Recognizing and defining causality is difficult. Causal claims have both a direct and a subjunctive complexity [Spirtes, 2000] - they are associated with claims about what did happen, or what did not happen, or has not happened yet, or what would have happened if some other circumstance had been otherwise. The following show some of the difficulties:

• *Example 1: Simultaneous Plant Death:* My rose bushes and my neighbor's rose bushes both die. Did the death of one cause the other to die? (Probably not, although the deaths are associated.)

• *Example 2: Drought:* There has been a drought. My rose bushes and my neighbor's rose bushes both die. Did the drought cause both rose bushes to die? (Most likely.)

• *Example 3: Traffic:* My friend calls me up on the telephone and asks me to drive over and visit her. While driving over, I ignore a stop sign and drive through an intersection. Another driver hits me. I die. Who caused my death? -- Me? -- The other driver? -- My friend? -- The traffic engineer who designed the intersection? -- Fate? (Based on an example suggested by Zadeh [2000].)

• *Example 4: Umbrellas:* A store owner doubles her advertising for umbrellas. Her sales increase by 20% What caused the increase? -- Advertising? -- Weather? -- Fashion? -- Chance?

• *Example 5: Poison:* (Chance increase without causation) Fred and Ted both want Jack dead. Fred poisons Jack's soup; and, Ted poisons his coffee. Each act increases Jack's chance of dying. Jack eats the soup but (feeling rather unwell) leaves the coffee, and dies later. Ted's act raised the chance of Jack's death but was not a cause of it.

Exactly what makes a causal relationship is open to varying definition. However, causal asymmetries often play a part [Hausman 1998]. Some claimed asymmetries are:

• *Time order:* Effects do not come before effects (at least as locally observed)

• *Probabilistic independence:* Causes of an event are probabilistically independent of another, while effects of a cause are probabilistically dependent on one another.

- *Counterfactual dependency:* Effects counterfactually depend on their causes, while causes do not counterfactually depend on their effects and effects of a common cause do not counterfactually depend on each other.
- *Over determination:* Effects over determine their causes, while causes rarely over determine their effects.
- *Fixity:* Causes are "fixed" no later than their effects
- *Connection dependency:* If one were to break the connection between cause and effect; only the effect might be affected.

2.3 Classical statistical dependence

Statistical independence:

Statistical dependence is interesting in this context because it is often confused with causality. Such reasoning is not correct. Two events E_1, E_2 may be statistical dependent because both have a common cause E_0. But this does not mean that E_1 is the cause of E_2.

For example, lack of rain (E_0) may cause my rose bush to die (E_1) as well as that of my neighbor (E_2). This does not mean that the dying of my rose has caused the dying of my neighbor's rose, or conversely. However, the two events E_1, E_2 are statistically dependent.

The general definition of statistical dependence is:

> Let A, B be two random variables that can take on values in the domains $\{a_1, a_2, \ldots, a_i\}$ and $\{b_1, b_2, \ldots, b_j\}$ respectively. Then A is said to be statistically independent of B iff
>
> $$\text{prob}(a_i|b_j) = \text{prob}(a_i) \text{ for all } b_j \text{ and for all } a_i.$$

The formula

$$\text{prob}(a_i|b_j) = \text{prob}(a_i)\,\text{prob}(b_j)$$

describes the joint probability of a_i AND b_j when A and B are independent random variables. Then follows the law of compound probabilities

$$\text{prob}(a_i,b_j) = \text{prob}(a_i)\,\text{prob}(b_j|a_i)$$

In the absence of causality, this is a symmetric measure. Namely,

$$\text{prob}(a_i,b_j) = \text{prob}(b_j,a_i)$$

Causality vs. statistical dependence:

A causal relationship between two events E_1 and E_2 will always give rise to a certain degree of statistical dependence between them. The converse is not true. A statistical dependence between two events may; but need not, indicate a causal relationship between them. We can tell if there is a positive correlation if

$$\text{prob}(a_i,b_j) > \text{prob}(a_i)\,\text{prob}(b_j)$$

However, all this tells us that it is an interesting relationship. It does not tell us if there is a causal relationship.

Following this reasoning, it is reasonable to suggest that association rules developed as the result of link analysis might be considered causal; if only because of a time sequence is involved. In some applications, such as communication fault analysis [Hatonen 1996], causality is assumed. In other potential applications, such as market basket analysis[3], the strength of time sequence causality is less apparent. For example, if someone buys milk on day_1 and dish soap on day_2, is there a causal relationship? Perhaps, some strength of implication function could be developed.

Some forms of experimental marketing might be appropriate. However, how widely it might be applied is unclear. For example, a food store could carry milk ($E_{1,m=1}$) one month and not carry dish soap. The second month the store could carry dish soap ($E_{2,m=2}$) and not milk. On the third month, it could carry both milk and dish soap ($E_{1,m=3}$) ($E_{2,m=3}$). That would determine both the independent and joint probabilities (setting aside seasonality issues). Then, if

$$\text{prob}(E_{1,m=3})\ \text{prob}(E_{2,m=3}) > \text{prob}(E_{1,m=1})\ \text{prob}(E_{2,m=2})$$

there would be some evidence that there might be a causal relationship as greater sales would occur when both bread and soap were present.

2.4 Probabilistic Causation

Probabilistic Causation designates a group of philosophical theories that aim to characterize the relationship between cause and effect using the tools of probability theory. A primary motivation is the desire for a theory of causation that does not presuppose physical determinism.

The success of quantum mechanics, and to a lesser extent, other theories using probability, brought some to question determinism. Some philosophers became interested in developing causation theories that do not presuppose determinism.

One notable feature has been a commitment to indeterminism, or rather, a commitment to the view that an adequate analysis of causation must apply equally to deterministic and indeterministic worlds. Mellor [1995] argues that indeterministic causation is consistent with the connotations of causation. Hausman [1998], on the other hand, defends the view that in indeterministic settings there is, strictly speaking, no indeterministic causation, but rather deterministic causation of probabilities.

Following Suppes [1970] and Lewis [Lewis 1996], the approach has been to replace the thought that causes are sufficient for, or determine,

3 Time sequence link analysis can be applied to market basket analysis when the customers can be recognized; for example through the use of supermarket customer "loyalty" cards or "cookies" in e-commerce.

their effects with the thought that a cause need only raise the probability of its effect. This shift of attention raises the issue of what kind of probability analysis, if any, is up to the job of underpinning indeterministic causation.

3. Representation

The representation constrains and supports the methods that can be used. Several representations have been proposed. Fully representing imprecision remains undone.

3.1 First order logic

Hobbs [2001] uses first-order logic to represent causal relationships. One difficulty with this approach is that the representation does not allow for any gray areas. For example, if an event occurred when the wind was blowing east, how could a wind blowing east-northeast be accounted for? The causality inferred may be incorrect due to the representation's rigidity.

Nor can first order logic deal with dependencies that are only *sometimes* true. For example, *sometimes* when the wind blows *hard*, a tree falls. This kind of *sometimes* event description can possibly be statistically described. Alternatively, a qualitative fuzzy measure might be applied.

Another problem is recognizing differing effect strengths. For example, if some events in the causal complex are more strongly tied to the effect? Also, it is not clear how a relationship such as the following would be represented: α causes β some of the time; β causes α some of the time; other times there is no causal relationship.

3.2 Probability and decision trees

Various forms of root initiated, tree-like graphs have been suggested [Shafer 1996]. A tree is a digraph starting from one vertice, the root. The vertices represent situations. Each edge represents a particular variable with a corresponding probability (branching). Among them are:

- Probability trees: Have a probability for every event in every situation, and hence a probability distribution and expected value for every variable in every situation. A probability tree is shown in *Fig 4.* Probability trees with zero probabilities can be used to represent deterministic events; an example of this can be seen in *Fig. 5.*

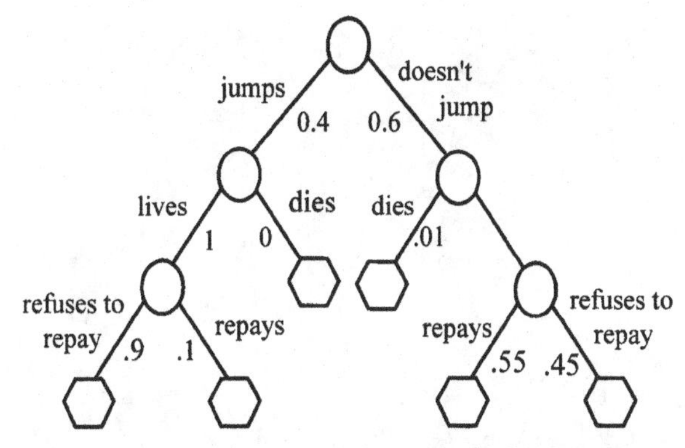

Fig. 4. Probability tree, dependent. (Based on: Shafer [1996], p71.)

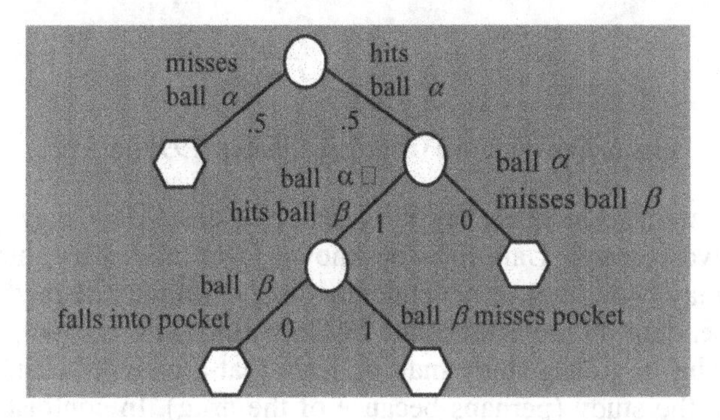

Fig. 5. Determinism in a probability tree. (Based on: Shafer [1996], p72.)

- Decision trees: Trees in which branching probabilities are supplied for some, while others are unknown. An example of a decision tree is provided in *Fig 6.* An often useful variant is Martingale trees.

Time ordering of the variables is represented via the levels in the tree. The higher a variable is in the tree, the earlier it is in time. This can become ambiguous for networked representations; i.e., when a node can have more than two parents and thus two competing paths (and their imbedded time sequences). By evaluating the expectation and probability changes among the nodes in the tree, one can decide whether the two variables are causally related.

There are various difficulties with this kind of tree. One of them is computational complexity. Another is the assumptions that need to be made about independence, such as the Markoff condition. In the context of large databases, learning the trees is computationally intractable.

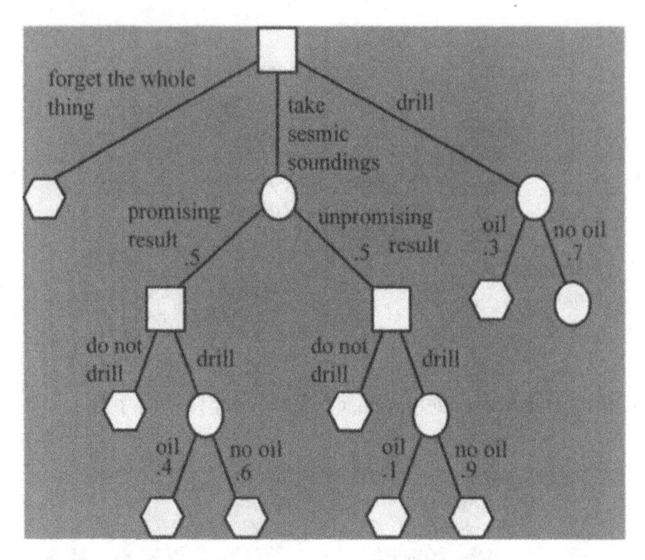

Fig. 6. Decision tree (based on: Shafer [1996], p 249).

Another significant difficulty is incomplete data. Data may be incomplete for two reasons. Data may be known to be necessary, but missing. Data also may be hidden. A dependency distinction may be made. Missing data is dependent on the actual state of the variables. For example, a missing data point in a drug study may indicate that a patient became too sick to continue the study (perhaps because of the drug). In contrast, if a variable is hidden, the absence of this data is independent of state. Both of these situations have approaches that may help. The reader is directed to Spirtes [2000] for a discussion.

3.3 Directed graphs

Some authors have suggested that *sometimes* it is possible to recognize causal relations through the use of directed graphs (digraphs) [Pearl 1991] [Spirtes 2000].

In a digraph, the vertices correspond to the variables and each directed edge corresponds to a causal influence. Diagraphs are not cyclic; the same node in the graph cannot be visited twice. An example is shown in *Fig. 7.* Pearl [2000] and Spirtes [2001] use a form of digraphs called DAGs for representing causal relationships.

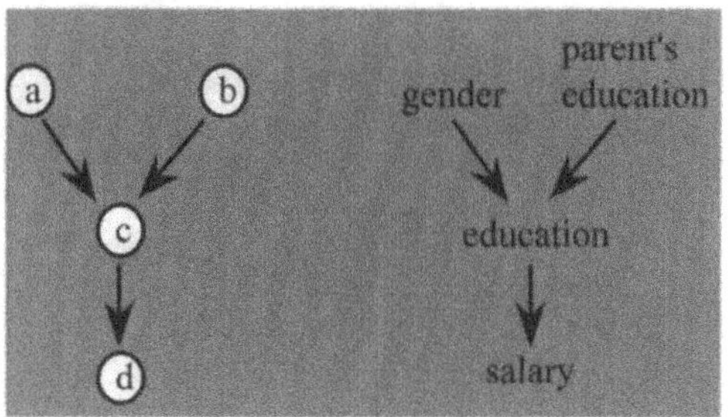

Fig. (a) An example digraph (DAG) (b) Example instantiating (a).

Sometimes, cycles exist. For example, a person's family medical history influences both whether they are depressive and whether they will have some diseases. Drinking alcohol combined with the genetic predisposition to certain disease influences whether the person has a particular disease; that then influence depression; that in turn may influence the person's drinking habits. *Fig. 8* shows an example of a cyclic digraph.

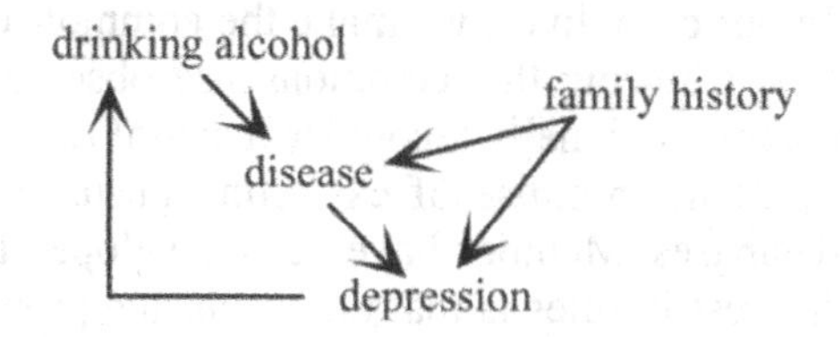

Fig. 8. Cyclic causal relationships.

Developing directed acyclic graphs from data is computationally expensive. The amount of work increases geometrically with the number of attributes. For constraint based methods, the reader is directed to Pearl [2000], Spirtes [2000], Silverstein [1998], and Cooper [1997]. For Bayesian discouvery, the reader is directed to Heckerman [1997] and Geiger [1995].

Quantitatively describing relationships between the nodes can be complex. One possibility is an extension of the random Markoff model; shown in *Fig. 9.* The state value is 1/0 as an event either happens or does not.

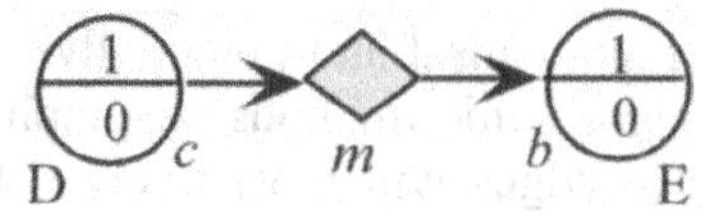

Fig. 9. Random Markoff model: c = P(D), m = the probability that when D is present, the causal mechanism brings about E, b = the probability that some other (unspecified) causal mechanism brings about E.

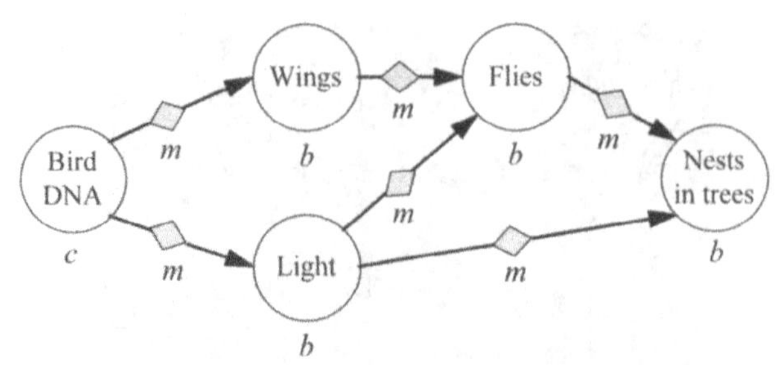

Fig. 10. Markoff model applied to a "bird" example [Rehder, 2002]

4. Epilogue

Causality occupies a central position in human commonsense reasoning. In particular, it plays an essential role in common sense human decision-making by providing a basis for choosing an action that is likely to lead to a desired result. In our daily lives, we make the commonsense observation that causality exists. Carrying this commonsense observation further, the concern is how to computationally recognize a causal relationship.

Data mining holds the promise of extracting unsuspected information from very large databases. Methods have been developed to build rules. In many ways, the interest in rules is that they offer the promise (or illusion) of causal, or at least, predictive relationships. However, the most common form of rules (association) only calculate a joint occurrence frequency; not causality. A fundamental question is determining whether or not recognizing an association can lead to recognizing a causal relationship.

An interesting question how to determine when causality can be said to be stronger or weaker. Either in the case where the causal strength may be different in two independent relationships; or, where in the case where two items each have a causal relationship on the other.

Causality is a central concept in many branches of science and philosophy. In a way, the term "causality" is like "truth" -- a word with many meanings and facets. Some of the definitions are extremely precise. Some of them involve a style of reasoning best be supported by fuzzy logic.

Defining and representing causal and potentially causal relationships is necessary to applying algorithmic methods. A graph consisting of a collection of simple directed edges will most likely not offer a sufficiently rich representation. Representations that embrace some aspects of imprecision are necessary.

A deep question is when anything can be said to cause anything else. And if it does, what is the nature of the causality? There is a strong motivation to attempt causality discouvery in association rules. The research concern is how to best approach the recognition of causality or non-causality in association rules. Or, if there is to recognize causality as long as association rules are the result of secondary analysis?

References

G. Chatin [1987] **Algorithmic Information Theory**, Cambridge University Press, Cambridge, United Kingdom

G. Chatin [1990] "A Random Walk In Arithmetic," New Scientist 125, n 1709 (March, 1990), 44-66

G. Cooper [1997] "A Simple Constraint-Based Algorithm for Efficiently Mining Observational For Causal Relationships" in *Data Mining and Knowledge Discouvery*, v 1, n 2, 203-224

D. Geiger, D. Heckerman [1995] "A Characterization Of The Dirichlet Distribution With Application To Learning Bayesian Networks," in *Proceedings of the 11th Conference on Uncertainty in AI*, Montreal, Quebec, 196-207, August

C. Glymour, G. Cooper, Eds. [1999] **Computation, Causation, and Discouvery,** AAAI Press, Menlo Park, California

C. Glymour [2001] **The Mind's Arrows, Bayes Nets And Graphical Causal Models In Psychology**, MIT Press, Cambridge, Massachusetts

P. Halpern [2000] **The Pursuit Of Destiny**, Perseus, Cambridge, Massachusetts

K. Hatonen, M. Klemettinen, H. Mannila, P. Ronkainen, H. Toivonen (1996) "Knowledge Discovery From Telecommunication Network Alarm Databases," *Conference On Data Engineering (ICDE'96) Proceedings*, New Orleans, 115-122

D. Hausman [1998] **Causal Asymmetries**, Cambridge University Press, Cambridge, United Kingdom

D. Heckerman, C. Meek, G. Cooper [1997] A Bayesian Approach To Causal Discovery, Microsoft Technical Report MSR-TR-97-05

J. Hobbs [2001] "Causality," *Proceedings, Common Sense 2001, Fifth Symposium on Logical Formalizations of Commonsense Reasoning*, New York University, New York, May, 145-155

G. Lakoff [1990] **Women, Fire, And Dangerous Things: What Categories Reveal About The Mind**, University of Chicago Press

L. Lederman, D. Teresi [1993] **The God Particle: If the Universe Is the Answer, What Is the Question?** Delta, New York

D. Lewis [1986] "Causation" and "Postscripts to Causation," in **Philosophical Papers**, Volume II, Oxford University Press, Oxford, 172-213

L. Mazlack [1987] "Machine Conceptualization Categories," *Proceedings 1987 IEEE Conference on Systems, Man, and Cybernetics,* 92-95

D. H. Mellor [1995] **The Facts of Causation**, Routledge, London

N. Packard, J. Chrutchfield, J. Farmer, R. Shaw [1980] "Geometry From A Time Series," Physical Review Letters, v 45, n 9, 712-716

J. Pearl, T. Verma [1991] "A Theory Of Inferred Causation," *Principles Of Knowledge Representation And Reasoning: Proceedings Of The Second International Conference,* J. Allen, R. Fikes, E. Sandewall, Morgan Kaufmann, 441-452

J. Pearl [2000] **Causality: Models, Reasoning, And Inference**, Cambridge University Press New York

B. Rehder [2002] "A Causal-Model Theory of the Conceptual Representation of Categories," FLAIRS 2002 Workshop on Causality, Pensacola, May

G. Shafer [1996] **The Art of Causal Conjecture**, MIT Press, Cambridge, Massachusetts

C. Silverstein, S. Brin, R. Motwani, J. Ullman [1998] "Scaleable Techniques For Mining Causal Structures," Proceedings 1998 *International Conference Very Large Data Bases*, NY, 594-605

P. Spirtes, C. Glymour, R. Scheines [2000] **Causation, Prediction, and Search**, second edition, MIT Press, Cambridge, Massachusetts

P. Spirtes [2001] "An Anytime Algorithm for Causal Inference," *Proceedings AI and Statistics 2001,* 213-221

P. Suppes [1970] **A Probabilistic Theory of Causality**, North-Holland Publishing Company, Amsterdam

L. Zadeh [2000] "Abstract Of A Lecture Presented At The Rolf Nevanilinna Colloquium, University of Helsinki," reported to: *Fuzzy Distribution List*, fuzzy-mail@dbai.tuwien.ac.at, August 24, 2000

Definability of Association Rules in Predicate Calculus

Jan Rauch

Department of Information and Knowledge Engineering
University of Economics, Prague
nám. W. Churchilla 4, 130 67 Praha 3, Czech Republic
rauch@vse.cz

Summary. Observational calculi are special logical calculi in which statements concerning observed data can be formulated. Their special case is predicate observational calculus. It can be obtained by modifications of classical predicate calculus - only finite models are allowed and generalised quantifiers are added. Association rules can be understood as special formulas of predicate observational calculi. Such association rules correspond to general relations of two Boolean attributes. A problem of the possibility to express association rule by the means of classical predicate calculus is investigated. A reasonable criterion of classical definability of association rules is presented.

Key words: Data mining, association rules, mathematical logic, observational calculi

1 Introduction

The goal of this chapter is to contribute to the theoretical foundations of data mining. We are interested in association rules of the form $\varphi \sim \psi$ where φ and ψ are derived Boolean attributes. Meaning of association rule $\varphi \sim \psi$ is that Boolean attributes φ and ψ are associated in a way corresponding to the symbol $\sim$ that is called 4ft-quantifier. The 4ft-quantifier makes possible to express various types of associations e.g. several types of implication or equivalency and also associations corresponding to statistical hypothesis tests.

Association rules of this form are introduced in [2]. Some more examples are e.g. in [7, 8]. To keep this chapter self-contained we will overview basic related notions in the next section.

Logical calculi formulae of which correspond to such association rules were defined and studied e.g. in [2, 4, 5, 6, 7]. It was shown that there are practically important theoretical properties of these calculi. Deduction rules of the form

$\frac{\varphi \sim \psi}{\varphi' \sim \psi'}$ where $\varphi \sim \psi$ and $\varphi' \sim \psi'$ are association rules are examples of such results [7].

Logical calculus of association rules can be understood as a special case of the monadic observational predicate calculus [2]. It can be obtained by modifications of classical predicate calculus such that *only finite models are allowed* and *4ft quantifiers are added.* We call this calculus *observational calculus of association rules.*

Observational calculus is a language formulae of which are statements concerning observed data. Various types of observational calculi are defined and studied [2]. The observational calculi are introduced in Sect. 3. Association rules as formulas of observational calculus are defined in Sect. 4.

The natural question is what association rules are classically definable. We say that the association rule is classically definable if it can be expressed by means of classical predicate calculus (i.e. predicates, variables, classical quantifiers $\forall$, $\exists$, Boolean connectives and the predicate of equality). The formal definition is in Sect. 5. The problem of definability in general monadic observational predicate calculi is solved by the Tharp's theorem, see Sect. 5.

Tharp's theorem is but too general from the point of view of association rules. We show, that there is a more intuitive criterion of classical definability of association rules. This criterion concerns 4ft-quantifiers. We need some theoretical results achieved in [2], see Sect. 6. The criterion of classical definability of association rules is proved in Sect. 7.

2 Association Rules

The association rule is an expression $\varphi \sim \psi$ where φ and ψ are Boolean attributes. The association rule $\varphi \sim \psi$ means that the Boolean attributes φ and ψ are associated in the way given by the symbol $\sim$. The symbol $\sim$ is called *4ft-quantifier.* Boolean attributes φ and ψ are derived from columns of an analysed data matrix $\mathcal{M}$. An example of the association rule is the expression

$$A(\alpha) \wedge D(\delta) \sim B(\beta) \wedge C(\gamma) .$$

The expression $A(\alpha)$ is a *basic Boolean attribute* The symbol α denotes a subset of all possible values of the attribute A (i.e. column of the data matrix $\mathcal{M}$). The basic Boolean attribute $A(\alpha)$ is true in row o of $\mathcal{M}$ if it is $a \in \alpha$ where a is the value of the attribute A in row o. Boolean attributes φ and ψ are derived from basic Boolean attributes using propositional connectives $\vee$, $\wedge$ and $\neg$ in the usual way.

The association rule $\varphi \sim \psi$ can be true or false in the analysed data matrix $\mathcal{M}$. It is verified on the basis of the four-fold contingency table of φ and ψ in $\mathcal{M}$, see Table 1. This table is denoted 4ft(φ, ψ, $\mathcal{M}$).

Here a is the number of the objects (i.e. the rows of $\mathcal{M}$) satisfying both φ and ψ, b is the number of the objects satisfying φ and not satisfying ψ, c is

Table 1. 4ft table 4ft(φ, ψ, $\mathcal{M}$) of φ and ψ in data matrix $\mathcal{M}$

$\mathcal{M}$	ψ	$\neg\psi$
φ	a	b
$\neg\varphi$	c	d

the number of objects not satisfying φ and satisfying ψ and d is the number of objects satisfying neither φ nor ψ . We write 4ft(φ, $\psi, \mathcal{M}$) $= \langle a, b, c, d\rangle$. We use the abbreviation "4ft" instead of "four-fold table". The notion *4ft table* is used for all possible tables $4ft(\varphi,\ \psi,\ \mathcal{M})$.

Definition 1. 4ft table *is a quadruple $\langle a, b, c, d\rangle$ of the integer non-negative numbers a, b, c, d such that $a + b + c + d > 0$.*

A condition concerning all 4ft tables is associated to each 4ft-quantifier. The association rule $\varphi \sim \psi$ is true in the analysed data matrix $\mathcal{M}$ if the condition associated to the 4ft-quantifier $\sim$ is satisfied for the 4ft table $4ft(\varphi,$ $\psi,\ \mathcal{M})$ of φ and ψ in $\mathcal{M}$. If this condition is not satisfied then the association rule $\varphi \sim \psi$ is false in the analysed data matrix $\mathcal{M}$.

This condition defines a $\{0, 1\}$ - function $Asf_\sim$ that is *called associated function of the 4ft-quantifier* $\sim$, see [2]. This function is defined for all 4ft tables such that

$$Asf_\sim = \begin{cases} 1 & \text{if the condition associated to} \sim \text{is satisfied} \\ 0 & \text{otherwise.} \end{cases}$$

Here are several examples of 4ft quantifiers.

The 4ft-quantifier $\Rightarrow_{p,Base}$ of *founded implication* for $0 < p \leq 1$ and $Base > 0$ [2] is defined by the condition $\frac{a}{a+b} \geq p \wedge a \geq Base$. The association rule $\varphi \Rightarrow_{p,Base} \psi$ means that at least $100p$ per cent of objects satisfying φ satisfy also ψ and that there are at least $Base$ objects of $\mathcal{M}$ satisfying both φ and ψ.

The 4ft-quantifier $\Rightarrow^!_{p,\alpha,Base}$ of *lower critical implication* for $0 < p \leq 1$, $0 < \alpha < 0.5$ and $Base > 0$ [2] is defined by the condition $\sum_{i=a}^{a+b} \binom{a+b}{i} p^i (1 - p)^{a+b-i} \leq \alpha \wedge a \geq Base$. This corresponds to the statistical test (on the level α) of the null hypothesis $H_0 : P(\psi|\varphi) \leq p$ against the alternative one $H_1 : P(\psi|\varphi) > p$. Here $P(\psi|\varphi)$ is the conditional probability of the validity of ψ under the condition φ.

The 4ft-quantifier $\Leftrightarrow_{p,Base}$ of *founded double implication* for $0 < p \leq 1$ and $Base > 0$ [3] is defined by the condition $\frac{a}{a+b+c} \geq p \wedge a \geq Base$. This means that at least $100p$ per cent of objects satisfying φ or ψ satisfy both φ and ψ and that there are at least $Base$ objects of $\mathcal{M}$ satisfying both φ and ψ.

The 4ft-quantifier $\equiv_{p,Base}$ of *founded equivalence* for $0 < p \leq 1$ and $Base > 0$ [3] is defined by the condition $\frac{a+d}{a+b+c+d} \geq p \wedge a \geq Base$. The association rule $\varphi \equiv_{p,Base} \psi$ means that φ and ψ have the same value (either *true* or *false*) for at least 100p per cent of all objects of $\mathcal{M}$ and that there are at least *Base* objects satisfying both φ and ψ.

The 4ft-quantifier $\sim^{+}_{p,Base}$ of *above average dependence* for $0 < p$ and $Base > 0$ [7] is defined by the condition $\frac{a}{a+b} \geq (1+p)\frac{a+c}{a+b+c+d} \wedge a \geq Base$. This means that among the objects satisfying φ is at least 100p per cent more objects satisfying ψ than among all objects and that there are at least *Base* objects satisfying both φ and ψ.

Further various 4ft-quantifiers are defined e.g. in [2, 3, 7, 8].

3 Observational Calculi

A mathematical theory related to the question "*Can computers formulate and justify scientific hypotheses?*" is developed in [2]. GUHA method as a tool for mechanising hypothesis formation is defined in relation to this theory. The procedure 4ft-Miner described in [8] is a GUHA procedure in the sense of [2].

Observational calculi are defined and studied in [2] as a language in which statements concerning observed data are formulated. We will use monadic observational predicate calculi to solve the question what predicate association rules can be logically equivalent expressed using classical quantifiers $\forall$ and $\exists$ and by the predicate of equality.

We will use the following notions introduced in [2].

Definition 2. *Observational semantic system and observational V-structures are defined as follows:*

1. **Semantic system** $\mathtt{S}= \langle \mathtt{Sent}, \mathtt{M}, \mathtt{V}, \mathtt{Val}\rangle$ *is given by a non–empty set* $\mathtt{Sent}$ *of sentences, a non–empty set* $\mathtt{M}$ *of models, non–empty set* $\mathtt{V}$ *of abstract values and by an evaluation function* $\mathtt{Val} : (\mathtt{Sent} \times \mathtt{M}) \rightarrow \mathtt{V}$. *If it is* $\varphi \in \mathtt{Sent}$ *and* $M \in \mathtt{M}$ *then* $\mathtt{Val}(\varphi, \mathtt{M})$ *is the value of* φ *in* $\mathtt{M}$.
2. **Semantic system** $\mathtt{S}= \langle \mathtt{Sent}, \mathtt{M}, \mathtt{V}, \mathtt{Val}\rangle$ *is an* **observational semantic system** *if* $\mathtt{Sent}$, $\mathtt{M}$ *and* $\mathtt{V}$ *are recursive sets and* $\mathtt{Val}$ *is a partially recursive function.*
3. *A* **type** *is a finite sequence* $\langle t_1, \ldots, t_n\rangle$ *of positive natural numbers. We write* $< 1^n >$ *instead of* $\langle\underbrace{1, 1, \ldots, 1}_{n-times}\rangle$.
4. *A* **V-structure** *of the type* $t = \langle t_1, \ldots, t_n\rangle$ *is a n+1-tuple*

$$\mathcal{M} = \langle M, f_1, \ldots, f_n\rangle,$$

where M *is a non-empty set and each* f_i $(i = 1, \ldots, n)$ *is a mapping from* M^{t_i} *into* V. *The set* M *is called the* **domain** *of* $\mathcal{M}$.
5. *If* $\mathcal{M}_1 = \langle M_1, f_1, \ldots, f_n\rangle$, $\mathcal{M}_2 = \langle M_2, g_1, \ldots, g_n\rangle$ *are the V-structures of the type* $t = \langle t_1, \ldots, t_n\rangle$ *then the one-one mapping* ζ *of* M_1 *onto* M_2 *is an* **isomorphism** *of* $\mathcal{M}_1$, $\mathcal{M}_2$ *if it preserves the structure, i.e. for each* i *and* $o_1, \ldots, o_{t_i} \in M_1$ *we have* $f_i(o_1, \ldots, o_{t_i}) = g_i(\zeta(o_1), \ldots, \zeta(o_{t_i}))$.

6. *Denote by* $\underline{M}_t^V$ *the set of all V-structures* $\mathcal{M}$ *of the type* t *such that the domain of* $\mathcal{M}$ *is a finite set of natural numbers. If* V *is a recursive set then the elements of* $\underline{M}_t^V$ *are called* **observational V-structures.**

Various observational semantic systems are defined in [2]. Observational predicate calculus (OPC for short) is one of them. It is defined by modifications of (classical) predicate calculus such that

- only finite models are admitted
- more quantifiers than $\forall$ and $\exists$ are used
- assumptions are made such that the closed formulae, models and the evaluation function form an observational semantic system

see definitions 3, 4 and 5.

Definition 3. *A* **predicate language** $\mathcal{L}$ *of type* $t = \langle t_1, \ldots, t_n \rangle$ *is defined in the following way.*

Symbols *of the language are:*

- **predicates** $P_1, \ldots, P_n$ *of arity* $t_1, \ldots, t_n$ *respectively*
- *an infinite sequence* $x_0, x_1, x_2, \ldots$ *of* **variables**
- **junctors** $\underline{0}$, $\underline{1}$ *(nullary),* $\neg$ *(unary) and* $\wedge, \vee, \rightarrow, \leftrightarrow$ *(binary), called falsehood, truth, negation, conjunction, disjunction, implication and equivalence.*
- **quantifiers** $q_0, q_1, q_2, \ldots$ *of types* $s_0, s_1, s_2, \ldots$ *respectively. The sequence of quantifiers is either infinite or finite (non-empty). The type of quantifier* q_i *is a sequence* $\langle 1^{s_i} \rangle$. *If there are infinitely many quantifiers then the function associating the type* s_i *with each* i *is recursive.*
- *A* **predicate language with equality** *contains an additional binary predicate* $=$ *(the equality predicate) distinct from* $P_1, \ldots, P_n$.

Formulae *are defined inductively as usual:*

- *Each expression* $P_i(u_1, \ldots, u_{t_i})$ *where* $u_1, \ldots, u_{t_i}$ *are variables is an* **atomic formula** *(and* $u_1 = u_2$ *is an atomic formula).*
- *Atomic formula is a formula,* $\underline{0}$ *and* $\underline{1}$ *are formulae. If* φ *and* ψ *are formulae, then* $\neg\varphi$, $\varphi \wedge \psi$, $\varphi \vee \psi$, $\varphi \rightarrow \psi$ *and* $\varphi \leftrightarrow \psi$ *are formulae.*
- *If* q_i *is a quantifier of the type* $\langle 1^{s_i} \rangle$, *if* u *is a variable and* $\varphi_1, \ldots, \varphi_{s_i}$ *are formulae then* $(q_i u)(\varphi_1, \ldots, \varphi_{s_i})$ *is a formula.*

Free and bound variables *are defined as usual. The induction step for* $(q_i u)(\varphi_1, \ldots, \varphi_s)$ *is as follows:*

- *A variable is free in* $(q_i u)(\varphi_1, \ldots, \varphi_s)$ *iff it is free in one of the formulae* $\varphi_1, \ldots, \varphi_s$ *and it is distinct from* u.
- *A variable is bound in* $(q_i u)(\varphi_1, \ldots, \varphi_s)$ *iff it is bound in one of the formulae* $\varphi_1, \ldots, \varphi_s$ *or it is* u.

Definition 4. Observational predicate calculus *OPC of the type* $t = \langle t_1, \ldots, t_n \rangle$ *is given by*

- *Predicate language $\mathcal{L}$ of the type t.*
- *Associated function Asf_{q_i} for each quantifier q_i of the language $\mathcal{L}$. Asf_{q_i} maps the set $\underline{M}_{s_i}^{\{0,1\}}$ of all models (i.e. V-structures) of the type s_i whose domain is a finite subset of the set of natural numbers into $\{0,1\}$ such that the following is satisfied:*
  - *Each Asf_{q_i} is invariant under isomorphism, i.e. if $\mathcal{M}_1, \mathcal{M}_2 \in M_{s_i}^{\{0,1\}}$ are isomorphic, then $Asf_{q_i}(\mathcal{M}_1) = Asf_{q_i}(\mathcal{M}_2)$.*
  - *$Asf_{q_i}(\mathcal{M})$ is a recursive function of two variables $q_i, \mathcal{M}$.*

Definition 5. (**Values of formulae**) *Let $\mathcal{P}$ be an OPC, let $\mathcal{M} = \langle M, f_1, \ldots, f_n\rangle$ be a model and let φ be a formula; write $FV(\varphi)$ for the set of free variables of φ. An* **$\mathcal{M}$-sequence** *for φ is a mapping ε of $FV(\varphi)$ into M. If the domain of ε is $\{u_1, \ldots u_n\}$ and if $\varepsilon(u_i) = m_i$ then we write $\varepsilon = \frac{u_1,\ldots,u_n}{m_1,\ldots,m_n}$. We define inductively $\|\varphi\|_{\mathcal{M}}[\varepsilon]$ - the $\mathcal{M}$-value of φ for ε.*

- $\|P_i(u_1, \ldots, u_k)\|_{\mathcal{M}}[\frac{u_1,\ldots,u_k}{m_1,\ldots,m_k}] = f_i(m_1, \ldots, m_k)$
- $\|u_1 = u_2)\|_{\mathcal{M}}[\frac{u_1,u_2}{m_1,m_2}] = 1$ *iff* $m_1 = m_2$
- $\|\underline{0}\|_{\mathcal{M}}[\emptyset] = 0,\ \|\underline{1}\|_{\mathcal{M}}[\emptyset] = 1,$
- $\|\neg\varphi\|_{\mathcal{M}}[\varepsilon] = 1 - \|\varphi\|_{\mathcal{M}}[\varepsilon]$
- *If $FV(\varphi) \subseteq domain(\varepsilon)$ then write ε/φ instead of restriction of ε to $FV(\varphi)$. Let ι be one of $\wedge$, $\vee$, $\rightarrow$, $\leftrightarrow$ and let Asf_ι be its associated function given by the usual truth table. Then $\|\varphi\ \iota\ \psi\|_{\mathcal{M}}[\varepsilon] = Asf_\iota(\|\varphi\|_{\mathcal{M}}[\varepsilon/\varphi], \|\psi\|_{\mathcal{M}}[\varepsilon/\psi])$.*
- *If $domain(\varepsilon) \supseteq FV(\varphi) - \{x\}$ and $x \notin domain(\varepsilon)$ then letting x vary over M we obtain an unary function $\|\varphi\|_{\mathcal{M}}^{\varepsilon}$ on M such that for $m \in M$ it is:*

$$\|\varphi\|_{\mathcal{M}}^{\varepsilon}(m) = \|\varphi\|_{\mathcal{M}}[(\varepsilon \cup \frac{x}{m})/\varphi] \ .$$

($\|\varphi\|_{\mathcal{M}}$ can be viewed as a k-ary function, k being the number of free variables of φ. Now all variables except x are fixed according to ε; x varies over M). We define: $\|(q_i x)(\varphi_1, \ldots, \varphi_k)\|_{\mathcal{M}}[\varepsilon] = Asf_q(\langle M, \|\varphi_1\|_{\mathcal{M}}^{\varepsilon}, \ldots, \|\varphi_k\|_{\mathcal{M}}^{\varepsilon}\rangle)$.

The following theorem is proved in [2].

Theorem 1. *Let $\mathcal{P}$ be an OPC of type t and let $\mathcal{S}$ be the semantic system whose sentences are closed formulas of $\mathcal{P}$, whose models are elements of $\underline{M}_t^{\{0,1\}}$ and whose evaluation function is defined by:*

$$\mathtt{Val}(\varphi, \mathcal{M}) = \|\varphi\|_{\mathcal{M}}[\emptyset] \ .$$

Then $\mathcal{S}$ is an observational semantic system.

Remark 1. Let $\mathcal{P}$ be an OPC of type t and let φ be its closed formula. Then we write only $\|\varphi\|_{\mathcal{M}}$ instead of $\|\varphi\|_{\mathcal{M}}[\emptyset]$.

We will also use the following notions defined in [2].

Definition 6. *An OPC is* **monadic** *if all its predicates are unary, i.e. if its type is* $t = \langle 1, \ldots, 1 \rangle$. *We write* **MOPC** *MOPC for "monadic observational predicate calculus". A MOPC whose only quantifiers are the classical quantifiers* $\forall, \exists$ *is called a* **classical MOPC** *or* **CMOPC**. *Similarly for MOPC with equality, in particular a CMOPC with equality.*

4 Association Rules in Observational Calculi

Let $\mathcal{P}_4$ be a MOPC of the type $\langle 1, 1, 1, 1 \rangle$ with unary predicates P_1, P_2, P_3, P_4 and with the quantifier $\Rightarrow_{p,Base}$ of the type $\langle 1, 1 \rangle$. Let x be the variable of $\mathcal{P}_4$. Then the closed formula

$$(\Rightarrow_{p,Base}, x)(P_1(x) \wedge P_4(x), P_2(x) \wedge P_3(x))$$

of MOPC $\mathcal{P}_4$ can be understood as the association rule

$$P_1 \wedge P_4 \Rightarrow_{p,Base} P_2 \wedge P_3$$

if we consider the associated function $Asf_{\Rightarrow_{p,Base}}$ of the quantifier $\Rightarrow_{p,Base}$ as the function mapping the set $\underline{M}^{\{0,1\}}_{\langle 1,1 \rangle}$ of all models $\mathcal{M} = \langle M, f_1, f_2 \rangle$ (see Fig. 1) of the type $\langle 1, 1 \rangle$ into $\{0, 1\}$ such that

$$Asf_{\Rightarrow_{p,Base}} = \begin{cases} 1 & if \ \frac{a}{a+b} \geq p \wedge a \geq Base \\ 0 & \text{otherwise.} \end{cases}$$

We suppose that a is the number of $o \in M$ such that both $f_1(o) = 1$, $f_2(o) = 1$ and that b is the number of $o \in M$ such that $f_1(o) = 1$ and $f_2(o) = 0$. These considerations lead to definition 7.

element of M	f_1	f_2
o_1	1	1
o_2	0	1
$\vdots$	$\vdots$	$\vdots$
o_n	0	0

Fig. 1. An example of the model $\mathcal{M} = \langle M, f_1, f_2 \rangle$

Definition 7. *Let* $\mathcal{P}$ *be a MOPC (with or without equality) with unary predicates* $P_1, \ldots, P_n$, $n \geq 2$. *Each formula*

$$(\sim x)(\varphi(x), \psi(x))$$

of $\mathcal{P}$ *where* $\sim$ *is a quantifier of the type* $\langle 1, 1 \rangle$, x *is a variable and* $\varphi(x)$, $\psi(x)$ *are open formulas built from the unary predicates, junctors and from the variable* x *is an* **association rule**. *We can write the association rule* $(\sim x)(\varphi(x), \psi(x))$ *also in the form* $\varphi \sim \psi$.

Definition 8. *Let $\mathcal{M} = \langle M; f, g\rangle$ be an observational $\{0,1\}$-structure of the type $\langle 1,1\rangle$ (i.e. $\mathcal{M} \in \underline{M}^{\{0,1\}}_{\langle 1,1\rangle}$ see definition 2). Then the the 4ft table*

$$T_{\mathcal{M}} = \langle a_{\mathcal{M}}, b_{\mathcal{M}}, c_{\mathcal{M}}, d_{\mathcal{M}}\rangle$$

of $\mathcal{M}$ is defined such that $a_{\mathcal{M}}$ is the number of $o \in M$ for which $f(o) = g(o) = 1$, $b_{\mathcal{M}}$ is the number of $o \in M$ for which $f(o) = 1$ and $g(o) = 0$ $c_{\mathcal{M}}$ is the number of $o \in M$ for which $f(o) = 0$ and $g(o) = 1$ and $d_{\mathcal{M}}$ is the number of $o \in M$ for which $f(o) = g(o) = 0$.

Let $\mathcal{N} = \langle M; f, g, h\rangle$ be an observational $\{0,1\}$-structure of the type $\langle 1,1,2\rangle$ Then the 4ft table $T_{\mathcal{N}}$ of $\mathcal{N}$ is defined as the 4ft table $T_{\mathcal{M}}$ of $\mathcal{M} = \langle M; f, g\rangle$.

Remark 2. The associated function $\mathrm{Asf}_\sim$ is invariant under isomorphism of observational structures (see definition 4). It means that if $\mathcal{M} = \langle M; f, g\rangle$ is an observational $\{0,1\}$-structure of the type $\langle 1,1\rangle$ then the value $\mathrm{Asf}_\sim(\mathcal{M}) = \mathrm{Asf}_\sim(\langle M; f, g\rangle)$ is fully determined by the 4ft table $T_{\mathcal{M}} = \langle a_{\mathcal{M}}, b_{\mathcal{M}}, c_{\mathcal{M}}, d_{\mathcal{M}}\rangle$. Thus we can write $\mathrm{Asf}_\sim(a_{\mathcal{M}}, b_{\mathcal{M}}, c_{\mathcal{M}}, d_{\mathcal{M}})$ instead of $\mathrm{Asf}_\sim(\langle M; f, g\rangle)$.

Remark 3. Let $\mathcal{P}$ be a MOPC as in definition 7. Let $\varphi \sim \psi$ be an associational rule, let $\mathcal{M}$ be a model of $\mathcal{P}$ and let M be a domain of $\mathcal{M}$. Then it is

$$\|\varphi \sim \psi\|_{\mathcal{M}} = \|(\sim x)(\varphi(x), \psi(x))\|_{\mathcal{M}}[\emptyset] = \mathrm{Asf}_\sim(\langle M, \|\varphi\|^{\emptyset}_{\mathcal{M}}, \|\psi\|^{\emptyset}_{\mathcal{M}}\rangle)\ .$$

Let us remember that both $\|\varphi\|^{\emptyset}_{\mathcal{M}}$ and $\|\psi\|^{\emptyset}_{\mathcal{M}}$ are unary functions defined on M, see the last point of definition 5. We denote the $\{0,1\}$-structure $\langle M, \|\varphi\|^{\emptyset}_{\mathcal{M}}, \|\psi\|^{\emptyset}_{\mathcal{M}}\rangle)$ by $\mathcal{M}_{\varphi,\psi}$. It means that

$$\|\varphi \sim \psi)\|_{\mathcal{M}} = \mathrm{Asf}_\sim(a_{\mathcal{M}_{\varphi,\psi}}, b_{\mathcal{M}_{\varphi,\psi}}, c_{\mathcal{M}_{\varphi,\psi}}, d_{\mathcal{M}_{\varphi,\psi}})\ .$$

We are interested in the question of what association rules are classically definable (i.e. they can be expressed by predicates, variables, classical quantifiers $\forall$, $\exists$, Boolean connectives and by the predicate of equality). We defined association rules of the form such as

$$A(a_1, a_7) \wedge D(d_2, d_3) \Rightarrow_{p,Base} B(b_2)$$

see Sect. 2. Informally speaking this association rule is equivalent to the association rule

$$(A(a_1) \vee A(a_7)) \ \wedge\ (D(d_2) \vee D(d_3)) \Rightarrow_{p,Base} B(b_2)$$

with Boolean attributes $A(a_1)$, $A(a_7)$, $D(d_2)$, $D(d_3)$ and $B(b_2)$.

Thus we can concentrate on the question of what association rules - i.e. closed formulas $(\sim x)(\varphi, \psi)$ of the MOPC are classically definable.

5 Classical Definability and Tharp's Theorem

We use the following two definitions and lemma from [2].

Definition 9. *Let $\mathcal{P}$ be an OPC. Suppose that φ and ψ are formulas such that $FV(\varphi) = FV(\psi)$ (φ and ψ have the same free variables). Then φ and ψ are said to be* **logically equivalent** *if* $\|\varphi\|_{\mathcal{M}} = \|\psi\|_{\mathcal{M}}$ *for each model $\mathcal{M}$.*

Remark that the equality $\|\varphi\|_{\mathcal{M}} = \|\psi\|_{\mathcal{M}}$ in the previous definition is generally the equality of functions. We will use it for the closed formulas. In this case it is the equality of two values.

Definition 10. *Let $\mathcal{P}$ be a MOPC (including MOPC with equality) of the type $\langle 1^n \rangle$ and let q be a quantifier of type $\langle 1^k \rangle$, $k \leq n$. Then q is* **definable** *in $\mathcal{P}$ if there is a sentence Φ of $\mathcal{P}$ not containing q such that the sentence*

$$(qx)(P_1(x), \ldots, P_n(x))$$

is logically equivalent to Φ.

Lemma 1. *Let $\mathcal{P}$ and q be as in definition 10. Then q is definable in $\mathcal{P}$ iff each sentence of $\mathcal{P}$ is logically equivalent to a sentence not containing the quantifier q.*

The following theorem concerns the problem of definability of quantifiers, see [2] (and [10] cited in [2]).

Theorem 2. *(Tharp) Let $\mathcal{P}^{=}$ be a CMOPC with equality and unary predicates $P_1, \ldots, P_n$ and let $\mathcal{P}'$ be its extension by adding a quantifier q of type $\langle 1^k \rangle$ ($k \leq n$). Then q is definable in $\mathcal{P}'$ iff there is a natural number m such that the following holds for $\varepsilon \in \{0, 1\}$ and each model M of type $\langle 1^k \rangle$:*

$$Asf_q(M) = \varepsilon \textit{ iff } (\exists M_0 \subseteq M)(M_0 \textit{ has } \leq m \textit{ elements}$$
$$\textit{and } (\forall M_1)(M_0 \subseteq M_1 \subseteq M \textit{ implies } Asf_q(M) = \varepsilon$$

We formalise the notion of classically definable association rules in the following definition.

Definition 11. *Let $\mathcal{P}^{=}$ be a CMOPC with equality and unary predicates $P_1, \ldots, P_n$ ($n \geq 2$), let $\mathcal{P}^{\sim}$ be its extension by adding a quantifier $\sim$ of type $\langle 1, 1 \rangle$. Then the association rule $\varphi \sim \psi$ is* **classically definable** *if the quantifier $\sim$ is definable in $\mathcal{P}^{\sim}$.*

The following lemma is an immediate consequence of lemma 1.

Lemma 2. *Let $\mathcal{P}^{\sim}$ and $\varphi \sim \psi$ be as in definition 11. Then the association rule $\varphi \sim \psi$ is classically definable iff it is logically equivalent to a sentence of $\mathcal{P}^{\sim}$ that contains only predicates, variables, classical quantifiers $\forall$, $\exists$, Boolean connectives and the predicate of equality.*

We search for another criterion of classical definability of association rules $\varphi \sim \psi$ than Tharp's theorem 2. Namely we are interested in a criterion closely related to the associated function $Asf_{\sim}$ of the 4ft-quantifier $\sim$.

Such criterion is presented in Sect. 7. It is based on the normal form theorem, see the next section.

6 Normal Form Theorem

We need the following notions and lemma from [2].

Definition 12.

1. *An* **n-ary card** *is a sequence $\langle u_i; 1 \leq i \leq n\rangle$ of n zeros and ones. Let $\mathcal{P}$ be a MOPC with predicates $P_1, \ldots, P_n$. If $\mathcal{M} = \langle M, p_1, \ldots, p_n\rangle$ is a model of $\mathcal{P}$ and if $o \in M$ then the* **$\mathcal{M}$-card of** *o is the tuple $C_{\mathcal{M}}(o) = \langle p_1(o), \ldots, p_n(o)\rangle$; it is evidently an n-ary card.*
2. *For each natural number $k > 0$, $\exists^k$ is a quantifier of the type $\langle 1\rangle$ whose associated function is defined as follows: For each finite model $\mathcal{M} = \langle M, f\rangle$ it is $Asf_{\exists^k}(\mathcal{M}) = 1$ iff there are at least k elements $o \in M$ such that $f(o) = 1$.*

Lemma 3. *Let k be a natural number and let $\mathcal{P}^{\,k}$ be the extension of CMOPC with equality $\mathcal{P}^{=}$ by adding $\exists^k$. Then $\exists^k$ is definable by the formula*

$$\phi : (\exists x_1, \ldots, x_k)(\bigwedge_{i \neq j, 1 \leq i,j \leq k} x_i \neq x_j \wedge \bigwedge_{1 \leq i \leq k} P_1(x_i))$$

Definition 13.

1. *Let $u = \langle u_1, \ldots, u_n\rangle$ be an n-card. Then the* **elementary conjunction given by** *u is the conjunction*

$$\kappa_u = \bigwedge_{i=1}^{n} \lambda_i$$

where for $i = 1, \ldots, n$ it holds:
   - *λ_i is $P_i(x)$ if $u_i = 1$*
   - *λ_i is $\neg P_i(x)$ if $u_i = 0$.*
2. *Each formula of the form $(\exists^k x)\kappa_{\mathrm{u}}$ where u is a card is called a* **canonical sentence** *for CMOPC's with equality.*

The further results concerning classical definability of association rules are based on the following normal form theorem proved in [2].

Theorem 3. *Let $\mathcal{P}^{=}$ be a CMOPC with equality and let $\mathcal{P}^{*}$ be the extension of $\mathcal{P}^{=}$ by adding the quantifiers $\exists^k$ (k is a natural number). Let Φ be a sentence from $\mathcal{P}^{=}$. Then there is a sentence Φ^* from $\mathcal{P}^*$ logically equivalent to Φ (in $\mathcal{P}^*$) and such that Φ^* is a Boolean combination of canonical sentences. (In particular, Φ^* contains neither the equality predicate nor any variable distinct from the canonical variable).*

7 Classical Definability of Association Rules

Tharp's theorem (see theorem 2) proved in [2] can be used as a criterion of classical definability of association rules. The main result of this section is theorem 5 that can be used as an alternative criterion of definability. This theorem shows that there is a relatively simple condition concerning the 4ft-quantifier that is equivalent to classical definability of corresponding association rules.

First we prove several lemmas and introduce some notions.

Lemma 4. *Let $\mathcal{P}_2$ be MOPC of the type $\langle 1,1\rangle$ with predicates P_1, P_2. Let $\mathcal{P}_n$ be MOPC of the type $\langle 1^n\rangle$ with predicates $P_1,\dots,P_n$. Let $\sim$ be a quantifier of the type $\langle 1,1\rangle$, let $\mathcal{P}_2'$ be extension of $\mathcal{P}_2$ by adding $\sim$ and let $\mathcal{P}_n'$ be extension of $\mathcal{P}_n$ by adding $\sim$. Then $\sim$ is definable in $\mathcal{P}_2'$ if and only if it is definable in $\mathcal{P}_n'$.*

Proof. (1) If $\sim$ is definable in $\mathcal{P}_2'$ then there is a sentence Φ of $\mathcal{P}_2$ (i.e. a sentence of $\mathcal{P}_2'$ not containing $\sim$) such that the sentence $(\sim x)(P_1(x),P_2(x))$ is logically equivalent to Φ, see definition 10. The sentence Φ is but also the sentence of $\mathcal{P}_n$ and thus the sentence of $\mathcal{P}_n'$ not containing $\sim$. It means that $\sim$ is definable in $\mathcal{P}_n'$

(2) Let $\sim$ is definable in $\mathcal{P}_n'$. It means that there is the sentence Φ of $\mathcal{P}_n$ (i.e. a sentence of $\mathcal{P}_n'$ not containing $\sim$) such that the sentence $(\sim x)(P_1(x),P_2(x))$ is logically equivalent to Φ. Let us construct a sentence Φ^* from Φsuch that we replace all occurrences of the atomic formulas $P_i(x)$ for $i\ge 3$ by the atomic formula $P_1(x)$. The sentence Φ^* is the sentence of the calculus $\mathcal{P}_2$ (i.e. the sentence of $\mathcal{P}_2'$ not containing $\sim$) that is logically equivalent to $(\sim x)(P_1(x),P_2(x))$. Thus $\sim$ is definable in $\mathcal{P}_2'$.

Definition 14. *Let $\mathcal{P}^=$ be CMOPC of the type $\langle 1,1\rangle$ with predicates P_1 and P_2 and with equality. Let $\mathcal{P}^*$ be the extension of $\mathcal{P}^=$ by adding all quantifiers $\exists^k$ (k is a natural number). Then we denote:*

- $\kappa_a = P_1(x)\wedge P_2(x)$
- $\kappa_b = P_1(x)\wedge\neg P_2(x)$
- $\kappa_c = \neg P_1(x)\wedge P_2(x)$
- $\kappa_d = \neg P_1(x)\wedge\neg P_2(x)$

Lemma 5. *Let $\mathcal{P}^=$ be CMOPC of the type $\langle 1,1\rangle$ with predicates P_1 and P_2 and with equality. Let $\mathcal{P}^*$ be the extension of $\mathcal{P}^=$ by adding all quantifiers $\exists^k$ (k is a natural number). Let Φ be a Boolean combination of canonical sentences of the calculus $\mathcal{P}^*$. Then Φ is logically equivalent to the formula*

$$\bigvee_{i=1}^{K}\varphi_a^{(i)}\wedge\varphi_b^{(i)}\wedge\varphi_c^{(i)}\wedge\varphi_d^{(i)}$$

where $K\ge 0$ and $\varphi_a^{(i)}$ is in one of the following formulas for each $i=1,\dots,K$:

- $(\exists^k x)\kappa_a$ *where* k *is natural number*
- $\neg(\exists^k x)\kappa_a$ *where* k *is natural number*
- $(\exists^k x)\kappa_a \wedge \neg(\exists^l x)\kappa_a$ *where* $0 < k < l$ *are natural numbers.*
- $\underline{1}$,

similarly for $\varphi_b^{(i)}, \varphi_c^{(i)}, \varphi_d^{(i)}$. *Let us remark that the value of an empty disjunction is* 0.

Proof. Φ *is a Boolean combination of canonical sentences of the calculus* $\mathcal{P}^*$ *thus we can write it in the form*

$$\bigvee_{i=1}^{K'} \bigwedge_{j=1}^{L_i} \psi_{i,j}$$

where $K' \geq 0$ *and each* $\psi_{i,j}$ *is the canonical sentence of the calculus* $\mathcal{P}^*$ *or a negation of such a canonical sentence. We create from each formula* $\bigwedge_{j=1}^{L_i} \psi_{i,j}$ $(i = 1, \ldots, K')$ *a new formula*

$$\varphi_a^{(i)} \wedge \varphi_b^{(i)} \wedge \varphi_c^{(i)} \wedge \varphi_d^{(i)}$$

in the required form.

We can suppose without loss of generality that

$$\bigwedge_{j=1}^{L_i} \psi_{i,j} = \bigwedge_{j=1}^{A} \psi_{a,j} \wedge \bigwedge_{j=1}^{B} \psi_{b,j} \wedge \bigwedge_{j=1}^{C} \psi_{c,j} \wedge \bigwedge_{j=1}^{D} \psi_{d,j}$$

where each formula $\psi_{a,j}$ $(j = 1, \ldots, A)$ *is equal to* $\exists^k \kappa_a$ *or to* $\neg(\exists^k \kappa_a)$ *where* k *is a natural number, analogously for* $\psi_{b,j}$, $\psi_{c,j}$ *and* $\psi_{d,j}$.

If $A = 0$ *then we define* $\varphi_a^{(i)} = \underline{1}$ *and* $\varphi_a^{(i)}$ *is logically equivalent to* $\bigwedge_{j=1}^{A} \psi_{a,j}$ *because the value of empty conjunction is* $\underline{1}$.

If $A = 1$ *then we define* $\varphi_a^{(i)} = \psi_{a,1}$ *and* $\varphi_a^{(i)}$ *is again logically equivalent to* $\bigwedge_{j=1}^{A} \psi_{a,j}$.

If $A \geq 2$ *then we can suppose without loss of generality that*

$$\bigwedge_{j=1}^{A} \psi_{a,j} = \bigwedge_{j=1}^{A_1} (\exists^{k_j} x)\kappa_a \wedge \bigwedge_{j=1}^{A_2} \neg(\exists^{k_j} x)\kappa_a \ .$$

If $A_1 > 0$ *then we define* $k = max\{k_j | j = 1, \ldots, A_1\}$ *and thus* $\bigwedge_{j=1}^{A_1} (\exists^{k_j} x)\kappa_a$ *is logically equivalent to* $(\exists^k x)\kappa_a$.

If $A_2 > 0$ *then we define* $l = min\{k_j | j = 1, \ldots, A_2\}$ *and thus* $\bigwedge_{j=1}^{A_2} \neg(\exists^{k_j} x)\kappa_a$ *is logically equivalent to* $\neg(\exists^l x)\kappa_a$.

In the case $A \geq 2$ *we define the formula* $\varphi_a^{(i)}$ *this way:*

- *if* $A_1 = 0$ *then* $\varphi_a^{(i)} = \neg(\exists^l x)\kappa_a$

- *if* $A_2 = 0$ *then* $\varphi_a^{(i)} = (\exists^k x)\kappa_a$
- *if* $A_1 > 0 \wedge A_2 > 0 \wedge k < l$ *then* $\varphi_a^{(i)} = (\exists^k x)\kappa_a \wedge \neg(\exists^l x)\kappa_a$
- *if* $A_1 > 0 \wedge A_2 > 0 \wedge k \geq l$ *then* $\varphi_a^{(i)} = \underline{0}$.

We defined the formula $\varphi_a^{(i)}$ *for all the possible cases (i.e.* $A = 0$, $A = 1$ *and* $A \geq 2$*) and in all cases is* $\bigwedge_{j=1}^{A} \psi_{a,j}$ *logically equivalent to* $\varphi_a^{(i)}$ *that is in the required form.*

We analogously create $\varphi_b^{(i)}$, $\varphi_c^{(i)}$ *and* $\varphi_d^{(i)}$ *thus they are equivalent to* $\bigwedge_{j=1}^{B} \psi_{b,j}$, $\bigwedge_{j=1}^{C} \psi_{c,j}$ *and* $\bigwedge_{j=1}^{D} \psi_{d,j}$ *respectively* ($i = 1, \ldots, K'$)*. Thus also the formulas*

$$\bigwedge_{j=1}^{L_i} \psi_{i,j} \quad \text{and} \quad \varphi_a^{(i)} \bigwedge \varphi_b^{(i)} \bigwedge \varphi_c^{(i)} \bigwedge \varphi_d^{(i)}$$

are logically equivalent. It means that also the formulas

$$\bigvee_{i=1}^{K'} \bigwedge_{j=1}^{L_i} \psi_{i,j} \text{ and } \bigvee_{i=1}^{K'} \varphi_a^{(i)} \wedge \varphi_b^{(i)} \wedge \varphi_c^{(i)} \wedge \varphi_d^{(i)}$$

are logically equivalent. Furthermore, all the $\varphi_a^{(i)}$, $\varphi_b^{(i)}$, $\wedge\varphi_c^{(i)}$ *and* $\varphi_d^{(i)}$ *are in the required form or equal to* $\underline{0}$*.*

Finally we omit all conjunctions $\varphi_a^{(i)} \wedge \varphi_b^{(i)} \wedge \varphi_c^{(i)} \wedge \varphi_d^{(i)}$ *with at least one member equal to* $\underline{0}$ *and we arrive at the required formula*

$$\bigvee_{i=1}^{K} \varphi_a^{(i)} \wedge \varphi_b^{(i)} \wedge \varphi_c^{(i)} \wedge \varphi_d^{(i)} .$$

Definition 15. *Let* $\mathcal{N}$ *be the set of all natural numbers. Then we define:*

- **The interval in** $\mathcal{N}^4$ *is the set*

  $$I = I_1 \times I_2 \times I_3 \times I_4$$

  such that it is for $i = 1, 2, 3, 4$ $I_j = \langle k, l)$ *or* $I_j = \langle k, \infty)$ *where* $0 \leq k < l$ *are natural numbers. The empty set* $\emptyset$ *is also the interval in* $\mathcal{N}^4$*.*
- *Let* $T = \langle a, b, c, d\rangle$ *be a 4ft table (see definition 1) and let* $I = I_1 \times I_2 \times I_3 \times I_4$ *be the interval in* $\mathcal{N}^4$ *. Then*

  $$T \in I \quad \text{iff} \quad a \in I_1 \wedge b \in I_2 \wedge c \in I_3 \wedge d \in I_4$$

Theorem 4. *Let* $\mathcal{P}^=$ *be CMOPC of the type* $\langle 1, 1\rangle$ *with equality and let* Φ *be a sentence of* $\mathcal{P}^=$*. Then there are* K *intervals* $I_1, \ldots, I_K$ *in* $\mathcal{N}^4$ *where* $K \geq 0$ *such that for each model* $\mathcal{M}$ *of the calculus* $\mathcal{P}^=$ *it is*

$$||\Phi||_{\mathcal{M}} = 1 \quad \text{iff} \quad T_{\mathcal{M}} \in \bigcup_{j=1}^{K} I_j$$

where $T_{\mathcal{M}} = \langle a_{\mathcal{M}}, b_{\mathcal{M}}, c_{\mathcal{M}}, d_{\mathcal{M}}\rangle$ *is the 4ft table of the model* $\mathcal{M}$*. (It is* $\bigcup_{j=1}^{0} I_j = \emptyset$*.)*

Proof. *Let $\mathcal{P}^*$ be the extension of the calculus $\mathcal{P}^=$ by adding the quantifiers $\exists^k$ $(k > 0)$. Then according to theorem 3 there is a sentence Φ^* of the calculus $\mathcal{P}^*$ such that Φ^* is the Boolean combination of the canonical sentences and Φ^* is equivalent to Φ.*

Furthermore, according to lemma 5 the formula Φ^* is equivalent to the formula

$$\bigvee_{i=1}^{K} \varphi_a^{(i)} \wedge \varphi_b^{(i)} \wedge \varphi_c^{(i)} \wedge \varphi_d^{(i)}$$

where $K \geq 0$ and $\varphi_a^{(i)}$ is in one of the following forms for each $i = 1, \ldots, K$:

- $(\exists^k x)\kappa_a$ *where k is natural number*
- $\neg(\exists^k x)\kappa_a$ *where k is natural number*
- $(\exists^k x)\kappa_a \wedge \neg(\exists^l x)\kappa_a$ *where $0 < k < l$ are natural numbers.*
- $\underline{1}$,

similarly for $\varphi_b^{(i)}$, $\varphi_c^{(i)}$, $\varphi_d^{(i)}$.

If $K = 0$ then $||\Phi||_{\mathcal{M}} = \underline{0}$ because the value of empty disjunction is $\underline{0}$. If $K = 0$ then also $\bigcup_{j=1}^{0} I_j = \emptyset$ and thus $T \notin \bigcup_{j=1}^{K} I_j$. It means that for $K = 0$ it is $||\Phi||_{\mathcal{M}} = 1$ iff $T_{\mathcal{M}} \in \bigcup_{j=1}^{K} I_j$.

If $K > 0$ then we create interval I_j for each $j = 1, \ldots, K$ in the following steps:

- *If $\varphi_a^{(j)} = (\exists^k x)\kappa_a$ then we define $I_a^{(j)} = \langle k, \infty)$.*
- *If $\varphi_a^{(j)} = \neg(\exists^k x)\kappa_a$ then we define $I_a^{(j)} = \langle 0, k)$.*
- *If $\varphi_a^{(j)} = (\exists^k x)\kappa_a \wedge \neg(\exists^l x)\kappa_a$ then we define $I_a^{(j)} = \langle k, l)$.*
- *If $\varphi_a^{(j)} = \underline{1}$ then we define $I_a^{(j)} = \langle 0, \infty)$.*
- *We analogously define $I_b^{(j)}$, $I_c^{(j)}$ and $I_d^{(j)}$ using $\varphi_b^{(j)}$, $\varphi_c^{(j)}$ and $\varphi_d^{(j)}$ respectively.*
- *We finally define $I_j = I_a^{(j)} \times I_b^{(j)} \times I_c^{(j)} \times I_d^{(j)}$*

Let us emphasize that the interval $I_a^{(j)}$ is defined such that if $||\varphi_a^{(j)}||_{\mathcal{M}} = 1$ then $a_{\mathcal{M}} \in I_a^{(j)}$ and if $||\varphi_a^{(j)}||_{\mathcal{M}} = 0$ then $a_{\mathcal{M}} \notin I_a^{(j)}$, similarly for $\varphi_b^{(j)}$, $\varphi_c^{(j)}$ and $\varphi_d^{(j)}$.

We prove that for each model $\mathcal{M}$ of the calculus $\mathcal{P}^=$ it is

$$||\Phi||_{\mathcal{M}} = 1 \quad \text{iff} \quad T_{\mathcal{M}} \in \bigcup_{j=1}^{K} I_j \,.$$

We use the fact that the formula Φ is equivalent to $\bigvee_{i=1}^{K} \varphi_a^{(i)} \wedge \varphi_b^{(i)} \wedge \varphi_c^{(i)} \wedge \varphi_d^{(i)}$ and we suppose that $T_{\mathcal{M}} = \langle a_{\mathcal{M}}, b_{\mathcal{M}}, c_{\mathcal{M}}, d_{\mathcal{M}} \rangle$.

Let $||\Phi||_{\mathcal{M}} = 1$. Thus there is $j \in \{1, \ldots, K\}$ such that $||\varphi_a^{(j)} \wedge \varphi_b^{(j)} \wedge \varphi_c^{(j)} \wedge \varphi_d^{(j)}||_{\mathcal{M}} = 1$ and it means that also $||\varphi_a^{(j)}||_{\mathcal{M}} = 1$. The interval $I_a^{(j)}$ is *constructed such that the fact* $||\varphi_a^{(j)}||_{\mathcal{M}} = 1$ *implies* $a_{\mathcal{M}} \in I_a^{(j)}$. *Analogously we get* $b_{\mathcal{M}} \in I_b^{(j)}$, $c_{\mathcal{M}} \in I_c^{(j)}$ *and* $d_{\mathcal{M}} \in I_d^{(j)}$. *It means that* $T_{\mathcal{M}} \in I_j$ *and also* $T_{\mathcal{M}} \in \bigcup_{j=1}^{K} I_j$.

Let $||\Phi||_{\mathcal{M}} = 0$. *Then it is for each* $j = 1, \ldots, K$ $||\varphi_a^{(j)} \wedge \varphi_b^{(j)} \wedge \varphi_c^{(j)} \wedge \varphi_d^{(j)}||_{\mathcal{M}} = 0$. *It means that also for each such* j *it is* $||\varphi_a^{(j)}||_{\mathcal{M}} = 0$ *or* $||\varphi_b^{(j)}||_{\mathcal{M}} = 0$ *or* $||\varphi_c^{(j)}||_{\mathcal{M}} = 0$ *or* $||\varphi_d^{(j)}||_{\mathcal{M}} = 0$. *If* $||\varphi_a^{(j)}||_{\mathcal{M}} = 0$ *then it must be* $a_{\mathcal{M}} \notin I_a^{(j)}$ *thus* $T_{\mathcal{M}} \notin I_j$.

Analogously we get $T_{\mathcal{M}} \notin I_j$ *for* $||\varphi_b^{(j)}||_{\mathcal{M}} = 0$, $||\varphi_c^{(j)}||_{\mathcal{M}} = 0$ *and* $||\varphi_d^{(j)}||_{\mathcal{M}} = 0$. *Thus* $T_{\mathcal{M}} \notin \bigcup_{j=1}^{K} I_j$. *This finishes the proof.*

Lemma 6. *Let* $\langle a, b, c, d\rangle$ *be a 4ft table. Then there is an observational structure* $\mathcal{M} = \langle M, f, g\rangle$ *of the type* $\langle 1, 1\rangle$ *such that it is*

$$T_{\mathcal{M}} = \langle a_{\mathcal{M}}, b_{\mathcal{M}}, c_{\mathcal{M}}, d_{\mathcal{M}}\rangle = \langle a, b, c, d\rangle$$

where $T_{\mathcal{M}}$ *is the 4ft table of* $\mathcal{M}$ *see definition 8.*

Proof. We construct $\mathcal{M} = \langle M, f, g\rangle$ *such that* M *has* $a+b+c+d$ *elements and the functions* f, g *are defined according to Fig. 2.*

element of M	f	g
$o_1, \ldots, o_a$	1	1
$o_{a+1}, \ldots, o_{a+b}$	1	0
$o_{a+b+1}, \ldots, o_{a+b+c}$	0	1
$o_{a+b+c+1}, \ldots, o_{a+b+c+d}$	0	0

Fig. 2. Structure $\mathcal{M}$ for which $T_{\mathcal{M}} = \langle a, b, c, d\rangle$

Theorem 5. *Let* $\mathcal{P}^=$ *be a CMOPC with equality of the type* $\langle 1^n\rangle$, *let* $\sim$ *be a quantifier of the type* $\langle 1, 1\rangle$ *and let* $\mathcal{P}'$ *be the extension of* $\mathcal{P}^=$ *by adding the quantifier* $\sim$. *Then* $\sim$ *is definable in* $\mathcal{P}'$ *if and only if there are* K *intervals* $I_1, \ldots, I_K$ *in* $\mathcal{N}^4$, $K \geq 0$ *such that it is for each 4ft table* $\langle a, b, c, d\rangle$

$$Asf_{\sim}(a, b, c, d) = 1 \ \text{ iff } \ \langle a, b, c, d\rangle \in \bigcup_{j=1}^{K} I_j \ .$$

Proof. According to lemma 4 we can restrict ourselves to the case when $\mathcal{P}^=$ *is of the type* $\langle 1, 1\rangle$.

First, let $\sim$ *is definable in* $\mathcal{P}'$. *Then there is a sentence* Φ *of the calculus* $\mathcal{P}^=$ *that is logically equivalent to the sentence* $(\sim x)(P_1(x), P_2(x))$. *It means that*

$$||\Phi||_{\mathcal{M}} = ||(\sim x)(P_1(x), P_2(x))||_{\mathcal{M}}$$

for each model $\mathcal{M}$ of $\mathcal{P}'$. Let us remark that

$$||(\sim x)(P_1(x), P_2(x))||_{\mathcal{M}} = Asf_{\sim}(a_{\mathcal{M}}, b_{\mathcal{M}}, c_{\mathcal{M}}, d_{\mathcal{M}})$$

where $T_{\mathcal{M}} = \langle a_{\mathcal{M}}, b_{\mathcal{M}}, c_{\mathcal{M}}, d_{\mathcal{M}}\rangle$ is the 4ft table of the model $\mathcal{M}$, see definition 8.

According to the theorem 4 there are K intervals $I_1, \ldots, I_K$ in $\mathcal{N}^4$ where $K \geq 0$ such that for each model $\mathcal{M}$ of the calculus $\mathcal{P}^=$ it is

$$||\Phi||_{\mathcal{M}} = 1 \quad \text{iff} \quad T_{\mathcal{M}} \in \bigcup_{j=1}^{K} I_j \ .$$

We show that for each 4ft table $\langle a, b, c, d\rangle$ it is

$$Asf_{\sim}(a, b, c, d) = 1 \quad \text{iff} \quad \langle a, b, c, d\rangle \in \bigcup_{j=1}^{K} I_j \ .$$

Let $\langle a, b, c, d\rangle$ be a 4ft table. There is according to lemma 6 model $\mathcal{M}_0$ such that $T_{\mathcal{M}_0} = \langle a, b, c, d\rangle$.

Let $Asf_{\sim}(a, b, c, d) = 1$. It means that

$$||\Phi||_{\mathcal{M}_0} = ||q(x)(P_1(x), P_2(x))||_{\mathcal{M}_0} = Asf_{\sim}(a, b, c, d) = 1$$

and thus $\langle a, b, c, d\rangle = T_{\mathcal{M}_0} \in \bigcup_{j=1}^{K} I_j$.

Let $Asf_{\sim}(a, b, c, d) = 0$. It means that

$$||\Phi||_{\mathcal{M}_0} = ||(\sim x)(P_1(x), P_2(x))||_{\mathcal{M}_0} = Asf_{\sim}(a, b, c, d) = 0$$

and thus $\langle a, b, c, d\rangle = T_{\mathcal{M}_0} \notin \bigcup_{j=1}^{K} I_j$.

We have proved the first part - if $\sim$ is definable in $\mathcal{P}'$ then there are K intervals $I_1, \ldots, I_K$ in $\mathcal{N}^4$, $K \geq 0$ such that it is for each 4ft table $\langle a, b, c, d\rangle$ $Asf_{\sim}(a, b, c, d) = 1$ iff $\langle a, b, c, d\rangle \in \bigcup_{j=1}^{K} I_j$.

Secondly, let us suppose that there are K intervals $I_1, \ldots, I_K$ in $\mathcal{N}^4$, $K \geq 0$ such that

$$Asf_{\sim}(a, b, c, d) = 1 \quad \text{iff} \quad \langle a, b, c, d\rangle \in \bigcup_{j=1}^{K} I_j \ .$$

We must prove that $\sim$ is definable $\mathcal{P}'$.

Let $K = 0$ then $\bigcup_{j=1}^{0} I_j = \emptyset$. It means that $Asf_{\sim}(a, b, c, d) = 0$ for each table $\langle a, b, c, d\rangle$. Thus for each model $\mathcal{M}$ it is $||(\sim x)(P_1(x), P_2(x))||_{\mathcal{M}} = 0$ and it means that $\sim$ is definable (e.g. by the formula $(\exists x)(x \neq x)$).

Let $K > 0$ then we denote $\mathcal{P}^*$ the extension of $\mathcal{P}^=$ by adding all the quantifiers $\exists^k$ for $k > 0$. We create for each $j = 1, \ldots, K$ the formula

$$\psi_j = \varphi_a^{(j)} \wedge \varphi_b^{(j)} \wedge \varphi_c^{(j)} \wedge \varphi_d^{(j)}$$

in the following way (see also lemma 5 and theorem 4). We suppose that $I_j = I_a \times I_b \times I_c \times I_d$.

- *If* $I_a = \langle 0, \infty)$ *then we define* $\varphi_a^{(j)} = \underline{1}$.
- *If* $I_a = \langle k, \infty)$ *then we define* $\varphi_a^{(j)} = (\exists^k x)\kappa_a$.
- *If* $I_a = \langle 0, k)$ *then we define* $\varphi_a^{(j)} = \neg(\exists^k x)\kappa_a$.
- *If* $I_a = \langle k, l)$ *then we define* $\varphi_a^{(j)} = (\exists^k x)\kappa_a \wedge \neg(\exists^l x)\kappa_a$.
- *We analogously define* $\varphi_b^{(j)}$, $\varphi_c^{(j)}$ *and* $\varphi_d^{(j)}$ *using* I_b, I_c *and* I_d *respectively.*

The formula ψ_j *is created such that for each model* $\mathcal{M}$ *of the calculus* $\mathcal{P}^*$ *it is* $||\psi_j||_{\mathcal{M}} = 1$ *if and only if* $T_{\mathcal{M}} \in I_j$. *Let us define*

$$\Phi = \bigvee_{j=1}^{K} \psi_j \, .$$

We show that the formula $(\sim x)(P_1(x), P_2(x))$ *is logically equivalent to* Φ *in the calculus* $\mathcal{P}^*$.

Let $||(\sim x)(P_1(x), P_2(x))||_{\mathcal{M}} = 1$. *It means that* $Asf_{\sim}(T_{\mathcal{M}}) = 1$, *we suppose* $Asf_{\sim}(a,b,c,d) = 1$ *iff* $\langle a,b,c,d\rangle \in \bigcup_{j=1}^{K} I_j$ *and thus it is* $T_{\mathcal{M}} \in \bigcup_{j=1}^{K} I_j$. *It implies that there is* $p \in \{1,\ldots,K\}$ *such that* $T_{\mathcal{M}} \in I_p$ *and therefore* $||\psi_p||_{\mathcal{M}} = 1$ *and also* $||\Phi||_{\mathcal{M}} = 1$.

Let $||\Phi||_{\mathcal{M}} = 1$. *Thus there is* $p \in \{1,\ldots,K\}$ *such that* $||\psi_p||_{\mathcal{M}} = 1$ *that implies* $T_{\mathcal{M}} \in I_p$ *and also* $T_{\mathcal{M}} \in \bigcup_{j=1}^{K} I_j$. *According to the supposition it means* $Asf_{\sim}(T_{\mathcal{M}}) = 1$ *and thus* $||(\sim x)(P_1(x), P_2(x))||_{\mathcal{M}} = 1$.

The quantifier $\exists^k$ *is for each* $k > 0$ *definable in the extension of* $\mathcal{P}^=$ *by adding* $\exists^k$ *(see lemma 3). It means that there is a formula* Φ^* *of the calculus* $\mathcal{P}^=$ *that is logically equivalent to the formula* Φ. *It also means that* Φ^* *is logically equivalent to the formula* $(\sim x)(P_1(x), P_2(x))$.

It finishes the proof.

Remark 4. The theorem just proved concerns quantifiers of the type $\langle 1, 1\rangle$. Let us remark that it can be generalised for quantifiers of general type $\langle 1^k\rangle$. The generalised criterion uses intervals in $\mathcal{N}^{2^k}$ instead of intervals in $\mathcal{N}^4$.

Remark 5. The theorem 5 can be used to prove that the 4ft-quantifier $\Rightarrow_{p,Base}$ of founded implication is not classically definable. The detailed proof does not fall with the scope of this chapter. It is done in details in [4]. It is also proved in [4] that most of 4ft-quantifiers implemented in the 4ft-Miner procedure [8] are not classically definable. We can suppose that the same is true for all the 4ft-quantifiers implemented in the 4ft-Miner procedure.

Acknowledgement: The work described here has been supported by the grant 201/05/0325 of the Czech Science Foundation and by the project IGA 17/04 of University of Economics, Prague.

References

1. Aggraval R, et al (1996) Fast Discovery of Association Rules. In: Fayyad UM, et al (eds). Advances in Knowledge Discovery and Data Mining. AAAI Press, Menlo Park California
2. Hájek P, Havránek T (1978) Mechanising Hypothesis Formation - Mathematical Foundations for a General Theory. Springer, Berlin Heidelberg New York (see also `http://www.cs.cas.cz/~hajek/guhabook/`)
3. Hájek P, Havránek T, Chytil M (1983) GUHA Method. Academia, Prague (in Czech)
4. Rauch J (1986) Logical Foundations of Hypothesis Formation from Databases. PhD Thesis, Mathematical Institute of the Czechoslovak Academy of Sciences, Prague (in Czech)
5. Rauch J (1997) Logical Calculi for Knowledge Discovery in Databases. In: Zytkow J, Komorowski J (eds) Principles of Data Mining and Knowledge Discovery. Springer, Berlin Heidelberg New York
6. Rauch J(1998) Contribution to Logical Foundations of KDD. Assoc. Prof. Thesis, Faculty of Informatics and Statistics, University of Economics, Prague (in Czech)
7. Rauch J (2005) Logic of Association Rules. Applied Intelligence 22, 9-28.
8. Rauch J, Šimůnek M (2005) An Alternative Approach to Mining Association Rules. In: Lin T Y, Ohsuga S, Liau C J, Tsumoto S, and Hu X (eds) Foundations of Data Mining and Knowledge Discovery, Springer-Verlag, 2005, pp. 219 - 238.
9. Šimůnek M (2003) Academic KDD Project LISp-Miner. In Abraham A et al (eds) Advances in Soft Computing - Intelligent Systems Design and Applications, Springer, Berlin Heidelberg New York
10. Tharp L H (1973) The characterisation of monadic logic. Journal of Symbolic Logic 38: 481–488

A Measurement-Theoretic Foundation of Rule Interestingness Evaluation

Yiyu Yao, Yaohua Chen, and Xuedong Yang

Department of Computer Science, University of Regina
Regina, Saskatchewan, Canada S4S 0A2
E-mail: {yyao, chen115y, yang}@cs.uregina.ca

Summary. Many measures have been proposed and studied extensively in data mining for evaluating the interestingness (or usefulness) of discovered rules. They are usually defined based on structural characteristics or statistical information about the rules. The meaningfulness of each measure was interpreted based either on intuitive arguments or mathematical properties. There does not exist a framework in which one is able to represent the user judgment explicitly, precisely, and formally. Since the usefulness of discovered rules must be eventually judged by users, a framework that takes user preference or judgment into consideration will be very valuable. The objective of this paper is to propose such a framework based on the notion of user preference. The results are useful in establishing a measurement-theoretic foundation of rule interestingness evaluation.

Key words: KDD, Rule Interestingness, Evaluation, Measurement Theory, User Preference

1 Introduction

With rapidly increasing capabilities of accessing, collecting, and storing data, knowledge discovery in databases (KDD) has emerged as a new area of research in computer science. The objective of KDD systems is to extract implicitly hidden, previously unknown, and potentially useful information and knowledge from databases [7, 10]. A core task of the KDD field, called data mining, is the application of specific machine learning algorithms, knowledge representations, statistical methods, and other data analysis techniques for knowledge extraction and abstraction. The discovered knowledge is often expressed in terms of a set of rules. They represent relationships, such as correlation, association, and causation, among concepts [48]. For example, the well-known association rules deal with relationships among sale items [1, 3]. Some fundamental tasks of data mining process in KDD are the discovery, interpretation, and evaluation of those relationships.

There are many types of rules embedded in a large database [46]. Furthermore, the number of rules is typically huge and only a small portion of rules is actually useful [36]. An important problem in data mining is the evaluation of the *interestingness* of the mined rules and filtering out useless rules [36]. Many measures have been proposed and studied to quantify the interestingness (or usefulness) of rules [11, 15, 35, 36, 48]. The results lead to an in-depth understanding of different aspects of rules. It is recognized that each measure reflects a certain characteristic of rules. In addition, many studies investigate and compare rule interestingness measures based on intuitive arguments or some mathematical properties. There is a lack of a well-accepted framework for examining the issues of rule interestingness in a systematic and unified manner.

We argue that measurement theory can be used to establish a solid foundation for rule interestingness evaluation. The theory provides necessary concepts and methodologies for the representation, classification, characterization, and interpretation of user judgment of the usefulness of rules. A measure of rule interestingness is viewed as a quantitative representation of user judgment. The meaningfulness of a measure is determined by the users' perception of the usefulness of rules.

Existing studies of rule interestingness evaluation can be viewed as measure-centered approaches. Measures are used as primitive notions to quantify the interestingness of rules. In contrast, our method is a user-centered approach. User judgment, expressed by a user preference relation on a set of rules, is used as a primitive notion to model rule interestingness. Measures are treated as a derived notion that provides a quantitative representation of user judgment.

The rest of this chapter is organized as follows. In the next section, we introduce the basic notion of evaluation and related issues. A critical review of existing measures of rules interestingness is presented, which reveals some limitations with existing studies. The third section provides motivations to the current study. The fourth section presents an overview of measurement theory. The fifth section applies measurement theory to build a framework of rule interestingness evaluation. Finally, the conclusion in the sixth section gives the summary of this chapter and discusses the future research.

2 Introduction of Evaluation

The discussion of the basic notion of evaluation is aimed at improving our understanding to the rule interestingness evaluation methodologies.

2.1 What is the Evaluation?

Many approaches define the term of evaluation based on specific views [13, 32], such as qualitative assessments and detailed statistical analysis. Suchman analyzes various definitions of evaluation with regard to the conceptual and operational approaches [38]. Simply speaking, the evaluation can be defined as the determination of the results, which are attained by some activity for accomplishing valued goals or objectives. The practice of evaluation can in fact be applied to many processes and research areas, such as the systematic collection of information of programs, personnel and products for reducing uncertainties, improving effectiveness, and making decisions [28].

Three basic components of an evaluation are summarized by Geisler [13]. The first component is the subjects for evaluation, which is what or whom needs to be evaluated. For the discovered rules, the subjects for evaluation are the properties or characteristics of each rule such as the association relationship between sale items and a type of business profit. The formulation of the subjects is always done in the first step of the evaluation procedure. The more the subjects are distinguished precisely, the better the framework and measurement can be produced.

The users who are interested in and willing to perform the evaluation are considered as the second component of an evaluation. Knowing who will participate in judging or who will benefit from the evaluation will help to clarify why the evaluation is performed and which measures or methods of evaluation should be used. Since the qualities of objects or events must be eventually judged by users, an evaluation needs to consider the user judgment. The users can be humans, organizations, or even systems. Different types of participants may have different purposes of conducting an evaluation and lead to different results of an evaluation.

The processes for evaluation and concrete measures are the evaluation's third component. Clarification of the criteria for the measures and designing the implementation for the evaluation are the key points in this component. One must consider the first two components, the subjects and the users, and then develop the processes and measurements of an evaluation. As Suchman points out, an evaluation can be constructed for different purposes, by different methods, with different criteria with respect to different users and subjects [38].

2.2 How to Do the Evaluation?

According to the definition of evaluation, the procedure of evaluation can be simply and generally described as follows [13, 38]:

- Identification of the subjects to be evaluated.
- Collection of data for the evaluation.
- Users analyze and measure those data to summarize their judgments based on the criteria and conduct the process of the evaluation for decision making.

The real procedures of an evaluation can be very complicated and might be iterative [38]. Furthermore, identifying and accurately measuring or quantifying the properties of subjects is very difficult to achieve. More often than not, an approximate approach can be accepted by general users. In the processes of an evaluation, it is very important that users determine an appropriate evaluation as the means of measuring.

2.3 Measurement of Evaluation

During the procedure of an evaluation, the measurement always plays a crucial role and the measurement theory provides the necessary concepts and methodologies for the evaluation. The subjects of measurement in measurement theory are about estimating the attributes or properties of empirical objects or events, such as weight, color, or intelligence [29]. The measurement can be performed by assigning numbers to the objects or events in order that the properties or attributes can be represented as numerical properties [17]. In other words, the properties of the quantity are able to faithfully reflect the properties of objects or events to be evaluated.

2.4 Subjectivity of Evaluation

From the discussion of the definition and procedure of evaluation, it is recognized that evaluation is an inherently subjective process [38]. The steps, methods, and measures used in an evaluation depend on the users who participate in the evaluation. The selection of the criteria and measures reflects the principles and underlying beliefs of the users [13].

Mackie argues that subjective values are commonly used when one evaluates objects, actions, or events [24]. Objectivity is only related to the objective measures and implementation of the measurement. People always judge the subjects with their subjective interests. Different people have different judgments on the same object, action, or event because they always stand on their own interests or standards of evaluations. In other words, the objective measurement is relative to personal standards of evaluations. In this regard, there are no absolutely objective evaluations, only relatively objective evaluations for human beings.

Nevertheless, these standards of evaluations can be derived from human being's subjective interests. In fact, the user preference is indeed realized as a very important issue for an evaluation to occur [13]. It can be described as the user's discrimination on two different objects rationally [23]. The users can simply describe their preference as "they act upon their interests and desires they have" [6]. In measurement and decision theories, user preferences are used to present the user judgments or user interests and can be viewed as the standards of an evaluation [12, 23, 29, 33]. The user preference or judgment should be considered in the process of an evaluation.

3 Rule Evaluation

As an active research area in data mining, rule evaluation has been considered by many authors from different perspectives. We present a critical review of the studies on rule evaluation in order to observe their difficulties. This leads to a new direction for future research.

3.1 A Model of Data Mining based on Granular Computing

In an information table, objects can be described by the conjunctions of attribute-value pairs [49]. The rows of the table represent the objects, the columns denote a set of attributes, and each cell is the value of an object on an attribute. In the model of granular computing, the objects in an information table are viewed as the universe and the information table can be expressed by a quadruple [45, 49]:

$$T = (U, At, \{V_a \mid a \in At\}, \{I_a \mid a \in At\}), \tag{1}$$

where U is a finite nonempty set of objects, At is a finite nonempty set of attributes, V_a is nonempty set of values for $a \in At$, and I_a is a function to map from U to V_a, that is, $I_a : U \rightarrow V_a$.

With respect to the notion of tables, we define a decision logic language [31]. In this language, an atomic formula is a pair (a, v), where $a \in At$ and $v \in V_a$. If ϕ and ψ are formulas, then $\neg\phi$, $\phi \wedge \psi$, $\phi \vee \psi$, $\phi \rightarrow \psi$, and $\phi \equiv \psi$ are also formulas. The set of objects that satisfy a formula ϕ are denoted by $m(\phi)$. Thus, given an atomic formula (a, v), the corresponding set of objects can be $m(a, v) = \{x \in U \mid I_a(x) = v\}$. The following properties hold:

(1) $m(\neg\phi) = \neg m(\phi)$,
(2) $m(\phi \wedge \psi) = m(\phi) \cap m(\psi)$,
(3) $m(\phi \vee \psi) = m(\phi) \cup m(\psi)$,
(4) $m(\phi \rightarrow \psi) = \neg m(\phi) \cup m(\psi)$,
(5) $m(\phi \equiv \psi) = (m(\phi) \cap m(\psi)) \cup (\neg m(\phi) \cap \neg m(\psi))$.

The formula ϕ can be viewed as the description of the set of objects $m(\phi)$.

In formal concept analysis, every concept consists of the intention and the extension [41, 42]. A set of objects is referred to as the extension, and the corresponding set of attributes as the intention of a concept. Therefore, a formula ϕ can represent the intention of a concept and a subset of objects $m(\phi)$ can be the extension of the concept. The pair $(\phi, m(\phi))$ is denoted as a concept.

One of the important functions of data mining of KDD is to find the strong relationships between concepts [48]. A rule can be represented as $\phi \Rightarrow \psi$, where ϕ and ψ are intensions of two concepts [45]. The symbol $\Rightarrow$ in the rules are interpreted based on the types of knowledge and rules can be classified according to the interpretations of $\Rightarrow$. In other words, different kinds of rules represent different types of knowledge extracted from a large database. Furthermore, based on the extensions $m(\phi)$, $m(\psi)$, and $m(\phi \wedge \psi)$, various quantitative measures can be used for the rules evaluation. A systematic analysis of quantitative measures associated with rules is given by Yao and Zhong [48].

3.2 A Critical Review of Existing Studies

Studies related to rule evaluation can be divided into two classes. One class, the majority of studies, deals with the applications of quantitative measures to reduce the size of search space of rules in the mining process, to filter out mined but non-useful rules, or to evaluate the effectiveness of a data mining system. The other class, only a small portion of studies, is devoted solely to the investigations of rule evaluation on its own. We summarize the main results from the following different points of views.

The roles of rule evaluation

It is generally accepted that KDD is an interactive and iterative process consisting of many phases [7, 14, 22, 26, 37, 53]. Fayyad *et al.* presented a KDD process consisting of the following steps: developing and understanding of the application domain, creating a target data set, data cleaning and preprocessing, data reduction and projection, choosing the data mining task, choosing the data mining algorithm(s), data mining, interpreting mined patterns, and consolidating, and acting on, the discovered knowledge [7, 8]. Rule evaluation plays different roles in different phases of the KDD process.

From the existing studies, one can observe that rule evaluation plays at least three different types of roles. In the data mining phase, quantitative measures can be used to reduce the size of search space. An example is the use of the well known *support* measure, which reduces the number of item sets need to be examined [1]. In the phase of interpreting mined patterns, rule evaluation plays a role in selecting the useful or interesting rules from the set of discovered rules [35, 36]. For example, the *confidence* measure of association rules is used to select only strongly associated item sets [1]. In fact, many measures associated with rules are used for such a purpose [48]. Finally, in the phase of consolidating and acting on discovered knowledge, rule evaluation can be used to quantify the usefulness and effectiveness of discovered rules. Many measures such as cost, classification error, and classification accuracy play such a role [11]. Rule evaluation in this regard is related to the evaluation of a data mining system.

The process-based approach captures the procedural aspects of KDD. Recently, Yao proposed a conceptual formulation of KDD in a three-layered framework [46]. They are the philosophy level, technique level, and application level. The philosophy level focuses on formal characterization, description, representation, and classification of knowledge embedded in a database without reference to mining algorithms. It provides answers to the question: What is the knowledge embedded in a database? The technique level concentrates on data mining algorithms without reference to specific applications. It provides answers to the question: How to discover knowledge embedded in a database? The application level focuses on the use of discovered knowledge with respect to particular domains. It provides answers to the question: How to apply the discovered knowledge?

With respect to the three-layered framework, rule evaluation plays the similar roles. In the philosophy level, quantitative measures can be used to characterize and classify different types of rules. In the technique level, measures can be used to reduce search space. In the application level, measures can be used to quantify the utility, profit, effectiveness, or actionability of discovered rules.

Subjective vs. objective measures

Silberschatz and Tuzhilin suggested that measures can be classified into two categories consisting of objective measures and subjective measures [35]. Objective measures depend only on the structure of rules and the underlying data used in the discovery process. Subjective measures also depend on the user who examines the rules [35]. In comparison, there are limited studies on subjective measures. For example, Silberschatz and Tuzhilin proposed a subjective measure of rule interestingness based on the notion of unexpectedness and in terms of a user belief system [35, 36].

Statistical, structural vs. semantic measures

Many measures, such as support, confidence, independence, classification error, etc., are defined based on statistical characteristics of rules. A systematic analysis of such measures can be performed by using a 2×2 contingency table induced by a rule [48, 50].

The structural characteristics of rules have been considered in many measures. For example, information, such as the size of disjunct (rule), attribute interestingness, the asymmetry of classification rules, etc., can be used [11]. These measures reflect the simplicity, easiness of understanding, or applicability of rules.

Although statistical and structural information provides an effective indicator of the potential effectiveness of a rule, its usefulness is limited. One needs to consider the semantic aspect of rules or explanations of rules [52]. Semantics centered approaches are application and user dependent. In addition to statistical information, one incorporates other domain specific knowledge such as user interest, utility, value, profit, actionability, and so on. Two examples of semantic-based approaches are discussed below.

Profit-based or utility-based mining is one example of a special kind of constraint-based mining, taking into account both statistical significance and profit significance [18, 39, 40]. Doyle discusses the importance and usefulness of the basic notions of economic rationality, such as utility functions, and suggests that economic rationality should play as large a role as logical rationality in rule reasoning [4]. For instance, one would not be interested in a frequent association that does not generate enough profit. The profit-based measures allow the user to prune the rules with high statistical significance, but low profit or high risk. For example, Barber and Hamilton propose the notion of share measure which considers the contribution, in terms of profit, of an item in an item set [2].

Actionable rule mining is another example of dealing with profit-driven actions required by business decision making [19, 20, 21]. A rule is referred to as actionable if the user can do something about it. For example, a user may be able to change the non-desirable/non-profitable patterns to desirable/profitable patterns.

Measures defined by statistical and structural information may be viewed as objective measures. They are user, application and domain independent. For example, a pattern is deemed interesting if it has certain statistical properties. These measures may be useful in the philosophical level of the three-layered framework. Different classes of rules can be identified based on statistical characteristics, such as peculiarity rules (low support and high confidence), exception rules (low support and high confidence, but complement to other high support and high confidence rules), and outlier patterns (far away from the statistical mean) [51].

Semantic-based measures involve the user interpretation of domain specific notions such as profit and actionability. They may be viewed as subjective measures. Such measures are useful in the application level of the three-layered framework. The usefulness of rules are measured and interpreted based on domain specific notions.

Single rule vs. multiple rules

Rule evaluation can also be divided into measures for a single rule and measures for a set of rules. Furthermore, a measure for a set of rules can be obtained from measures for single rules. For example, conditional probability can be used as a measure for a single classification rule, conditional entropy can be used as a measure for a set of classification rules [47]. The latter is defined in terms of the former.

Measures for multiple rules concentrate on properties of a set of rules. They are normally expressed as some kind of average. Hilderman and Hamilton examined many measures for multiple rules known as the summarization of a database [15].

Axiomatic approaches

Instead of focusing on rules, the axiomatic approaches study the required properties of quantitative measures.

Suppose that the discovered knowledge is represented in terms of rules of the form, $E \Rightarrow H$, and is paraphrased as "if E then H". Piatetsky-Shapiro [30] suggests that a quantitative measure of rule $E \Rightarrow H$ may be computed as a function of $support(E)$, $support(H)$, $support(E \wedge H)$, rule complexity, and possibly other parameters such as the mutual distribution of E and H or the size of E and H. For the evaluation of rules, Piatetsky-Shapiro [30] introduces three axioms. Major and Mangano [25] add the fourth axioms. Klösgen [16] studies a special class of measures that are characterized by two quantities, $confidence(E \Rightarrow H)$ and $support(E)$. The $support(H \wedge E)$ is obtained by $confidence(E \Rightarrow H)support(E)$. Suppose $support(E, H)$ is a measure associated with rule $E \Rightarrow H$. The version of the four axioms given by Klösgen [16] is:

(i). $Q(E, H) = 0$ if E and H are statistically independent,
(ii). $Q(E, H)$ monotonically increases in $confidence(E \Rightarrow H)$ for a fixed $support(E)$,
(iii). $Q(E, H)$ monotonically decreases in $support(E)$ for a fixed $support(E \wedge H)$,
(iv). $Q(E, H)$ monotonically increases in $support(E)$ for a fixed $confidence(E \Rightarrow H) > support(H)$.

The axiomatic approach is widely used in many other disciplines.

An axiomatic study of measures for multiple rules has been given by Hilderman and Hamilton [15].

3.3 A Direction for Future Research

From the previous discussions, one can make several useful observations. Studies on rule evaluations can be classified in several ways. Each of them provides a different view. Most studies on rule evaluation concentrate on specific measures and each measure reflects certain aspects of rules. Quantitative measures are typically interpreted by using intuitively defined notions, such as novelty, usefulness, and non-trivialness, unexpectedness, and so on. Therefore, there is a need for a unified framework that enables us to define, interpret, and compare different measures.

A very interesting research direction for rule evaluation is the study of its foundations. Several issues should be considered. One needs to link the meaningfulness of a measure to its usage. In theory, it may not be meaningful to argue which measure is better without reference to its roles and usage. It is also necessary to build a framework in which various notions of rule evaluation can be formally and precisely defined and interpreted. The study of rule evaluation needs to be connected to the study of foundations of data mining.

4 Overview of Measurement Theory

For completeness, we give a brief review of the basic notions of measurement theory that are pertinent to our discussion. The contents of this section draw heavily from Krantz *et al.* [17], Roberts [33] and French [12].

When measuring an attribute of a class of objects or events, we may associate numbers with the individual objects so that the properties of the attribute are faithfully represented as numerical properties [17, 29]. The properties are usually described by certain qualitative relations and operations. Consider an example discussed by Krantz *et al.* [17]. Suppose we are measuring the lengths of a set U of straight, rigid rods. One important property of length can be described by a qualitative relation "longer than". Such a relation can be obtained by first placing two rods, say a and b, side by side and adjusting them so that they coincide at one end, and then observing whether a extends beyond b at the other end. We say that a is longer than b, denoted by $a \succ b$, if a extends beyond b. In this case, we would like to assign numbers $f(a)$ and $f(b)$ with $f(a) > f(b)$ to reflect the results of the comparison. That is, we require the numbers assigned to the individual rods satisfy the condition: for all $a, b \in U$,

$$a \succ b \Longleftrightarrow f(a) > f(b). \tag{2}$$

In other words, the qualitative relation "longer than", $\succ$, in the empirical system is faithfully reflected by the quantitative relation "greater than", $>$, in the numerical system. Another property of length is that we can concatenate two or more rods by putting them end to end in a straight line, and compare the length of this set with that of another set. The concatenation of a and b can be written as $a \circ b$. In order to reflect such a property, we require the numbers assigned to the individual rods be additive with respect to concatenation. That is, in addition to condition (2), the numbers assigned must also satisfy the following condition: for all $a, b \in U$,

$$f(a \circ b) = f(a) + f(b). \tag{3}$$

Thus, concatenation $\circ$ in the empirical system is preserved by addition $+$ in the numerical system. Many other properties of length comparison and of concatenation of rods can be similarly formulated. For instance, $\succ$ should be transitive, and $\circ$ should be commutative and associative. The numbers assigned must reflect these properties as well. This simple example clearly illustrates the basic ideas of measurement theory, which is primarily concerned with choosing consistent quantitative representations of qualitative systems.

Based on the description of the basic notions of measurement theory in the above example, some basic concepts and notations are introduced and the formal definitions and formulations of the theory are reviewed.

4.1 Relational Systems

Suppose U is a set. The Cartesian product of U with U, denoted $U \times U$, is a set of all ordered pairs (a, b) so that $a, b \in U$. A binary relation R on a set U, simply denote (U, R), is a subset of the Cartesian product $U \times U$. For $a, b \in U$, if a is related to b under R, we write aRb or $(a, b) \in R$. For example, consider the binary relation "less than" ($<$) relation on real numbers. An ordered pair (a, b) is in the binary relation if and only if $a < b$. Similarly, "greater than" and "equals" also can be defined as the binary relations on real numbers.

With the set U, a function $f : U \rightarrow U$ can in fact also be thought of as a binary relation (U, R). A function $f : U^n \rightarrow U$ can be an (n+1)-ary relation (U, R). The functions from U into U is called binary operations, or just operations for short. For example, for addition ($+$), given a pair of real numbers a and b, there exists a third real number c so that $a + b = c$.

A *relational system* (*structure*) is a set of one or more relations (operations) on an arbitrary set. That is, a relational system is an ordered $(p + q + 1)$-tuple $\mathcal{A} = (U, R_1, \ldots, R_p, \circ_1, \ldots, \circ_q)$, where U is a set, $R_1, \ldots, R_p$ are (not necessarily binary) relations on U, and $\circ_1, \ldots, \circ_q$ are binary operations on U. If the binary operations are considered as a special type of relations, a relational system can be simply denoted as a $(p+1)$-tuple $\mathcal{A} = (U, R_1, \ldots, R_p)$. For convenience, we separate the operations from other relations.

If U is the set (or a subset) of real numbers, such a relational system is called as a numerical relational systems. As illustrated by the example of rigid rods, for measuring the property of length, we can start with an observed or empirical system $\mathcal{A}$ and seek a mapping into a numerical relational system $\mathcal{B}$ which preserves or faithfully reflects all the properties of the relations and operations in $\mathcal{A}$.

4.2 Axioms of the Empirical System

Based on the definitions of the relations and operations in the relation systems, we should describe the valid use or properties of these relations and operations in order to find the appropriate corresponding numerical systems. Many properties are common to well-defined relations. The consistency properties to be preserved are known as *axioms*. For example, if U is a set of real numbers and R is the relation of "equality" on U, R is reflexive, symmetric, and transitive. However, if U is the set of people in the real world and R is the relation "father of" on U, R is irreflexive, asymmetric, and nontransitive.

The set of axioms characterizing the relations in an empirical system should be complete so that every consistency property for the relations that is required is either in the list or deducible from those in the list [12, 17, 33].

4.3 Homomorphism of Relational Systems

Consider two relational systems, an empirical (a qualitative) system $\mathcal{A} = (U, R_1, \ldots, R_p, \circ_1, \ldots, \circ_q)$, and a numerical system $\mathcal{B} = (V, R'_1, \ldots, R'_p, \circ'_1, \ldots,$

$\circ'_q$). A function $f : U \rightarrow V$ is called a *homomorphism* from $\mathcal{A}$ to $\mathcal{B}$ if, for all $a_1, \ldots, a_{r_i} \in \mathcal{A}$,

$$R_i(a_1, \ldots, a_{r_i}) \Longleftrightarrow R'_i(f(a_1), \ldots, f(a_{r_i})), \qquad i = 1, \ldots, p,$$

and for all $a, b \in \mathcal{A}$,

$$f(a \circ_j b) = f(a) \circ'_j f(b), \qquad j = 1, \ldots, q.$$

The empirical system for the earlier example is denoted by $(U, \succ, \circ)$, where U is the set of rigid rods and their finite concatenations, $\succ$ is the binary relation "longer than" and $\circ$ is the concatenation operation. The numerical relation system is $(\Re, >, +)$, where $\Re$ is the set of real numbers, $>$ is the usual "greater than" relation and $+$ is the arithmetic operation of addition. The numerical assignment $f(\cdot)$ is a homomorphism which maps U into $\Re$, $\succ$ into $>$, and $\circ$ into $+$ in such a way that $>$ preserves the properties of $\succ$, and $+$ preserves the properties of $\circ$ as stated by conditions (2) and (3).

In general, a measurement has been performed if a homomorphism can be assigned from an empirical (observed) relational system $\mathcal{A}$ to a numerical relational system $\mathcal{B}$. The homomorphism is said to give a *representation*, and the triple $(\mathcal{A}, \mathcal{B}, f)$ of the empirical relational system $\mathcal{A}$, the numerical relational system $\mathcal{B}$, and the function f is called a *scale* or *measure*. Sometimes, a homomorphism from an empirical relational system into the set of real numbers is referred alone as a scale (measure).

With given numerical scales (measures), new scales or measures defined in terms of the old ones are called derived scales or derived measures. For example, density d can be defined in terms of mass m and volume v as $d = m/v$. The density d is the derived scale (measure), and the mass m and volume v are called as primitive scales (measures).

4.4 Procedure of Measurement

Generally, there are three fundamental steps in measurement theory [12, 17, 33]. Suppose we are seeking a quantitative representation of an empirical system. The first step, naturally, is to define the relations and operations to be represented. The axioms of the empirical system are determined. The next step is to choose a numerical system. The final step is to construct an appropriate homomorphism. A *representation theorem* asserts that if a given empirical system satisfies certain axioms, then a homomorphism into the chosen numerical system can be constructed.

The next question concerns the uniqueness of the scale. A uniqueness theorem is generally obtained by identifying a set of *admissible transformations*. If $f(\cdot)$ is a scale representing an empirical system and if $\lambda(\cdot)$ is an admissible transformation, then $\lambda(f(\cdot))$ is also a scale representing the same empirical system.

If the truth (falsity) of a numerical statement involving a scale or measure remains unchanged under all admissible transformations, we say that it is quantitatively meaningful. A numerical statement may be quantitatively meaningful, but qualitatively meaningless. In order for a quantitative statement to be qualitatively meaningful, it must reflect or model a meaningful statement in the empirical system.

Examples of the discussed view of measurement theory include the axiomatization of probability and expected utility theory [27, 34], the axiomatization of possibility functions [5] and the axiomatization of belief functions [43].

5 Application of Measurement Theory to Rule Evaluation

Given a database, in theory, there exists a set of rules embedded in it, independent of whether one has an algorithm to mine them. For a particular application, the user may only be interested in a certain type of rules. Therefore, the key issue of rules evaluation is in fact the measurement of rules' usefulness or interestingness expressed by a user preference relation. According to the procedure of measurement, for rule evaluation, we follow the three steps to seek a quantitative representation of an empirical system.

5.1 User Preference Relations

In the measurement theory, the user judgment or user preference can be modeled as a kind of binary relation, called user preference relation [33]. If the user prefers a rule to another rule, then we can say that one rule is more useful or interesting than the other rule.

Assume we are given a set of discovered rules. Let $\mathbf{R}$ be the set of rules. Since the usefulness or interestingness of rules should be finally judged by users, we focus on user preference relation as a binary relation on the set of discovered rules. Given two rules $r', r'' \in \mathbf{R}$, if a user judges r' to be more useful than r'', we say that the user prefers r' to r'' and denote it by $r' \succ r''$. That is,

$$r' \succ r'' \Leftrightarrow \text{the user prefers } r' \text{ to } r''. \tag{4}$$

In the absence of strict preference, i.e., if both $\neg(r' \succ r'')$ and $\neg(r' \succ r'')$ hold, we say that r' and r'' are indifferent. An indifference relation $\sim$ on $\mathbf{R}$ can be defined as follows:

$$r' \sim r'' \Leftrightarrow (\neg(r' \succ r''), \neg(r'' \succ r')). \tag{5}$$

The empirical relational system can be defined as following:

Definition 1. *Given a set of discovered rules* $\mathbf{R}$ *and user preference* $\succ$, *the pair* $(\mathbf{R}, \succ)$ *is called the (empirical) relational system of the set of discovered rules.*

The user judgment on rules can be formally described by a user preference relation $\succ$ on $\mathbf{R}$. In our formulation, we treat the user preference relation $\succ$ as a primitive notion. At this stage, we will not attempt to define and interpret a user preference relation using other notions.

5.2 Axioms of User Preference Relations

The next issue is to identify the desired properties of a preference relation so that it can be measured quantitatively. Such consistency properties are known as axioms. We consider the following two axioms:

- **Asymmetry**:
  $r' \succ r'' \Rightarrow \neg(r'' \succ r')$,
- **Negative transitivity**:
  $(\neg(r' \succ r''), \neg(r'' \succ r''')) \Rightarrow \neg(r' \succ r''')$.

The first axiom requires that a user cannot prefer r' to r'' and at the same time prefers r'' to r'. In other words, the result of a user preference on two different discovered rules is not contradictive. In fact, this axiom ensures the user preference or user judgment is rational. The second is the negative transitivity axiom, which means that if a user does not prefer r' to r'', nor r'' to r''', the user should not prefer r' to r'''.

If a preference relation is a weak order, it is transitive, i.e., $r' \succ r''$ and $r'' \succ r'''$ imply $r' \succ r'''$. It seems reasonable that a user preference relation should satisfy these two axioms.

A few additional properties of a weak order are summarized in the following lemma [33].

Lemma 1. *Suppose a preference relation $\succ$ on a finite set of rules $\mathbf{R}$ is a weak order. Then,*

- *the relation $\sim$ is an equivalence relation,*
- *exactly one of $r' \succ r''$, $r'' \succ r'$ and $r' \sim r''$ holds for every $r', r'' \in \mathbf{R}$.*
- *the relation $\succ'$ on $\mathbf{R}/\sim$ defined by $X \succ' Y \Leftrightarrow \exists r', r''(r' \succ r'', r' \in X, r'' \in Y)$, is a linear order, where X and Y are elements of $\mathbf{R}/\sim$.*

A linear order is a weak order in which any two different elements are comparable. This lemma implies that if $\succ$ is a weak order, the indifference relation $\sim$ divides the set of rules into disjoint subsets.

5.3 Homomorphism based on Real-valued Function

In the measurement-theoretic terminology, the requirement of a weak order indeed suggests the use of an ordinal scale (homomorphism) for the measurement of user preference of rules, as shown by the following representation theorem [33]. That is, we can find a real-valued function u as a measure.

Theorem 1. *Suppose* $\mathbf{R}$ *is a finite non-empty set of rules and* $\succ$ *is a relation on* $\mathbf{R}$. *There exists a real-valued function* $u : \mathbf{R} \longrightarrow \Re$ *satisfying the condition,*

$$r' \succ r'' \Leftrightarrow u(r') > u(r'') \tag{6}$$

if and only if $\succ$ *is a weak order. Moreover,* u *is defined up to a strictly monotonic increasing transformation.*

The numbers $u(r'), u(r''), \ldots$ as ordered by $>$ reflect the order of $r', r'', \ldots$ under $\succ$. The function u is referred to as an order-preserving utility function. It quantifies a user preference relation and provides a measurement of user judgments. According to Theorem 1, the axioms of a weak order are the conditions which allow the measurement. Thus, to see if we can measure a user's preference to the extent of producing an ordinal utility function, we just check if this preference satisfies the conditions of asymmetry and negative transitivity. A rational user's judgment must allow the measurement in terms of a quantitative utility function. On the other hand, another interpretation treats the axioms as testable conditions. Whether can measure the user judgments depends on whether the user preference relation is a weak order [44].

5.4 Ordinal Measurement of Rules Interestingness

In the above discussion, only the asymmetry and negative transitivity axioms must be satisfied. This implies that the ordinal scale is used for the measurement of user preference. For the ordinal scale, it is meaningful to examine or compare the order induced by the utility function.

The main ideas can be illustrated by a simple example. Suppose a user preference relation $\succ$ on a set of rules $\mathbf{R} = \{r_1, r_2, r_3, r_4\}$ is specified by the following weak order:

$$r_3 \succ r_1, \quad r_4 \succ r_1, \quad r_3 \succ r_2, \quad r_4 \succ r_2, \quad r_4 \succ r_3.$$

This relation $\succ$ satisfies the asymmetry and negative transitivity conditions (axioms). We can find three equivalence classes $\{r_4\}$, $\{r_3\}$, and $\{r_1, r_2\}$. In turn, they can be arranged as three levels:

$$\{r_4\} \succ' \{r_3\} \succ' \{r_1, r_2\}.$$

Obviously, we can defined the utility function u_1 as follows:

$$u_1(r_1) = 0, \quad u_1(r_2) = 0, \quad u_1(r_3) = 1, \quad u_1(r_4) = 2.$$

Another utility function u_2 also may also be used:

$$u_2(r_1) = 5, \quad u_2(r_2) = 5, \quad u_2(r_3) = 6, \quad u_2(r_4) = 7.$$

The two utility functions preserve the same order for any pair of rules, although they use different values.

Based on the formal model of measurement on rules interestingness, we can study different types of user preference relations. In order to do so, we need to impose more axioms on the user preference relation. The axioms on user preference relations can be easily interpreted and related to domain specific notions.

Luce and Suppes discuss the user preference and the closely related areas of utility and subjective probability from the mathematical psychology point of view [23]. The utility is defined as a type of property of any object, whereby it tends to produce benefit, advantage, pleasure, good, and happiness, or to prevent the happening of mischief, pain, evil, or unhappiness. In other words, utility is a type of subjective measure, not objective measure. The utility of an item depends on the user preference and differs among the individuals. In the theory of decision making, utility is viewed as essential elements of a user preference on a set of decision choices or candidates [9, 12].

6 Conclusion

A critical review of rule evaluation suggests that we can study the topic from different points of views. Each view leads to different perspectives and different issues. It is recognized that there is a need for a unified framework for rule evaluation, in which various notions can be defined and interpreted formally and precisely.

Measurement theory is used to establish a solid foundation for rule evaluation. Fundamental issues are discussed based on the user preference of rules. Conditions on a user preference relation are discussed so that one can obtain a quantitative measure that reflects the user-preferred ordering of rules.

The proposed framework provides a solid basis for future research. We will investigate additional qualitative properties on the user preference relation. Furthermore, we will identify the qualitative properties on user preference relations that justify the use of many existing measures.

References

1. R. Agrawal, T. Imielinski and A. Swami, "Mining association rules between sets of items in massive databases", *Proceedings of the 1993 ACM SIGMOD International Conference on Management of Data*, 207-216, 1993.
2. B. Barber and H. Hamilton, "Extracting share frequent itemsets with infrequent subsets", *Data Mining and Knowledge Discovery*, **7**, 153-185, 2003.
3. M.S. Chen, J. Han and P.S. Yu, "Data mining: an overview from database perspective", *IEEE Transactions on Knowledge and Data Engineering*, **8**, 866-833, 1996.
4. J. Doyle, "Rationality and its role in reasoning", *Computational Intelligence*, **8**, 376-409, 1992.

5. D. Dubois, "Belief structures, possibility theory and decomposable confidence measures on finite sets", *Computers and Artificial Intelligence*, **5**, 403-416, 1986.
6. P.A. Facione, D. Scherer and T. Attig, *Values and Society: An Introduction to Ethics and Social Philosophy*, Prentice-Hall, Inc., New Jersey, 1978.
7. U.M. Fayyad, G. Piatetsky-Shapiro and P. Smyth, "From data mining to knowledge discovery: an overview", in: U.M. Fayyad, G. Piatetsky-Shapiro, P. Smyth and R. Uthurusamy (Eds.), *Advances in Knowledge Discovery and Data Mining*, AAAI/MIT Press, 1-34, 1996.
8. U.M. Fayyad, G. Piatetsky-Shapiro and P. Smyth, "From data mining to knowledge discovery in databases", *AI Magazine*, **17**, 37-54, 1996.
9. P.C. Fishburn, *Utility Theory for Decision Making*, John Wiley & Sons, New York, 1970.
10. W. Frawely, G. Piatetsky-Shapiro and C. Matheus, "Knowledge discovery in database: an overview", *Knowledge Discovery in Database*, AAAI/MIT Press, 1-27, 1991.
11. A.A. Freitas, "On rule interestingness measures", *Knowledge-Based Systems*, **12**, 309-315, 1999.
12. S. French, *Decision Theory: An Introduction to the Mathematics of Rationality*, Ellis Horwood Limited, Chichester, West Sussex, England, 1988.
13. E. Geisler, *Creating Value with Science and Technology*, Quorum Books, London, 2001.
14. J. Han and M. Kamber, *Data mining: Concept and Techniques*, Morgan Kaufmann, Palo Alto, CA, 2000.
15. R.J. Hilderman and H.J. Hamilton, *Knowledge Discovery and Measures of Interest*, Kluwer Academic Publishers, Boston, 2001.
16. W. Klösgen, "Explora: a multipattern and multistrategy discovery assistant", in: U.M. Fayyad, G. Piatetsky-Shapiro, P. Smyth and R. Uthurusamy (Eds.), *Advances in Knowledge Discovery and Data Mining*, AAAI/MIT Press, 249-271, 1996.
17. D.H. Krantz, R.D. Luce, P. Suppes and A. Tversky, *Foundations of Measurement*, Academic Press, New York, 1971.
18. T.Y. Lin, Y.Y. Yao and E. Louie, "Value added association rules", *Advances in Knowledge Discovery and Data Mining, Proceedings of 6th Pacific-Asia Conference (PAKDD'02)*, 328-333, 2002.
19. C. Ling, T. Chen, Q. Yang and J. Chen, "Mining optimal actions for profitable CRM", *Proceedings of the 2002 IEEE International Conference on Data Mining (ICDM'02)*, 767-770, 2002.
20. B. Liu, W. Hsu and S. Chen, "Using general impressions to analyze discovered classification rules", *Proceedings of the 3rd International Conference on Knowledge Discovery and Data Mining (KDD'97)*, 31-36, 1997.
21. B. Liu, W. Hsu and Y. Ma, "Identifying non-actionable association rules", *Proceedings of the 7th ACM SIGKDD International Conference on Knowledge Discovery and Data Mining*, 329-334, 2001.
22. C. Liu, N. Zhong and S. Ohsuga," A multi-agent based architecture for distributed KDD process", *Foundations of Intelligent Systems, Proceedings of 12th International Symposium (ISMIS'00)*, 591-600, 2000
23. R.D. Luce and P. Suppes, "Preference, utility, and subjective probability", in: R.D. Luce, R.R. Bush and E. Galanter (Eds.), *Handbook of Mathematical Psychology, John Wiley and Sons, Inc., New York*, 249-410, 1965.

24. J.L. Machie, *Ethics: Inventing Right and Wrong*, Penguin Books Ltd., Harmondsworth, 1977.
25. J. Major and J. Mangano, "Selecting among rules induced from a hurricane database", *Journal of Intelligent Information Systems*, **4**, 1995.
26. H. Mannila, "Methods and problems in data mining", *Database Theory, Proceedings of 6th International Conference (ICDT'97)*, 41-55, 1997.
27. J. Neumann and O. Morgenstern, *Theory of Games and Economic Behavior*, Princeton University Press, 1944.
28. M.Q. Patton, *Practical Evaluation*, Sage Publications, Newbury Park, 1982.
29. J. Pfanzagl, *Theory of Measurement*, John Wiley & Sons, New York, 1968.
30. G. Piatetsky-Shapiro, "Discovery, analysis, and presentation of strong rules", in: G. Piatetsky-Shapiro and W.J. Frawley (Eds.), *Knowledge Discovery in Databases*, AAAI/MIT Press, 229-238, 1991.
31. Z. Pawlak, *Rough Sets, Theoretical Aspects of Reasoning about Data*, Kluwer Academic Publishers, Dordrecht, 1991.
32. D. Reith, "Evaluation, a function of practice", in: J. Lishman, (Ed.), *Evaluation, 2nd Edition*, Kingsley Publishers, London, 23-39, 1988.
33. F. Roberts, *Measurement Theory*, Addison Wesley, Massachusetts, 1979.
34. L.J. Savage, *The Foundations of Statistics*, Dover, New York, 1972.
35. A. Silberschatz and A. Tuzhilin, "On subjective measures of interestingness in knowledge discovery", *Proceedings of the 1st International Conference on Knowledge Discovery and Data Mining (KDD'95)*, 275-281, 1995.
36. A. Silberschatz and A. Tuzhilin, "What makes patterns interesting in knowledge discovery systems", *IEEE Transactions on Knowledge and Data Engineering*, **8**, 970-974, 1996.
37. E. Simoudis, "Reality check for data mining", *IEEE Expert*, **11**, 26-33, 1996.
38. E.A. Suchman, *Evaluation Research*, Russell Sage Foundation, USA, 1967.
39. K. Wang and Y. He, "User-defined association mining", *Knowledge Discovery and Data Mining, Proceedings of 5th Pacific-Asia Conference (PAKDD'01)*, 387-399, 2001.
40. K. Wang, S. Zhou and J. Han, "Profit mining: from patterns to actions", *Advances in Database Technology, Proceedings of 8th International Conference on Extending Database Technology (EDBT'02)*, 70-87, 2002.
41. R. Wille, "Concept lattices and concept knowledge systems", *Computers Mathematics with Applications*, **23**, 493-515, 1992.
42. R. Wille, "Restructuring lattice theory: an approach based on hierarchies of concepts", in: Ivan Rival (Ed.), *Ordered sets*, Reidel, Dordecht-Boston, 445-470, 1982.
43. S.K.M. Wong, Y.Y. Yao, P. Bollmann and H.C. Bürger, "Axiomatization of qualitative belief structure", *IEEE Transaction on Systems, Man, and Cybernetics*, **21**, 726-734, 1991.
44. Y.Y. Yao, "Measuring retrieval performance based on user preference of documents", *Journal of the American Society for Information Science*, **46**, 133-145, 1995.
45. Y.Y. Yao, "Modeling data mining with granular computing", *Proceedings of the 25th Annual International Computer Software and Applications Conference*, 638-643, 2001.
46. Y.Y. Yao, "A step towards the foundations of data mining", *Data Mining and Knowledge Discovery: Theory, Tools, and Technology V, The International Society for Optical Engineering*, 254-263, 2003.

47. Y.Y. Yao, "Information-theoretic measures for knowledge discovery and data mining", in: Karmeshu (Ed.), *Entropy Measures, Maximum Entropy and Emerging Applications*, Springer, Berlin, 115-136, 2003.
48. Y.Y. Yao and N. Zhong, "An analysis of quantitative measures associated with rules", *Methodologies for Knowledge Discovery and Data Mining, Proceedings of the 3rd Pacific-Asia Conference on Knowledge Discovery and Data Mining (PAKDD'99)*, 479-488, 1999.
49. Y.Y. Yao. and N. Zhong, "Potential applications of granular computing in knowledge discovery and data mining", *Proceedings of World Multiconference on Systemics, Cybernetics and Informatics*, 573-580, 1999.
50. Y.Y. Yao and C.J. Liau, "A generalized decision logic language for granular computing", *FUZZ-IEEE on Computational Intelligence*, 1092-1097, 2002.
51. Y.Y. Yao, N. Zhong and M. Ohshima, "An analysis of Peculiarity oriented multidatabase mining", *IEEE Transactions on Knowledge and Data Engineering*, **15**, 952-960, 2003.
52. Y.Y. Yao, Y. Zhao and R.B. Maguire, "Explanation-oriented association mining using a combination of unsupervised and supervised learning algorithms", *Advances in Artificial Intelligence, Proceedings of the 16th Conference of the Canadian Society for Computational Studies of Intelligence (AI'03)*, 527-532, 2003.
53. N. Zhong, C. Liu, and S. Ohsuga, "Dynamically organizing KDD processes", *Journal of Pattern Recognition and Artificial Intelligence*, **15**, 451-473, 2001.

Statistical Independence as Linear Dependence in a Contingency Table

Shusaku Tsumoto

Department of Medical Informatics,
Shimane University, School of Medicine
89-1 Enya-cho Izumo 693-8501 Japan
tsumoto@computer.org

Summary. A contingency table summarizes the conditional frequencies of two attributes and shows how these two attributes are dependent on each other. Thus, this table is a fundamental tool for pattern discovery with conditional probabilities, such as rule discovery. In this paper, a contingency table is interpreted from the viewpoint of granular computing. The first important observation is that a contingency table compares two attributes with respect to the number of equivalence classes. The second important observation is that matrix algebra is a key point of analysis of this table. Especially, the degree of independence, rank plays a very important role in extracting a probabilistic model from a given contingency table.

Key words: Statistical Independence, Linear Independence, Contingency Table, Matrix Theory

1 Introduction

Independence (dependence) is a very important concept in data mining, especially for feature selection. In rough sets[4], if two attribute-value pairs, say $[c = 0]$ and $[d = 0]$ are independent, their supporting sets, denoted by C and D do not have a overlapping region ($C \cap D = \phi$), which means that one attribute independent to a given target concept may not appear in the classification rule for the concept. This idea is also frequently used in other rule discovery methods: let us consider deterministic rules, described as *if-then* rules, which can be viewed as classic propositions ($C \rightarrow D$). From the set-theoretical point of view, a set of examples supporting the conditional part of a deterministic rule, denoted by C, is a subset of a set whose examples belong to the consequence part, denoted by D. That is, the relation $C \subseteq D$ holds and deterministic rules are supported only by positive examples in a dataset[8].

When such a subset relation is not satisfied, indeterministic rules can be defined as if-then rules with probabilistic information[7]. From the set-theoretical point of view, C is not a subset, but closely overlapped with D. That is, the relations $C \cap D \neq \phi$ and $|C \cap D|/|C| \geq \delta$ will hold in this case.[1] Thus, probabilistic rules are supported by a large number of positive examples and a small number of negative examples.

On the other hand, in a probabilistic context, independence of two attributes means that one attribute (a_1) will not influence the occurrence of the other attribute (a_2), which is formulated as $p(a_2|a_1) = p(a_2)$.

Although independence is a very important concept, it has not been fully and formally investigated as a relation between two attributes.

In this paper, a contingency table of categorical attributes is focused on from the viewpoint of granular computing. The first important observation is that a contingency table compares two attributes with respect to information granularity. Since the number of values of a given categorical attribute corresponds to the number of equivalence classes, a given contingency table compares the characteristics of information granules: $n \times n$ table compares two attributes with the same granularity, while a $m \times n(m \geq n)$ table can be viewed as comparison of two partitions, which have m and n equivalence classes.

The second important observation is that matrix algebra is a key point of analysis of this table. A contingency table can be viewed as a matrix and several operations and ideas of matrix theory are introduced into the analysis of the contingency table. Especially, the degree of independence, rank plays a very important role in extracting a probabilistic model from a given contingency table. When the rank of the given table is equal to 1.0, one attribute in the table are statistically independent of the other attributes. When the rank is equal to n, which is the number of values of at least one attribute, then two attributes are dependent. Otherwise, the row or columns of contingency table are partially independent, which gives very interesting statistical models of these two attributes.

The paper is organized as follows: Section 2 discusses the characteristics of contingency tables. Section 3 shows the definitions of statistical measures used for contingency tables and their assumptions. Section 4 discusses the rank of the corresponding matrix of a contingency table when the given table is two way. Section 5 extends the above idea into a multi-way contingency table. Section 6 presents an approach to statistical evaluation of a rough set model. Finally, Section 7 concludes this paper.

This paper is a preliminary study on the concept of independence in statistical analysis, and the following discussions are very intuitive. Also, for simplicity of discussion, it is assumed that a contingency table gives a square matrix ($n$xn).

---

[1]The threshold δ is the degree of the closeness of overlapping sets, which will be given by domain experts. For more information, please refer to Section 3.

2 Contingency Table from Rough Sets

2.1 Accuracy and Coverage

In the subsequent sections, the following notations is adopted, which is introduced in [6]. Let U denote a nonempty, finite set called the universe and A denote a nonempty, finite set of attributes, i.e., $a : U \to V_a$ for $a \in A$, where V_a is called the domain of a, respectively. Then, a decision table is defined as an information system, $IS = (U, A \cup \{\mathcal{D}\})$, where $\{\mathcal{D}\}$ is a set of given decision attributes. The atomic formulas over $B \subseteq A \cup \{\mathcal{D}\}$ and V are expressions of the form $[a = v]$, called descriptors over B, where $a \in B$ and $v \in V_a$. The set $F(B, V)$ of formulas over B is the least set containing all atomic formulas over B and closed with respect to disjunction, conjunction and negation. For each $f \in F(B, V)$, f_A denote the meaning of f in A, i.e., the set of all objects in U with property f, defined inductively as follows.

1. If f is of the form $[a = v]$ then, $f_A = \{s \in U | a(s) = v\}$
2. $(f \wedge g)_A = f_A \cap g_A$; $(f \vee g)_A = f_A \vee g_A$; $(\neg f)_A = U - f_A$

By using this framework, classification accuracy and coverage, or true positive rate is defined as follows.

Definition 1.
Let R and D denote a formula in $F(B, V)$ and a set of objects whose decision attribute is given as $\lceil$, respectively. Classification accuracy and coverage(true positive rate) for $R \to \mathcal{D}$ is defined as:

$$\alpha_R(D) = \frac{|R_A \cap D|}{|R_A|}(= P(D|R)), \text{ and } \kappa_R(D) = \frac{|R_A \cap D|}{|D|}(= P(R|D)),$$

where $|A|$ denotes the cardinality of a set A, $\alpha_R(D)$ denotes a classification accuracy of R as to classification of $\mathcal{D}$, and $\kappa_R(D)$ denotes a coverage, or a true positive rate of R to $\mathcal{D}$, respectively.

2.2 Two-way Contingency Table

From the viewpoint of information systems, a contingency table summarizes the relation between two attributes with respect to frequencies. This viewpoint has already been discussed in [9, 10]. However, this study focuses on more statistical interpretation of this table.

Definition 2. *Let R_1 and R_2 denote binary attributes in an attribute space A. A contingency table is a table of a set of the meaning of the following formulas: $|[R_1 = 0]_A|$,$|[R_1 = 1]_A|$, $|[R_2 = 0]_A|$,$|[R_2 = 1]_A|$, $|[R_1 = 0 \wedge R_2 = 0]_A|$,$|[R_1 = 0 \wedge R_2 = 1]_A|$, $|[R_1 = 1 \wedge R_2 = 0]_A|$,$|[R_1 = 1 \wedge R_2 = 1]_A|$, $|[R_1 = 0 \vee R_1 = 1]_A|(= |U|)$. This table is arranged into the form shown in Table 1, where: $|[R_1 = 0]_A| = x_{11} + x_{21} = x_{\cdot 1}$, $|[R_1 = 1]_A| = x_{12} + x_{22} = x_{\cdot 2}$,*

Table 1. Two way Contingency Table

	$R_1 = 0$	$R_1 = 1$	
$R_2 = 0$	x_{11}	x_{12}	$x_{1\cdot}$
$R_2 = 1$	x_{21}	x_{22}	$x_{2\cdot}$
	$x_{\cdot 1}$	$x_{\cdot 2}$	$x_{\cdot\cdot}$ $(= \|U\| = N)$

$|[R_2 = 0]_A| = x_{11} + x_{12} = x_{1\cdot}$, $|[R_2 = 1]_A| = x_{21} + x_{22} = x_{2\cdot}$, $|[R_1 = 0 \wedge R_2 = 0]_A| = x_{11}$, $|[R_1 = 0 \wedge R_2 = 1]_A| = x_{21}$, $|[R_1 = 1 \wedge R_2 = 0]_A| = x_{12}$, $|[R_1 = 1 \wedge R_2 = 1]_A| = x_{22}$, $|[R_1 = 0 \vee R_1 = 1]_A| = x_{\cdot 1} + x_{\cdot 2} = x_{\cdot\cdot} (= |U|)$.

From this table, accuracy and coverage for $[R_1 = 0] \to [R_2 = 0]$ are defined as:

$$\alpha_{[R_1=0]}([R_2 = 0]) = \frac{|[R_1 = 0 \wedge R_2 = 0]_A|}{|[R_1 = 0]_A|} = \frac{x_{11}}{x_{\cdot 1}},$$

$$\text{and}$$

$$\kappa_{[R_1=0]}([R_2 = 0]) = \frac{|[R_1 = 0 \wedge R_2 = 0]_A|}{|[R_2 = 0]_A|} = \frac{x_{11}}{x_{1\cdot}}.$$

Example.

Let us consider an information table shown in Table 2. The relationship be-

Table 2. A Small Dataset

a	b	c	d	e
1	0	0	0	1
0	0	1	1	1
0	1	2	2	0
1	1	1	2	1
0	0	2	1	0

tween b and e can be examined by using the corresponding contingency table as follows. First, the frequencies of four elementary relations are counted, called *marginal distributions*: $[b = 0]$, $[b = 1]$, $[e = 0]$, and $[e = 1]$. Then, the frequencies of four kinds of conjunction are counted: $[b = 0] \wedge [e = 0]$, $[b = 0] \wedge [e = 1]$, $[b = 1] \wedge [e = 0]$, and $[b = 1] \wedge [e = 1]$. Then, the following contingency table is obtained (Table 3). From this table, accuracy and coverage for $[b = 0] \to [e = 0]$ are obtained as $1/(1 + 2) = 1/3$ and $1/(1 + 1) = 1/2$.

One of the important observations from granular computing is that a contingency table shows the relations between two attributes with respect to

Table 3. Corresponding Contingency Table

	b=0	b=1	
e=0	1	1	2
e=1	2	1	3
	3	2	5

intersection of their supporting sets. For example, in Table 3, both b and e have two different partitions of the universe and the table gives the relation between b and e with respect to the intersection of supporting sets. It is easy to see that this idea can be extended into $n - way$ contingency tables, which can be viewed as $n \times n$-matrix.

2.3 Multi-way Contingency Table

Two-way contingency table can be extended into a contingency table for multinominal attributes.

Definition 3. *Let R_1 and R_2 denote multinominal attributes in an attribute space A which have m and n values. A contingency tables is a table of a set of the meaning of the following formulas: $|[R_1 = A_j]_A|$, $|[R_2 = B_i]_A|$, $|[R_1 = A_j \wedge R_2 = B_i]_A|$, and $|U|$ $(i = 1, 2, 3, \cdots, n$ and $j = 1, 2, 3, \cdots, m)$. This table is arranged into the form shown in Table 1, where: $|[R_1 = A_j]_A| = \sum_{i=1}^{m} x_{1i} = x_{\cdot j}$, $|[R_2 = B_i]_A| = \sum_{j=1}^{n} x_{ji} = x_{i\cdot}$, $|[R_1 = A_j \wedge R_2 = B_i]_A| = x_{ij}$, $|U| = N = x_{\cdot\cdot}$ $(i = 1, 2, 3, \cdots, n$ and $j = 1, 2, 3, \cdots, m)$.*

Table 4. Contingency Table ($n \times m$)

	A_1	A_2	$\cdots$	A_n	Sum
B_1	x_{11}	x_{12}	$\cdots$	x_{1n}	$x_{1\cdot}$
B_2	x_{21}	x_{22}	$\cdots$	x_{2n}	$x_{2\cdot}$
$\cdots$	$\cdots$	$\cdots$	$\cdots$	$\cdots$	$\cdots$
B_m	x_{m1}	x_{m2}	$\cdots$	x_{mn}	$x_{m\cdot}$
Sum	$x_{\cdot 1}$	$x_{\cdot 2}$	$\cdots$	$x_{\cdot n}$	$x_{\cdot\cdot} = \lvert U \rvert = N$

3 Chi-square Test

One of the most important tests which check the independence of two attributes is chi-square test. Although it is frequently used, the fundamental issues of this test have never been fully examined from the viewpoint of granular computing and data mining. Here, the meaning of this test from this view will be reinterpreted.

Especially, it is notable that this test statistic measures the square distance between joint probabilities of given two attributes a_1 and a_2, $p(a_1, a_2)$ and product sum of both probabilities $p(a_1)p(a_2)$: $(p(a_1, a_2) - p(a_1)p(a_2))^2$.

3.1 Definitions

The chi-square test is based on the following theorem[5].

Theorem 1. *In a general case, when a contingency table shown in Table 4 is given, the test statistic:*

$$\chi^2 = \sum_{i,j=1}^{n,m} \frac{(x_{ij} - x_{i\cdot}x_{\cdot j}/N)^2}{x_{i\cdot}x_{\cdot j}/N} \tag{1}$$

follows chi-square distribution with the freedom of $(n-1)(m-1)$.

In the case of binary attributes shown in Table 1, this test statistic can be transformed into the following simple formula and it follows the chi-square distribution with the freedom of one.

$$\chi^2 = \frac{N(x_{11}x_{22} - x_{12}x_{21})^2}{x_{1\cdot}x_{2\cdot}x_{\cdot 1}x_{\cdot 2}} \tag{2}$$

One of the core ideas of the chi-square test is that the test statistic measures the square of difference between the real value and the expected value of one column. In the example shown in Table 4, $(x_{11} - x_{\cdot 1}x_{1\cdot}/N)^2$ measures the difference between x_{11} and the expected value of this column $x_{\cdot 1} \times x_{1\cdot}/N$ where $x_{1\cdot}/N$ is a marginal distribution of B_1.

Another core idea is that $x_{\cdot 1}x_{1\cdot}/N$ is equivalent to the variance of marginal distributions if they follow multinominal distributions. [2] Thus, the chi-square test statistic is equivalent to tthe otal sum of the ratio of the square distance s^2 to the corresponding variance σ^2. Actually, the theorem above comes from more general theorem as a corollary if a given multinominal distribution converges into a normal distribution.

Theorem 2. *If* x_1,x_2,$\cdots$, x_n *are randomly selected from the population following a normal distribution* $N(m, \sigma^2)$, *the formula*

$$y = \frac{1}{\sigma^2} \sum_{i=1}^{n} (x_i - m)^2$$

follows the χ^2 *distribution with the freedom of* $(n-1)$.

In the subsequent sections, it is assumed that all these above statistical assumptions will hold.

---

[2]If the probabilities p and q come from the multinominal distribution, Npq is equal to variance.

3.2 Chi-square Test and Statistical Independence

Since x_{ij}/N, a_i/N and b_j/N are equal to $p(A_i, B_j)$, $p(A_i)$ and $p(B_j)$, the equation 1 will become:

$$\chi^2 = \sum_{i,j=1}^{n,m} \frac{(p(A_i, B_j) - p(A_i)p(B_j))^2}{p(A_i)p(B_j)} \tag{3}$$

Therefore, a chi-square test statistic exactly checks the degree of statistical independence. When all the combinations of i and m satisfy $p(A_i, B_j) = p(A_i)p(B_j)$, that is, when χ^2 is equal to 0, two attributes A and B are completely statistical independent. If χ^2 is strongly deviated from 0, this test statistic suggests that two attributes are strongly dependent.

It is notable that $(ad - bc)^2$ appears on the nominator in the equation (2). This factor, $ad - bc$ exactly corresponds to the determinant of a matrix when the contingency table(Fig. 1) is regarded as a matrix. According to the matrix theory, if $ad - bc$ is equal to 1.0, then the rank of a given matrix is equal to 1.0. Thus, this observation suggests that the rank can be used to check the independence of two attributes. For example, in Table 2, b and e are dependent because $ad - bc = 1 \times 1 - 1 \times 2 = -1 < 0$.

4 Rank of Contingency Table (two-way)

4.1 Preliminaries

Definition 4. *A corresponding matrix $C_{T_{a,b}}$ is defined as a matrix the element of which are equal to the value of the corresponding contingency table $T_{a,b}$ of two attributes a and b, except for marginal values.*

Definition 5. *The rank of a table is defined as the rank of its corresponding matrix. The maximum value of the rank is equal to the size of (square) matrix, denoted by r.*

Example.

Let the table given in Table 3 be defined as $T_{b,e}$. Then, $C_{T_{b,e}}$ is:

$$\begin{pmatrix} 1 & 1 \\ 2 & 1 \end{pmatrix}$$

Since the determinant of $C_{T_{b,e}}$ $det(C_{T_{b,e}})$ is not equal to 0, the rank of $C_{T_{b,e}}$ is equal to 2. It is the maximum value ($r = 2$), so b and e are statistically dependent.

4.2 Independence when the table is two-way

From the results in linear algebra, several results are obtained. (The proofs is omitted.) First, it is assumed that a contingency table is given as Table 1. Then the corresponding matrix $(C_{T_{R_1,R_2}})$ is given as:

$$\begin{pmatrix} x_{11} & x_{12} \\ x_{21} & x_{22} \end{pmatrix},$$

Proposition 1. *The determinant of* $det(C_{T_{R_1,R_2}})$ *is equal to* $|x_{11}x_{22} - x_{12}x_{21}|$.

Proposition 2. *The rank will be:*

$$rank = \begin{cases} 2, & if\ det(C_{T_{R_1,R_2}}) \neq 0 \\ 1, & if\ det(C_{T_{R_1,R_2}}) = 0 \end{cases}$$

If the rank of $det(C_{T_{b,e}})$ is equal to 1, according to the theorems of the linear algebra, it is obtained that one row or column will be represented by the other column. That is,

Proposition 3. *Let* r_1 *and* r_2 *denote the rows of the corresponding matrix of a given two-way table,* $C_{T_{b,e}}$. *That is,*

$$r_1 = (x_{11}, x_{12}), r_2 = (x_{21}, x_{22})$$

Then, r_1 *can be represented by* r_2*:*

$$r_1 = kr_2,$$

where k *is given as:*

$$k = \frac{x_{11}}{x_{21}} = \frac{x_{12}}{x_{22}} = \frac{x_{1\cdot}}{x_{2\cdot}}$$

What is the meaning of Proposition 3? Since $x_{11}/x_{\cdot\cdot}$, $x_{21}/x_{\cdot\cdot}$, $x_{1\cdot/x_{\cdot\cdot}}$ and $x_{2\cdot/x_{\cdot\cdot}}$ are equal to $p([R_1 = 0, R_2 = 0])$, $p([R_1 = 0, R_2 = 1])$, $p([R_2 = 0])$ and $p([R_2 = 1])$, respectively,

$$k = \frac{p([R_1 = 0, R_2 = 0])}{p([R_1 = 0, R_2 = 1])} = \frac{p([R_2 = 0])}{p([R_2 = 1])}$$

After several steps of computation, finally the following relation is obtained:

$$p([R_1 = 0], [R_2 = 0]) = p([R_1 = 0])p([R_2 = 0]),$$

which is equivalent to statistical independence. That is,

Theorem 3. *If the rank of the corresponding matrix is 1, then two attributes in a given contingency table are statistically independent. Thus,*

$$rank = \begin{cases} 2, & dependent \\ 1, & statistical\ independent \end{cases}$$

4.3 Other Characteristics

Since the meanings of attributes are very clear in binary cases, the following result is obtained.

Theorem 4. *The distribution of cells in a given contingency table will be roughly evaluated by the determinant of the corresponding matrix. That is,*

$$\begin{cases} det(C_{T_{R_1,R_2}}) > 0: & x_{11} \text{ and } x_{22} \text{ are dominant,} \\ det(C_{T_{R_1,R_2}}) < 0: & x_{12} \text{ and } x_{21} \text{ are dominant,} \end{cases}$$

where dominant means that these values play important roles in determining the dependencies of the two attributes.

5 Rank of Contingency Table (Multi-way)

In the case of a general square matrix, the results in the two-way contingency table can be extended. Especially, it is very important to observe that conventional statistical independence is only supported when the rank of the corresponding is equal to 1. Let us consider the contingency table of c and d in Table 2, which is obtained as follows. Thus, the corresponding matrix of

Table 5. Contingency Table for c and d

	c=0	c=1	c=2	
d=0	1	0	0	1
d=1	0	1	1	2
d=2	0	1	1	2
	1	2	2	5

this table is:

$$\begin{pmatrix} 1 & 0 & 0 \\ 0 & 1 & 1 \\ 0 & 1 & 1 \end{pmatrix},$$

whose determinant is equal to 0. It is clear that its rank is 2. It is interesting to see that if the case of $[d = 0]$ is removed, then the rank of the corresponding matrix is equal to 1 and two rows are equal. Thus, if the value space of d into $\{1, 2\}$ is restricted, then c and d are statistically independent. This relation is called *contextual independence*[1], which is related with conditional independence.

However, another type of weak independence is observed: let us consider the contingency table of a and c. The table is obtained as Table 6:

Table 6. Contingency Table for a and c

	a=0	a=1	
c=0	0	1	1
c=1	1	1	2
c=2	2	0	2
	3	2	5

Its corresponding matrix is:

$$\begin{pmatrix} 0 & 1 \\ 1 & 1 \\ 2 & 0 \end{pmatrix},$$

Since the corresponding matrix is not square, the determinant is not defined. But it is easy to see that the rank of this matrix is two. In this case, even any attribute-value pair removed from the table will not generate statistical independence.

However, interesting relations can be found. Let r_1, r_2 and r_3 denote the first, second and third row, respectively. That is, $r_1 = (0,1)$, $r_2 = (1,1)$ and $r_3 = (2,0)$.

Then, it is easy to see that $r_3 = 2(r_2 - r_1)$, which can be viewed as:

$$p([c = 2]) = 2(p([c = 1]) - p([c = 0])),$$

which gives a probabilistic model between attribute-value pairs. According to linear algebra, when we have a $m \times n(m \geq n)$ or $n \times m$ corresponding matrix, the rank of the corresponding matrix is less than n. Thus:

Theorem 5. *If the corresponding matrix of a given contingency table is not square and of the form $m \times n(m \geq n)$, then its rank is less than n. Especially, the row $r_{n+1}, r_{n+2}, \cdots, r_m$ can be represented by:*

$$r_k = \sum_{i=1}^{r} k_i r_i (n + 1 \leq k \leq m),$$

where k_i and r denote the constant and the rank of the corresponding matrix, respectively. This can be interpreted as:

$$p([R_1 = A_k]) = \sum_{i=1}^{r} k_i p([R_1 = A_i])$$

Finally, the relation between rank and independence in a multi-way contingency table is obtained.

Theorem 6. *Let the corresponding matrix of a given contingency table be a square $n \times n$ matrix. If the rank of the corresponding matrix is 1, then two attributes in a given contingency table are statistically independent. If the rank* *of the corresponding matrix is n , then two attributes in a given contingency table are dependent. Otherwise, two attributes are contextual dependent, which means that several conditional probabilities can be represented by a linear combination of conditional probabilities. Thus,*

$$rank = \begin{cases} n & \textit{dependent} \\ 2, \cdots, n-1 & \textit{contextual independent} \\ 1 & \textit{statistical independent} \end{cases}$$

6 Rough Set Approximations and Contingency Tables

The important idea in rough sets is that real-world concepts can be captured by two approximations: lower and upper approximations [4]. Although these ideas are deterministic, they can be extended into naive probabilistic models if we set up thresholds as shown in Ziarko's variable precision rough set model(VPRS)[11].

From the ideas of Ziarko's VPRS, Tsumoto shows that lower and upper approximations of a target concept correspond to sets of examples which satisfy the following conditions[8]:

Lower Approximation of D: $\cup\{R_A|\alpha_R(D) = 1.0\}$,
Upper Approximation of D: $\cup\{R_A|\kappa_R(D) = 1.0\}$,

where R is a disjunctive or conjunctive formula. Thus, if we assume that all the attributes are binary, we can construct contingency tables corresponding to these two approximations as follows.

6.1 Lower approximation

From the definition of accuracy shown in Section 2, the contingency table for lower approximation is obtained if c is set to 0. That is, the following contingency table corresponds to the lower approximation of $R_2 = 0$. In this

Table 7. Contingency Table for Lower Approximation

	$R_1 = 0$	$R_1 = 1$	
$R_2 = 0$	x_{11}	x_{12}	$x_{1\cdot}$
$R_2 = 1$	0	x_{22}	x_{22}
	x_{11}	$x_{\cdot 2}$	$x_{\cdot\cdot}$ $(= \lvert U \rvert = N)$

case, the test statistic is simplified into:

$$\chi^2 = \frac{N(ad)^2}{a(b+d)(a+b)d} \tag{4}$$

6.2 Upper approximation

From the definition of coverage shown in Section 3, the contingency table for lower approximation is obtained if c is set to 0. That is, the following contingency table corresponds to the lower approximation of $R_2 = 0$. In this

Table 8. Contingency Table for Upper Approximation

	$R_1 = 0$	$R_1 = 1$	
$R_2 = 0$	x_{11}	0	x_{11}
$R_2 = 1$	x_{21}	x_{22}	$x_{2\cdot}$
	$x_{\cdot 1}$	x_{22}	$x_{\cdot\cdot}$ $(= \|U\| = N)$

case, the test statistic is simplified into:

$$\chi^2 = \frac{N(ad)^2}{(a+c)da(c+d)} \tag{5}$$

It is notable that the corresponding matrix of two approximations are corresponding to triangular matrices. In the case of a lower approximation, the corresponding matrix is equivalent to the upper triangular matrix. On the other hand, the corresponding matrix of an upper approximation is equivalent to the lower triangular matrix.

Furthermore, when the lower approximation is equal to the upper one of a given target concept, the corresponding matrix is of the form of a diagonal matrix.

7 Conclusion

In this paper, a contingency table is interpreted from the viewpoint of granular computing and statistical independence. From the correspondence between contingency table and matrix, the following observations are obtained: in the case of statistical independence, the rank of the corresponding matrix of a given contingency table is equal to 1. That is, all the rows of contingency table can be described by one row with the coefficient given by a marginal distribution. If the rank is maximum, then two attributes are dependent. Otherwise, some probabilistic structure can be found within attribute -value pairs in a given attribute. Thus, matrix algebra is a key point of the analysis of a contingency table and the degree of independence, rank plays a very important role in extracting a probabilistic model.

This paper is a preliminary study on the formal studies on contingency tables, and the discussions are very intuitive, not mathematically rigor. More formal analysis will appear in the future work.

References

1. Butz, C.J. Exploiting contextual independencies in web search and user profiling, *Proceedings of World Congress on Computational Intelligence (WCCI'2002)*, CD-ROM, 2002.
2. Polkowski, L. and Skowron, A.(Eds.) *Rough Sets and Knowledge Discovery 1*, Physica Verlag, Heidelberg, 1998.
3. Polkowski, L. and Skowron, A.(Eds.) *Rough Sets and Knowledge Discovery 2*, Physica Verlag, Heidelberg, 1998.
4. Pawlak, Z., *Rough Sets*. Kluwer Academic Publishers, Dordrecht, 1991.
5. Rao, C.R. *Linear Statistical Inference and Its Applications, 2nd Edition*, John Wiley & Sons, New York, 1973.
6. Skowron, A. and Grzymala-Busse, J. From rough set theory to evidence theory. In: Yager, R., Fedrizzi, M. and Kacprzyk, J.(eds.) *Advances in the Dempster-Shafer Theory of Evidence*, pp.193-236, John Wiley & Sons, New York, 1994.
7. Tsumoto S and Tanaka H: Automated Discovery of Medical Expert System Rules from Clinical Databases based on Rough Sets. In: *Proceedings of the Second International Conference on Knowledge Discovery and Data Mining 96*, AAAI Press, Palo Alto CA, pp.63-69, 1996.
8. Tsumoto, S. Knowledge discovery in clinical databases and evaluation of discovered knowledge in outpatient clinic. *Information Sciences*, **124**, 125-137, 2000.
9. Yao, Y.Y. and Wong, S.K.M., A decision theoretic framework for approximating concepts, *International Journal of Man-machine Studies*, **37**, 793-809, 1992.
10. Yao, Y.Y. and Zhong, N., An analysis of quantitative measures associated with rules, N. Zhong and L. Zhou (Eds.), *Methodologies for Knowledge Discovery and Data Mining, Proceedings of the Third Pacific-Asia Conference on Knowledge Discovery and Data Mining*, LNAI **1574**, Springer, Berlin, pp. 479-488, 1999.
11. Ziarko, W., Variable Precision Rough Set Model. *Journal of Computer and System Sciences*, 46, 39-59, 1993.

Foundations of Classification

J. T. Yao Y. Y. Yao and Y. Zhao

Department of Computer Science, University of Regina
Regina, Saskatchewan, Canada S4S 0A2
{jtyao, yyao, yanzhao}@cs.uregina.ca

Summary. Classification is one of the main tasks in machine learning, data mining, and pattern recognition. A granular computing model is suggested for learning two basic issues of concept formation and concept relationship identification. A classification problem can be considered as a search for suitable granules organized under a partial order. The structures of search space, solutions to a consistent classification problem, and the structures of solution space are discussed. A classification rule induction method is proposed. Instead of searching for a suitable partition, we concentrate on the search for a suitable covering of the given universe. This method is more general than partition-based methods. For the design of covering granule selection heuristics, several measures on granules are suggested.

1 Introduction

Classification is one of the main tasks in machine learning, data mining, and pattern recognition [3, 10, 12]. It deals with classifying labelled objects. Knowledge for classification can be expressed in different forms, such as classification rules, discriminant functions, and decision trees. Extensive research has been done on the construction of classification models.

Mainstream research in classification focus on classification algorithms and their experimental evaluations. By comparison, less attention has been paid to the study of fundamental concepts such as structures of search space, solution to a consistent classification problem, as well as the structures of a solution space. For this reason, we present a granular computing based framework for a systematic study of these fundamental issues.

Granular computing is an umbrella term to cover any theories, methodologies, techniques, and tools that make use of granules in problem solving [25, 27, 33, 34]. A *granule* is a subset of the universe. A family of granules that contains every object in the universe is called a *granulation* of the universe. The granulation of a given universe involves dividing the universe into subsets or grouping individual objects into clusters. There are many fundamental issues in granular computing, such as the granulation of a given universe, the descriptions of granules, the relationships between granules, and the computation of granules.

Data mining, especially rule-based mining, can be molded in two steps, namely, the formation of concepts and the identification of relationship between concepts. Formal concept analysis may be considered as a concrete model of granular computing. It deals with the characterization of a concept by a unit of thoughts consisting the intension and the extension of the concept [4, 23]. From the standing point of granular computing, the concept of a granule may be exemplified by a set of instances, i.e., the extension; the concept of a granule may be described or labelled by a name, i.e., the intension. Once concepts are constructed and described, one can develop computational methods using granules [27]. In particular, one may study relationships between concepts in terms of their intensions and extensions, such as sub-concepts and super-concepts, disjoint and overlap concepts, and partial sub-concepts. These relationships can be conveniently expressed in the form of rules and associated quantitative measures indicating the strength of rules. By combining the results from formal concept analysis and granular computing, knowledge discovery and data mining, especially rule mining, can be viewed as a process of forming concepts and finding relationships between concepts in terms of intensions and extensions [28, 30, 32].

The organization of this chapter is as follows. In Section 2, we first present the fundamental concepts of granular computing which serve as the basis of classification problems. Measures associated with granules for classification will be studied in Section 3. In Section 4, we will examine the search spaces of classification rules. In Section 5, we remodel the ID3 and PRISM classification algorithms from the viewpoint of granular computing. We also propose the kLR algorithm and a granule network algorithm to complete the study of the methodology in granular computing model.

2 Fundamentals of a Granular Computing Model for Classification

This section provides an overview of the granular computing model [28, 30].

2.1 Information tables

Information tables are used in granular computing models. An information table provides a convenient way to describe a finite set of objects called a universe by a finite set of attributes [14, 33]. It represents all available information and knowledge. That is, objects are only perceived, observed, or measured by using a finite number of properties.

Definition 1. *An information table is the following tuple:*

$$S = (U, At, \mathcal{L}, \{V_a \mid a \in At\}, \{I_a \mid a \in At\}),$$

where

U is a finite nonempty set of objects,
At is a finite nonempty set of attributes,
$\mathcal{L}$ is a language defined by using attributes in At,
V_a is a nonempty set of values of $a \in At$,
$I_a : U \to V_a$ is an information function that maps an object of U to exactly one possible value of attribute a in V_a.

We can easily extend the information function I_a to an information function on a subset of attributes. For a subset $A \subseteq At$, the values of an object x on A is denoted by $I_A(x)$, where $I_A(x) = \bigwedge_{a \in A} I_a(x)$.

Definition 2. *In the language $\mathcal{L}$, an atomic formula is given by $a = v$, where $a \in At$ and $v \in V_a$. If ϕ and ψ are formulas, then so are $\neg\phi$, $\phi \wedge \psi$, and $\phi \vee \psi$.*

The semantics of the language $\mathcal{L}$ can be defined in the Tarski's style through the notions of a model and the satisfiability of the formulas.

Definition 3. *Given the model as an information table S, the satisfiability of a formula ϕ by an object x, written $x \models_S \phi$, or in short $x \models \phi$ if S is understood, is defined by the following conditions:*

$$\begin{aligned}
&(1)\ \ x \models a = v \text{ iff } I_a(x) = v,\\
&(2)\ \ x \models \neg\phi \text{ iff } \text{ not } x \models \phi,\\
&(3)\ \ x \models \phi \wedge \psi \text{ iff } x \models \phi \text{ and } x \models \psi,\\
&(4)\ \ x \models \phi \vee \psi \text{ iff } x \models \phi \text{ or } x \models \psi.
\end{aligned}$$

Definition 4. *Given a formula π, the set $m_S(\phi)$, defined by*

$$m_S(\phi) = \{x \in U \mid x \models \phi\}, \tag{1}$$

is called the meaning of the formula ϕ in S. If S is understood, we simply write $m(\phi)$.

A formula ϕ can be viewed as the description of the set of objects $m(\phi)$, and the meaning $m(\phi)$ of a formula is the set of all objects having the property expressed by ϕ. Thus, a connection between formulas of $\mathcal{L}$ and subsets of U is established.

2.2 Concept formulation

To formalize data mining, we have to analyze the concepts first. There are two aspects of a concept, the intension and the extension [4, 23]. The intension of a concept consists of all properties or attributes that are valid for all objects to which the concept applies. The intension of a concept is its meaning, or its complete definition. The extension of a concept is the set of objects or entities which are instances of the concept. The extension of a concept is a collection, or a set, of things to which the concept applies. A concept is thus described jointly by its intension and extension, i.e., a set of properties and a set of objects. The intension of a concept can be expressed by a formula, or an expression, of a certain language, while the extension of a concept is presented as a set of objects that satisfies the formula. This formulation enables us to study formal concepts in a logic setting in terms of intensions and also in a set-theoretic setting in terms of extensions.

With the introduction of language $\mathcal{L}$, we have a formal description of concepts. A concept which is definable in an information table is a pair of $(\phi, m(\phi))$, where $\phi \in \mathcal{L}$. More specifically, ϕ is a description of $m(\phi)$ in S, i.e. the intension of concept $(\phi, m(\phi))$, and $m(\phi)$ is the set of objects satisfying ϕ, i.e. the extension of concept $(\phi, m(\phi))$. We say a formula has meaning if it has an associated subset of objects; we also say a subset of objects is definable if it is associated with at least one formula.

Definition 5. *A subset $X \subseteq U$ is called a definable granule in an information table S if there exists at least one formula ϕ such that $m(\phi) = X$.*

By using the language $\mathcal{L}$, we can define various granules. For an atomic formula $a = v$, we obtain a granule $m(a = v)$. If $m(\phi)$ and $m(\psi)$ are granules corresponding to formulas ϕ and ψ, we obtain granules $m(\phi) \cap m(\psi) = m(\phi \wedge \psi)$ and $m(\phi) \cup m(\psi) = m(\phi \vee \psi)$.

Object	**height**	**hair**	**eyes**	**class**
o_1	short	blond	blue	+
o_2	short	blond	brown	-
o_3	tall	red	blue	+
o_4	tall	dark	blue	-
o_5	tall	dark	blue	-
o_6	tall	blond	blue	+
o_7	tall	dark	brown	-
o_8	short	blond	brown	-

Table 1. An information table

To illustrate the idea developed so far, consider an information table given by Table 1, which is adopted from Quinlan [15]. The following expressions are some of the formulas of the language $\mathcal{L}$:

$$\begin{aligned}&\textbf{height} = \text{tall},\\&\textbf{hair} = \text{dark},\\&\textbf{height} = \text{tall} \wedge \textbf{hair} = \text{dark},\\&\textbf{height} = \text{tall} \vee \textbf{hair} = \text{dark}.\end{aligned}$$

The meanings of the above formulas are given by:

$$\begin{aligned}&m(\textbf{height} = \text{tall}) = \{o_3, o_4, o_5, o_6, o_7\},\\&m(\textbf{hair} = \text{dark}) = \{o_4, o_5, o_7\},\\&m(\textbf{height} = \text{tall} \wedge \textbf{hair} = \text{dark}) = \{o_4, o_5, o_7\},\\&m(\textbf{height} = \text{tall} \vee \textbf{hair} = \text{dark}) = \{o_3, o_4, o_5, o_6, o_7\}.\end{aligned}$$

By pairing intensions and extensions, we can obtain formal concepts, such as ($\textbf{height} = \text{tall}, \{o_3, o_4, o_5, o_6, o_7\}$), ($\textbf{hair} = \text{dark}, \{o_4, o_5, o_7\}$), ($\textbf{height} = \text{tall} \wedge \textbf{hair} = \text{dark}, \{o_4, o_5, o_7\}$) and ($\textbf{height} = \text{tall} \vee \textbf{hair} = \text{dark}, \{o_3, o_4, o_5, o_6, o_7\}$). The involved granules such as $\{o_3, o_4, o_5, o_6, o_7\}$, $\{o_4, o_5, o_7\}$ and $\{o_3, o_4, o_5, o_6, o_7\}$) are definable granules.

In the case where we can precisely describe a subset of objects X, the description may not be unique. That is, there may exist two formulas such that $m(\phi) = m(\psi) = X$. For example,

$$\begin{aligned}&\textbf{hair} = dark,\\&\textbf{height} = tall \wedge \textbf{hair} = dark,\end{aligned}$$

have the same meaning set $\{o_4, o_5, o_7\}$. Another two formulas

$$\begin{aligned}&\textbf{class} = +,\\&\textbf{hair} = \text{red} \vee (\textbf{hair} = \text{blond} \wedge \textbf{eyes} = \text{blue}),\end{aligned}$$

have the same meaning set $\{o_1, o_3, o_6\}$.

In many classification algorithms, one is only interested in formulas of a certain form. Suppose we restrict the connectives of language $\mathcal{L}$ to only the conjunction connective $\wedge$. Each formula is a conjunction of atomic formulas and such a formula is referred to as a conjunctor.

Definition 6. *A subset $X \subseteq U$ is a conjunctively definable granule in an information table S if there exists a conjunctor ϕ such that $m(\phi) = X$.*

The notion of definability of subsets in an information table is essential to data analysis. In fact, definable granules are the basic logic units that can be described and discussed, upon which other notions can be developed.

2.3 Granulations as partitions and coverings

Partitions and coverings are two simple and commonly used granulations of the universe.

Definition 7. *A partition of a finite universe U is a collection of non-empty, and pairwise disjoint subsets of U whose union is U. Each subset in a partition is also called a block or an equivalence class.*

When U is a finite set, a partition $\pi = \{X_i \mid 1 \leq i \leq m\}$ of U consists of a finite number m of blocks. In this case, the conditions for a partition can be simply stated by:

$$\begin{aligned} &\text{(i).} \quad \text{for all } i,\ X_i \neq \emptyset, \\ &\text{(ii).} \quad \text{for all } i \neq j,\ X_i \cap X_j = \emptyset, \\ &\text{(iii).} \quad \bigcup \{X_i \mid 1 \leq i \leq m\} = U. \end{aligned}$$

There is a one-to-one correspondence between the partitions of U and the equivalence relations (i.e., reflexive, symmetric, and transitive relations) on U. Each equivalence class of the equivalence relation is a block of the corresponding partition. In this paper, we use partitions and equivalence relations, and blocks and equivalence classes interchangeably.

Definition 8. *A covering of a finite universe U is a collection of non-empty subsets of U whose union is U. The subsets in a covering are called covering granules.*

When U is a finite set, a covering $\tau = \{X_i \mid 1 \leq i \leq m\}$ of U consists of a finite number m of covering granules. In this case, the conditions for a covering can be simply stated by:

$$\begin{aligned} &\text{(i).} \quad \text{for all } i, X_i \neq \emptyset, \\ &\text{(ii).} \quad \bigcup \{X_i \mid 1 \leq i \leq m\} = U. \end{aligned}$$

According to the definition, a partition consists of disjoint subsets of the universe, and a covering consists of possibly overlapping subsets. Partitions are a special case of coverings.

Definition 9. *A covering τ of U is said to be a non-redundant covering if the collection of subsets derived by deleting any one of the granules from τ is not a covering.*

One can obtain a finer partition by further dividing the equivalence classes of a partition. Similarly, one can obtain a finer covering by further decomposing the granules of the covering.

Definition 10. *A partition π_1 is a refinement of another partition π_2, or equivalently, π_2 is a coarsening of π_1, denoted by $\pi_1 \preceq \pi_2$, if every block of π_1 is contained in some block of π_2. A covering τ_1 is a refinement of another covering τ_2, or equivalently, τ_2 is a coarsening of τ_1, denoted by $\tau_1 \preceq \tau_2$, if every granule of τ_1 is contained in some granule of τ_2.*

The refinement relation is a partial ordering of the set of all partitions, namely, it is reflexive, antisymmetric and transitive. This naturally defines a refinement order on the set of all partitions, and thus form a partition lattice, denoted as $\Pi(U)$. Likewise, a refinement order on the set of all covering forms a covering lattice, denoted as $\mathcal{T}(U)$.

Based on the refinement relation, we can construct multi-level granulations of the universe [29]. Given two partitions π_1 and π_2, their meet, $\pi_1 \wedge \pi_2$, is the finest partition of π_1 and π_2, their join, $\pi_1 \vee \pi_2$, is the coarsest partition of π_1 and π_2. The equivalence classes of a meet are all nonempty intersections of an equivalence class from π_1 and an equivalence class from π_2. The equivalence classes of a join are all nonempty unions of an equivalence class from π_1 and an equivalence class from π_2.

Since a partition is a covering, we use the same symbol to denote the refinement relation on partitions and refinement relation on covering. For a covering τ and a partition π, if $\tau \preceq \pi$, we say that τ is a refinement of π, which indicates that every granule of τ is contained in some granule of π.

Definition 11. *A partition is called a definable partition (π_D) in an information table S if every equivalence class is a definable granule. A covering is called a definable covering (τ_D) in an information table S if every covering granule is a definable granule.*

For example, in information Table 1 $\{\{o_1, o_2, o_6, o_8\}, \{o_3, o_4, o_5, o_7\}\}$ is a definable partition/covering, since the granule $\{o_1, o_2, o_6, o_8\}$ can be defined by the formula **hair**=blond, and the granule $\{o_3, o_4, o_5, o_7\}$ can be defined by the formula $\neg$**hair**=blond. We can also justify that another partition $\{\{o_1, o_2, o_3, o_4\}, \{o_5, o_6, o_7, o_8\}\}$ is not a definable partition.

If partitions π_1 and π_2 are definable, $\pi_1 \wedge \pi_2$ and $\pi_1 \vee \pi_2$ are definable partitions. The family of all definable partitions forms a partition lattice $\Pi_D(U)$, which is a sub-lattice of $\Pi(U)$. Likewise, if two coverings τ_1 and τ_2 are definable, $\tau_1 \wedge \tau_2$ and $\tau_1 \vee \tau_2$ are definable coverings. The family of all definable coverings forms a covering lattice $\mathcal{T}_D(U)$, which is a sub-lattice of $\mathcal{T}(U)$.

Definition 12. *A partition is called a conjunctively definable partition (π_{CD}) if every equivalence class is a conjunctively definable granule. A covering is called a conjunctively definable covering (τ_{CD}) if every covering granule is a conjunctively definable granule.*

For example, in Table 1 $\{\{o_1, o_2, o_8\}, \{o_3, o_4, o_5, o_6\}, \{o_7\}\}$ is a conjunctively definable partition or covering since the granule $\{o_1, o_2, o_8\}$ can be defined by the conjunctor **height**=short, the granule $\{o_3, o_4, o_5, o_6\}$ can be defined by the conjunctor **height**=tall$\wedge$**eyes**=blue, and the granule $\{o_7\}$ can be defined by the conjunctor **hair**=dark$\wedge$**eyes**=brown. Note, the join of these three formulas cannot form a tree structure.

The family of conjunctively definable partitions forms a definable partition lattice $\Pi_{CD}(U)$, which is a sub-lattice of $\Pi_D(U)$. The family of conjunctively definable coverings forms a definable covering lattice $\mathcal{T}_{CD}(U)$, which is a sub-lattice of $\mathcal{T}_D(U)$.

Definition 13. *A partition is called a tree definable partition (π_{AD}) if every equivalence class is a conjunctively definable granule, and all the equivalence classes form a tree structure.*

A partition $\Pi_{AD}(U)$ is defined by At, or a subset of At. For a subset A of attributes, we can define an equivalence relation E_A as follows:

$$\begin{aligned} xE_A y &\iff \bigwedge a \in A, I_a(x) = I_a(y) \\ &\iff I_A(x) = I_A(y). \end{aligned} \tag{2}$$

For the empty set, we obtain the coarsest partition $\{U\}$. For a nonempty subset of attributes, the induced partition is conjunctively definable. The family of partitions defined by all subsets of attributes forms a definable partition lattice $\Pi_{\mathrm{AD}}(U)$, which is a sub-lattice of $\Pi_{\mathrm{CD}}(U)$.

For the information in Table 1, we obtain the following partitions with respect to subsets of the attributes:

$$\begin{aligned} \pi_{\emptyset} &= \{U\}, \\ \pi_{\mathbf{height}} &= \{\{o_1, o_2, o_8\}, \{o_3, o_4, o_5, o_6, o_7\}\}, \\ \pi_{\mathbf{hair}} &= \{\{o_1, o_2, o_6, o_8\}, \{o_3\}, \{o_4, o_5, o_7\}\}, \\ \pi_{\mathbf{eyes}} &= \{\{o_1, o_3, o_4, o_5, o_6\}, \{o_2, o_7, o_8\}\}, \\ \pi_{\mathbf{height} \wedge \mathbf{hair}} &= \{\{o_1, o_2, o_8\}, \{o_3\}, \{o_4, o_5, o_7\}, \{o_6\}\}, \\ \pi_{\mathbf{height} \wedge \mathbf{eyes}} &= \{\{o_1\}, \{o_2, o_8\}, \{o_3, o_4, o_5, o_6\}, \{o_7\}\}, \\ \pi_{\mathbf{hair} \wedge \mathbf{eyes}} &= \{\{o_1, o_6\}, \{o_2, o_8\}, \{o_3\}, \{o_4, o_5\}, \{o_7\}\}, \\ \pi_{\mathbf{height} \wedge \mathbf{hair} \wedge \mathbf{eyes}} &= \{\{o_1\}, \{o_2, o_8\}, \{o_3\}, \{o_4, o_5\}, \{o_6\}, \{o_7\}\}. \end{aligned}$$

Since each subset defines a different partition, the partition lattice has the same structure as the lattice defined by the power set of the three attributes **height**, **hair**, and **eyes**.

All the notions developed in this section can be defined relative to a particular subset $A \subseteq At$ of attributes. A subset $X \subseteq U$ is called a definable granule with respect to a subset of attributes $A \subseteq At$ if there exists at least one formula ϕ over A such that $m(\phi) = X$. A partition π is called a definable partition with respect to a subset of attributes A if every equivalence class is a definable granule with respect to A. Let $\Pi_{D(A)}(U)$, $\Pi_{CD(A)}(U)$, and $\Pi_{AD(A)}(U)$ denote the partition (semi-) lattices with respect to a subset of attributes $A \subseteq At$, respectively. We have the following connection between partition (semi-) partition lattices and they provide a formal framework of classification problems.

$$\begin{aligned}
\Pi_{AD}(U) &\subseteq \Pi_{CD}(U) \subseteq \Pi_{D}(U) \subseteq \Pi(U),\\
\Pi_{AD(A)}(U) &\subseteq \Pi_{CD(A)}(U) \subseteq \Pi_{D(A)}(U) \subseteq \Pi(U),\\
\mathcal{T}_{CD}(U) &\subseteq \mathcal{T}_{D}(U) \subseteq \mathcal{T}(U),\\
\mathcal{T}_{CD(A)}(U) &\subseteq \mathcal{T}_{D(A)}(U) \subseteq \mathcal{T}(U).
\end{aligned}$$

3 Measures associated with granules

We introduce and review three types of quantitative measures associated with granules, measures of a single granule, measures of relationships between a pair of granules [28, 32], and measures of relationships between a granule and a family of granules, as well as a pair of family of granules.

The only measure of a single granule $m(\phi)$ of a formula ϕ is the *generality*, defined as:

$$G(\phi) = \frac{|m(\phi)|}{|U|}, \tag{3}$$

which indicates the relative size of the granule $m(\phi)$. A concept defined by the formula ϕ is more general if it covers more instances of the universe. The quantity may be viewed as the probability of a randomly selected object satisfying ϕ.

Given two formulas ϕ and ψ, we introduce a symbol $\Rightarrow$ to connect ϕ and ψ in the form of $\phi \Rightarrow \psi$. It may be intuitively interpreted as a rule which enables us to infer information about ψ from ϕ. The strength of $\phi \Rightarrow \psi$ can be quantified by two related measures [20, 28].

The *confidence* or *absolute support* of ψ provided by ϕ is the quantity:

$$AS(\phi \Rightarrow \psi) = \frac{|m(\phi \wedge \psi)|}{|m(\phi)|} = \frac{|m(\phi) \cap m(\psi)|}{|m(\phi)|}. \tag{4}$$

It may be viewed as the conditional probability of a randomly selected object satisfying ψ given that the object satisfies ϕ. In set-theoretic terms, it is the degree to which $m(\phi)$ is included in $m(\psi)$. Thus, AS is a measure of the correctness or the precision of the inference. A rule with the maximum absolute support 1 is a certain rule. The *coverage* ψ provided by ϕ is the quantity:

$$CV(\phi \Rightarrow \psi) = \frac{|m(\phi \wedge \psi)|}{|m(\psi)|} = \frac{|m(\phi) \cap m(\psi)|}{|m(\psi)|}. \tag{5}$$

It may be viewed as the conditional probability of a randomly selected object satisfying ϕ given that the object satisfies ψ. Thus, CV is a measure of the applicability or recall of the inference. Obviously, we can infer more information about ψ from ϕ if we have both a high absolute support and a high coverage. In general, there is a trade-off between support and coverage.

Consider now a family of formulas $\Psi = \{\psi_1, \ldots, \psi_n\}$ which induces a partition $\pi(\Psi) = \{m(\psi_1), \ldots, m(\psi_n)\}$ of the universe. Let $\phi \Rightarrow \Psi$ denote the inference relation between ϕ and Ψ. In this case, we obtain the following probability distribution in terms of $\phi \Rightarrow \psi_i$'s:

$$P(\Psi \mid \phi) = \left(P(\psi_1 \mid \phi) = \frac{|m(\phi) \cap m(\psi_1)|}{|m(\phi)|}, \ldots, P(\psi_n \mid \phi) = \frac{|m(\phi) \cap m(\psi_n)|}{|m(\phi)|} \right).$$

The conditional entropy $H(\Psi \mid \phi)$ defined by:

$$H(\Psi \mid \phi) = -\sum_{i=1}^{n} P(\psi_i \mid \phi) \log P(\psi_i \mid \phi), \tag{6}$$

provides a measure that is inversely related to the strength of the inference $\phi \Rightarrow \Psi$. If $P(\psi_{i_0} \mid \phi) = 1$ for one formula ψ_{i_0} and $P(\psi_i \mid \phi) = 0$ for all $i \neq i_0$, the entropy reaches the minimum value 0. In this case, if an object satisfies ϕ, one can identify one equivalence class of $\pi(\Psi)$ to which the object belongs without uncertainty. When $P(\psi_1 \mid \phi) = \ldots = (\psi_n \mid \phi) = 1/n$, the entropy reaches the maximum value $\log n$. In this case, we are in a state of total uncertainty. Knowing that an object satisfies the formula ϕ does not help in identifying an equivalence class of $\pi(\Psi)$ to which the object belongs.

Suppose another family of formulas $\Phi = \{\phi_1, \ldots, \phi_m\}$ define a partition $\pi(\Phi) = \{m(\phi_1), \ldots, m(\phi_m)\}$. The same symbol $\Rightarrow$ is also used to connect two families of formulas that define two partitions of the universe, namely, $\Phi \Rightarrow \Psi$. The strength of this connection can be measured by the conditional entropy:

$$\begin{aligned} H(\Psi \mid \Phi) &= \sum_{j=1}^{m} P(\phi_j) H(\Psi \mid \phi_j) \\ &= -\sum_{j=1}^{m} \sum_{i=1}^{n} P(\psi_i \wedge \phi_j) \log P(\psi_i \mid \phi_j), \end{aligned} \tag{7}$$

where $P(\phi_j) = G(\phi_j)$. In fact, this is the most commonly used measure for selecting attributes in the construction of decision tree for classification [15].

The measures discussed so far quantified two levels of relationships, i.e., granule level and granulation level. As we will show in the following section, by focusing on different levels, one may obtain different methods for the induction of classification rules.

4 Induction of Classification Rules by Searching Granules

For classification tasks, it is assumed that each object is associated with a unique class label. Objects can be divided into classes which form a granulation of the universe. We further assume that information about objects are given by an information table as defined in Section 2. Without loss of generality, we assume that there is a unique attribute **class** taking class labels as its value. The set of attributes is expressed as $At = C \cup \{\mathbf{class}\}$, where C is the set of attributes used to describe the objects. The goal is to find classification rules of the form, $\phi \Longrightarrow \mathbf{class} = c_i$, where ϕ is a formula over C and c_i is a class label.

Let $\pi_{\mathbf{class}} \in \Pi(U)$ denote the partition induced by the attribute **class**. An information table with a set of attributes $At = C \cup \{\mathbf{class}\}$ is said to provide a consistent classification if all objects with the same description over C have the same class label, namely, if $I_C(x) = I_C(y)$, then $I_{\mathbf{class}}(x) = I_{\mathbf{class}}(y)$. Using the concept of a partition lattice, we define the consistent classification problem as follows.

Definition 14. *An information table with a set of attributes $At = C \cup \{\mathbf{class}\}$ is a consistent classification problem if and only if there exists a partition $\pi \in \Pi_{D(C)}(U)$ such that $\pi \preceq \pi_{\mathbf{class}}$, or a covering $\tau \in T_{D(C)}(U)$ such that $\tau \preceq \pi_{\mathbf{class}}$.*

It can be easily verified that a consistent classification problem can be considered as a search for a definable partition $\pi \in \Pi_D(U)$, or more generally, a conjunctively definable partition $\pi \in \Pi_{CD}(U)$, or a tree definable partition $\pi \in \Pi_{AD}(U)$ such that $\pi \preceq \pi_{\mathbf{class}}$. For the induction of classification rules, the partition $\pi_{AD(C)}(U)$ is not very interesting. In fact, one is more interested in finding a subset $A \subset C$ of attributes such that $\pi_{AD(A)}(U)$ that also produces the correct classification. Similarly, a consistent classification problem can also be considered as a search for a conjunctively definable covering τ such that $\tau \preceq \pi_{\mathbf{class}}$. This leads to different kinds of solutions to the classification problem.

Definition 15. *A partition solution to a consistent classification problem is a conjunctively definable partition π such that $\pi \preceq \pi_{\mathbf{class}}$. A covering solution to a consistent classification problem is a conjunctively definable covering τ such that $\tau \preceq \pi_{\mathbf{class}}$.*

Let X denote a block in a partition or a covering granule of the universe, and let $des(X)$ denote its description using language $\mathcal{L}$. If $X \subseteq m(\mathbf{class} = c_i)$, we can construct a classification rule: $des(X) \Rightarrow \mathbf{class} = c_i$. For a partition or a covering, we can construct a family of classification rules. The main difference between a partition solution and a covering solution is that an object is only classified by one rule in a partition-based solution, while an object may be classified by more than one rule in a covering-based solution.

Consider the consistent classification problem of Table 1. We have the partition by **class**, a conjunctively defined partition π, and a conjunctively defined covering τ:

$$\begin{aligned} \pi_{\mathbf{class}}: &\quad \{\{o_1, o_3, o_6\}, \{o_2, o_4, o_5, o_7, o_8\}\}, \\ \pi: &\quad \{\{o_1, o_6\}, \{o_2, o_8\}, \{o_3\}, \{o_4, o_5, o_7\}\}, \\ \tau: &\quad \{\{o_1, o_6\}, \{o_2, o_7, o_8\}, \{o_3\}, \{o_4, o_5, o_7\}\}. \end{aligned}$$

Clearly, $\pi \preceq \pi_{\mathbf{class}}$ and $\tau \preceq \pi_{\mathbf{class}}$. So $\pi \preceq \pi_{\mathbf{class}}$ and $\tau \preceq \pi_{\mathbf{class}}$ are solutions of consistent classification problem of Table 1. A set of classification rules of π is:

(r1) **hair** = blond $\wedge$ **eyes** = blue $\Longrightarrow$ **class** = +,
(r2) **hair** = blond $\wedge$ **eyes** = brown $\Longrightarrow$ **class** = −,
(r3) **hair** = red $\Longrightarrow$ **class** = +,
(r4) **hair** = dark $\Longrightarrow$ **class** = −.

A set of classification rules of τ consists of (r1), (r3), (r4) and part of (r2):

(r2′) **eyes** = brown $\Longrightarrow$ **class** = −.

The first set of rules is in fact obtained by the ID3 learning algorithm [16], and the second set is by the PRISM algorithm [2]. In comparison, rule (r2′) is shorter than (r2). Object o_7 is classified by (r4) in the partition solution, while it is classified by two rules (r2′) and (r4) in the covering solution.

The left hand side of a rule is a formula whose meaning is a block of the solution. For example, for the first rule, we have $m($**hair** = blond $\wedge$ **eyes** = blue$) = \{o_1, o_6\}$.

We can re-express many fundamental notions of classification in terms of partitions.

Depending on the particular lattice used, one can easily establish properties of the family of solutions. Let $\Pi_\alpha(U)$, where $\alpha = \mathrm{AD}(C), \mathrm{CD}(C), \mathrm{D}(C)$, denote a (semi-) lattice of definable partitions. Let $\Pi_\alpha^S(U)$ be the corresponding set of all solution partitions. We have:

(i) For $\alpha = \mathrm{AD}(C), \mathrm{CD}(C), \mathrm{D}(C)$, if $\pi' \in \Pi_\alpha(U)$, $\pi \in \Pi_\alpha^S(U)$ and $\pi' \preceq \pi$, then $\pi' \in \Pi_\alpha^S(U)$;
(ii) For $\alpha = \mathrm{AD}(C), \mathrm{CD}(C), \mathrm{D}(C)$, if $\pi', \pi \in \Pi_\alpha^S(U)$, then $\pi' \wedge \pi \in \Pi_\alpha^S(U)$;
(iii)For $\alpha = \mathrm{D}(C)$, if $\pi', \pi \in \Pi_\alpha^S(U)$, then $\pi' \vee \pi \in \Pi_\alpha^S(U)$;

It follows that the set of all solution partitions forms a definable lattice, a conjunctively definable lattice, or a tree definable lattice.

Mining classification rules can be formulated as a search for a partition from a partition lattice. A definable lattice provides the search space of potential solutions, and the partial order of the lattice provides the search direction. The standard search methods, such as depth-first search, breadth-first search, bounded depth-first search, and heuristic search, can be used to find a solution to the consistent classification problem. Depending on the required properties of rules, one may use different definable lattices that are introduced earlier. For example, by searching the conjunctively definable partition lattice $\Pi_{\mathrm{CD}(C)}(U)$, we can obtain classification rules whose left hand sides are only conjunctions of atomic formulas. By searching the lattice $\Pi_{\mathrm{AD}(C)}(U)$, one can obtain a similar solution that can form a tree structure. The well-known ID3 learning algorithm in fact searches $\Pi_{AD(C)}(U)$ for classification rules [15].

Definition 16. *For two solutions $\pi_1, \pi_2 \in \Pi_\alpha$ of a consistent classification problem, namely, $\pi_1 \preceq \pi_{\mathbf{class}}$ and $\pi_2 \preceq \pi_{\mathbf{class}}$, if $\pi_1 \preceq \pi_2$, we say that π_1 is a more specific solution than π_2, or equivalently, π_2 is a more general solution than π_1.*

Definition 17. *A solution $\pi \in \Pi_\alpha$ of a consistent classification problem is called the most general solution if there does not exist another solution $\pi' \in \Pi_\alpha$, $\pi \neq \pi'$, such that $\pi \preceq \pi' \preceq \pi_{\mathbf{class}}$.*

For a consistent classification problem, the partition defined by all attributes in C is the finest partition in Π_α. Thus, the most general solution always exists. However, the most general solution may not be unique. There may exist more than one the most general solutions.

In the information Table 1, consider three partitions from the lattice $\Pi_{\mathrm{CD}(C)}(U)$:

$$\begin{aligned}
\pi_1 &: \{\{o_1\}, \{o_2, o_8\}, \{o_3\}, \{o_4, o_5\}, \{o_6\}, \{o_7\}\},\\
\pi_2 &: \{\{o_1, o_6\}, \{o_2, o_8\}, \{o_3\}, \{o_4, o_5, o_7\}\},\\
\pi_3 &: \{\{o_1, o_6\}, \{o_2, o_7, o_8\}, \{o_3\}, \{o_4, o_5\}\}.
\end{aligned}$$

We have $\pi_1 \preceq \pi_2 \preceq \pi_{\mathbf{class}}$ and $\pi_1 \preceq \pi_3 \preceq \pi_{\mathbf{class}}$. Thus, π_1 is the more specific solution than both π_2 and π_3. In fact, both π_2 and π_3 are the most general solutions.

The roles of attributes are well-studied in the theory of rough sets [14], and can be re-expressed as follows:

Definition 18. *An attribute $a \in C$ is called a core attribute if $\pi_{C-\{a\}}$ is not a solution to the consistent classification problem.*

Definition 19. *An attribute $a \in C$ is called a superfluous attribute if $\pi_{C-\{a\}}$ is a solution to the consistent classification problem, namely, $\pi_{C-\{a\}} \preceq \pi_{\mathbf{class}}$.*

Definition 20. *A subset $A \subseteq C$ is called a reduct if π_A is a solution to the consistent classification problem and π_B is not a solution for any proper subset $B \subset A$.*

For a given consistent classification problem, there may exist more than one reduct.

In the information Table 1, attributes **hair** and **eyes** are core attributes. Attribute **height** is a superfluous attribute. The only reduct is the set of attributes $\{\mathbf{hair}, \mathbf{eyes}\}$.

5 The Studies on Classification Algorithms

With the concepts introduced so far, we can remodel some popular classification algorithms. We study various algorithms from a granular computing view and propose a more general and flexible granulation algorithm.

5.1 ID3

The ID3 [15] algorithm is probably the most popular algorithm in data mining. Many efforts have been made to extend the ID3 algorithm in order to get a better classification result. The C4.5 algorithm [17] is proposed by Quinlan himself and generates fuzzy decision tree [8].

The ID3-like learning algorithms can be formulated as a heuristic search of the semi- conjunctively definable lattice $\Pi_{\mathrm{AD}(C)}(U)$. The heuristic used for the ID3-like algorithms is based on an information-theoretic measure of dependency between the partition defined by **class** and another conjunctively definable partition with respect to the set of attributes C. Roughly speaking, the measure quantifies the degree to which a partition $\pi \in \Pi_{\mathrm{AD}(C)}(U)$ satisfies the condition $\pi \preceq \pi_{\mathbf{class}}$ of a solution partition.

Specifically, the direction of ID3 search is from the coarsest partitions of $\Pi_{\mathrm{AD}(C)}(U)$ to more refined partitions. The largest partitions in $\Pi_{\mathrm{AD}(C)}(U)$ are the partitions defined by single attributes in C. Using the information-theoretic measure, ID3 first selects a partition defined by a single attribute. If an equivalence class in the partition is not a conjunctively definable granule with respect to **class**, the equivalence class is further divided into smaller granules by using an additional attribute. The same information-theoretic measure is used for the selection of the new attribute. The smaller granules are conjunctively definable granules with respect to C. The search process continues until a partition $\pi \in \Pi_{\mathrm{AD}(C)}(U)$ is obtained such that $\pi \preceq \pi_{\mathbf{class}}$.

Figure 1 shows the learning algorithm of ID3.

Fig. 1. The learning algorithm of ID3

```
IF all cases in the training set belong to the same class
THEN Return the value of the class
ELSE
   (1) Select an attribute a to split the universe, which is with the
       maximum information gain.
   (3) Divide the training set into non empty subsets, one for each
       value of attribute a.
   (3) Return a tree with one branch for each subset, each branch
       having a descendant subtree or a class value produced by
       applying the algorithm recursively for each subset in turn.
```

Following the algorithm we start with the selection of the attribute **hair**. The first step of granulation is to partition the universe with values of **hair** as it is with the largest information gain. Since there are three values for **hair**, we obtain three granules for this partition. Elements of (**hair**=dark) and (**hair**=red) granules belong to the same class, we do not need to further decompose these two granules. As elements in granule (**hair**=blond) do not belong to same class, we granulate the new universe (**hair**=blond) with attribute **eyes**. We stop granulation when elements in the two new granules (**eyes**=blue) and (**eyes**=brown) are in the same class. The partition tree is shown in Figure 2 which is the familiar ID3 decision tree.

ID3 is a granulation oriented search algorithm. It searches a partition of a problem at one time. The top-down construction of a decision tree for classification searches for a partition solution to a consistent classification problem. The induction process can be briefly described as follows. Based on a measure of connection between two partitions, one selects an attribute to divide the universe into a partition [15]. If an equivalence class is not a subset of a user defined class, it is further divided by using another attribute. The process continues until one finds a decision tree that correctly classifies all objects. Each node of the decision tree is labelled by an attribute, and each branch is labelled by a value of the parent attribute.

Fig. 2. An example partition generated by ID3

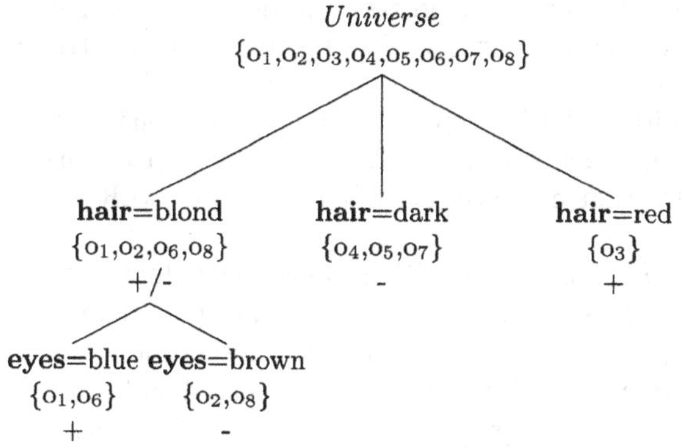

Universe
$\{o_1,o_2,o_3,o_4,o_5,o_6,o_7,o_8\}$

hair=blond $\{o_1,o_2,o_6,o_8\}$ +/-

hair=dark $\{o_4,o_5,o_7\}$ -

hair=red $\{o_3\}$ +

eyes=blue $\{o_1,o_6\}$ +

eyes=brown $\{o_2,o_8\}$ -

5.2 *k*LR

Algorithms for finding a reduct in the theory of rough sets can also be viewed as heuristic search of the partition lattice $\Pi_{\mathrm{AD}(C)}(U)$. Two directions of search can be carried, either from coarsening partitions to refinement partitions or from refinement partitions to coarsening partitions.

The smallest partition in $\Pi_{\mathrm{AD}(C)}(U)$ is π_C. By dropping an attribute a from C, one obtains a coarsening partition $\pi_{C-\{a\}}$. Typically, a certain fitness measure is used for the selection of the attribute. The process continues until no further attributes can be dropped. That is, we find a subset $A \subseteq C$ such that $\pi_A \preceq \pi_{\mathbf{class}}$ and $\neg(\pi_B \preceq \pi_{\mathbf{class}})$ for all proper subsets $B \subset A$. The resulting set of attributes A is a reduct.

Fig. 3. The learning algorithm of kLR

```
Let k = 0.
The k-level, k > 0, of the classification tree is built based on the
   (k-1)^th level described as follows:
if there is a node in (k - 1)^th level that does not consist of only
   elements of the same class then
   (1) Choose an attribute based on a certain criterion γ : At ⟶ ℜ;
   (2) Divide all the inconsistent nodes based on the selected
       attribute and produce the k^th level nodes, which are subsets
       of the inconsistent nodes;
   (3) Label the inconsistent nodes by the attribute name, and label
       the branches coming out from the inconsistent nodes by the
       values of the attribute.
```

The largest partition in $\Pi_{\mathrm{AD}(C)}(U)$ is $\pi_\emptyset$. By adding an attribute a, one obtains a refined partition π_a. The process continues until we have a partition satisfying the condition $\pi_A \preceq \pi_{\mathbf{class}}$. The resulting set of attributes A is a reduct.

The kLR algorithm [31] is proposed as one of the rough set-type search algorithms to find a reduct, which is a set of individually necessary and jointly sufficient attributes that correctly classify the objects. kLR is described in Figure 3.

Comparing the kLR algorithm with ID3-like algorithms, we note that an important feature of ID3-like algorithms is that when partitioning an inconsistent granule, a formula is chosen based on only information about the current granule. The criteria used by ID3-like methods is based on local optimization. In the decision tree, different granules at the same level may use different formulas, and moreover the same attribute may be used at different levels. The use of local optimal criteria makes it difficult to judge the overall quality of the partial decision tree during its construction process. The kLR algorithm may solve this difficulty by partitioning the inconsistent granules at the same level at the same time using the same formula. One can construct a kLR decision tree and evaluate its quality level by level. Normally, a kLR decision tree is different from the corresponding ID3 tree. However, the running example in Table 1 is too small to show the different resulting trees generated by ID3 and kLR.

5.3 PRISM

PRISM [2] is an algorithm proposed by Cendrowska in 1987. Instead of using the principle of generating decision trees which can be converted to decision rules, PRISM generates rules from the training set directly. More importantly, PRISM is a covering-based method. The PRISM algorithm can be formulated as a heuristic search of the conjunctively definable covering lattice $T_{CD(C)}(U)$. The heuristic used for PRISM is the conditional probability of a given class value given a formula. The algorithm is described in Figure 4.

Fig. 4. The learning algorithm of PRISM

```
For i=1 to n
repeat until all instances of class i have been removed
   (1) Calculate the probability of occurrence of class i for each
       attribute-value pair.
   (2) Select the attribute-value pair with maximum probability and
       create a subset of the training set comprising all instances
       with the selected combination.
   (3) Repeat (1) and (2) for this subset until it contains only
       instances of class i. The induced rule is then the conjunction
       of all the attribute-value pairs selected in creating this
       subset.
   (4) Remove all instances covered by this rule from training set.
```

Fig. 5. An example covering generated by PRISM

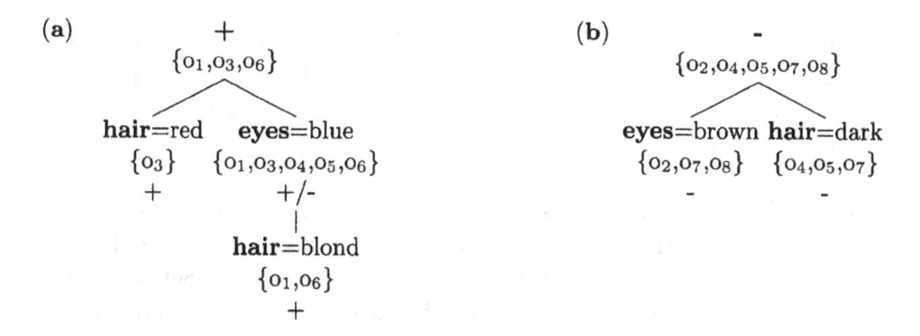

¿From granular computing point of view, PRISM is actually finding a covering of the universe. Let's still use the example of Table 1. There are two classes, + and -. For (**class** = +), the meaning set is $\{o_1, o_3, o_6\}$. The largest conditional probability of class + given all the attribute-value pairs is $P(+|\textbf{hair}=\text{red})$. We use this attribute-value pair to form a granule $\{o_3\}$. The second largest probability is $P(+|\textbf{eyes}=\text{blue})$. We use this attribute-value pair to form a second granule $\{o_1, o_3, o_4, o_5, o_6\}$, and further refine it by combining **eyes**= blue $\wedge$ **hair**= blond. The granule $\{o_1, o_6\}$ contains only (**class** = +). So far these two granules cover (**class** = +). We do the same for (**class** = −) and find two granules $\{o_2, o_7, o_8\}$ and $\{o_4, o_5, o_7\}$ which cover (**class** = −). The covering of universe has four granules. The covering is shown in Figure 5. For this particular example, PRISM provide shorter rules than ID3. This is consistent with Cendrowska's results and a recent review [1].

5.4 Granular computing approach

A granular computing approach [25, 26] is proposed as a granule network to extend the existing classification algorithms. In a granule network, each node is labelled by a subset of objects. The arc leading from a larger granule to a smaller granule is labelled by an atomic formula. In addition, a smaller granule is obtained by selecting those objects of the larger granule that satisfy the atomic formula. The family of the smallest granules thus forms a conjunctively definable covering of the universe.

We need to introduce the concepts of inactive and active granules for the implementation of this approach.

Definition 21. *A granule X is inactive if it meets one of the following two conditions:*

$$\text{(i).} \quad X \subseteq m(\textbf{class} = c_i) \text{ where } c_i \text{ is one possilbe value of } V_{\textbf{class}},$$

$$\text{(ii).} \quad X = \bigcup Y, \text{where each } Y \text{ is a child node of } X.$$

A granule X is active if it does not meet any of the above conditions.

Atomic formulas define *basic* granules, which serve as the basis for the granule network. The pair $(a = v, m(a = v))$ is called a basic concept. Each node in the granule network is a conjunction of some basic granules, and thus a conjunctively definable granule. The induction process of the granule network can be briefly described as follows. The whole universe U is selected as the root node at the initial stage. Evaluate and set the activity status of U. If U is active with respect to the conditions (i) and (ii), based on a measure of fitness, one selects a basic concept bc to cover a subset of the universe. Set the status for both the root node U and the new node. Based on a measure of activity, one of the active node is selected for further classification. This iterative process stops when a non-redundant covering solution is found. Figure 6 outlines an algorithm for the construction of a granule network [26].

The two importance issues of the algorithm is the evaluation of the fitness of each basic concept, and the the evaluation of the activity status of each active node. The algorithm is basically a heuristic search algorithm.

Fig. 6. An Algorithm for constructing a granule network

```
Construct the family of basic concepts with respect to atomic
   formulas:
            BC(U) = {(a = v, m(a = v)) | a ∈ C, v ∈ V_a}.
Set the granule network to GN = ({U}, ∅), which is a graph consisting
   of only one node and no arc.
Set the activity status of U.
While the set of inactive nodes is not a non-redundant covering
   solution of the consistent classification problem, do:
   (1) Select the active node N with the maximum value of activity.
   (2) Select the basic concept bc = (a = v, m(a = v)) with maximum value
       of fitness with respect to N.
   (3) Modify the granule network GN by adding the granule
       N ∩ m(a = v) as a new node, connecting it to N by arc, and
       labelling it by a = v.
   (4) Set the activity status of the new node.
   (5) Update the activity status of N.
```

The measures discussed in the last section can be used to define different fitness/activity functions. User can also use a measure or some measures to choose the basic concept and active node. The measure does not need to be fixed. In other words, in the process of granular computing classification, user can interactively decide what measure to be used. As a result, different measures can be used at the different levels of construction. In the rest of this section, we will use the running example, Table 1, to illustrate the basic ideas.

The initial node U is an active granule with respect to condition (i). Table 2 summarizes the measures of basic concepts with respect to U. There are three granules which are a subset of one of class values, i.e., $\{o_3\} \subseteq$ (**class** $= +$), $\{o_4, o_5, o_7\} \subseteq$ (**class** $= -$) and $\{o_2, o_7, o_8\} \subseteq$ (**class** $= -$). The values of entropy of these granules are minimum, i.e., 0. Therefore, these three granules

Formula	Granule	Generality	Confidence		Coverage		Entropy
			+	-	+	-	
height = short	$\{o_1, o_2, o_8\}$	3/8	1/3	2/3	1/3	2/5	0.92
height = tall	$\{o_3, o_4, o_5, o_6, o_7\}$	5/8	2/5	3/5	2/3	3/5	0.97
hair = blond	$\{o_1, o_2, o_6, o_8\}$	4/8	2/4	2/4	2/3	2/5	1.00
hair = red	$\{o_3\}$	1/8	1/1	0/1	1/3	0/5	0.00
hair = dark	$\{o_4, o_5, o_7\}$	3/8	0/3	3/3	0/3	3/5	0.00
eyes = blue	$\{o_1, o_3, o_4, o_5, o_6\}$	5/8	3/5	2/5	3/3	2/5	0.97
eyes = brown	$\{o_2, o_7, o_8\}$	3/8	0/3	3/3	0/3	3/5	0.00

Table 2. Basic granules and their measures for the selected active node U

are inactive. The generality of latter two granules are higher than the first granule. These three granules are added to the granule network one by one.

The union of the inactive granules $\{\{o_4, o_5, o_7\}, \{o_2, o_7, o_8\}, \{o_3\}$ cannot cover the universe. After adding these three granules to the granule network, the universe U is still the only active node (with respect to condition (i) and (ii)), therefore, it is selected. With the consideration of a non-redundant covering, we will not choose a granule that will not cover the universe even if other measures are in favor of this granule. Based on the fitness measures summarized in Table 3, **hair** = blond and **eyes** = blue contain the objects $\{o_1, o_6\}$ that can possibly form a non-redundant covering solution. **height** = tall and **eyes** = blue have the highest generality, confidence and coverage. **height** = short has the smallest entropy. One can make a comprehensive decision based on all these measures. For example, the basic concept (**eyes** = blue, $\{o_1, o_3, o_4, o_5, o_6\}$) is selected. The granule $\{o_1, o_3, o_4, o_5, o_6\}$ is added to the granule network, labelled by **eyes** = blue. The new node is active (with respect to condition (i)). By adding it to the granule network, U is no longer active (with respect to condition (ii)).

Formula	Granule	Generality	Confidence +	Confidence -	Coverage +	Coverage -	Entropy
height = short	$\{o_1, o_2, o_8\}$	3/8	1/3	2/3	1/3	2/5	0.92
height = tall	$\{o_3, o_4, o_5, o_6, o_7\}$	5/8	2/5	3/5	2/3	3/5	0.97
hair = blond	$\{o_1, o_2, o_6, o_8\}$	4/8	2/4	2/4	2/3	2/5	1.00
eyes = blue	$\{o_1, o_3, o_4, o_5, o_6\}$	5/8	3/5	2/5	3/3	2/5	0.97

Table 3. Basic granules and their measures for the selected active node U

The new node is the only active granule at this stage (with respect to condition (i)). Table 4 summarizes the measures of basic concepts with respect to it. Based on the fitness measures again, **hair** = blond contains all the objects of $\{o_1, o_6\}$ that can possibly form a non-redundant covering solution, it is also in favour of confidence, coverage and entropy. By adding

Formula	Granule	Generality	Confidence +	Confidence -	Coverage +	Coverage -	Entropy
$\wedge$**height** = short	$\{o_1\}$	1/5	1/1	0/1	1/3	0/2	0.00
$\wedge$**height** = tall	$\{o_3, o_4, o_5, o_6\}$	4/5	2/4	2/4	2/3	2/2	1.00
$\wedge$**hair** = blond	$\{o_1, o_6\}$	2/5	2/2	0/2	2/3	0/2	0.00
$\wedge$**hair** = red	$\{o_3\}$	1/5	1/1	0/1	1/3	0/2	0.00
$\wedge$**hair** = dark	$\{o_4, o_5\}$	2/5	0/2	2/2	0/3	2/2	0.00

Table 4. Basic granules and their measures for the selected active node $m($**eye** = blue)

the concept (**hair** = blond, $\{o_1, o_6\}$) to the granule network, we can get another inactive granule, and the union of all inactive granules forms a non-redundant covering solution of the consistent classification problem. It is $\tau = \{\{o_4, o_5, o_7\}, \{o_2, o_7, o_8\}, \{o_3\}, \{o_1, o_6\}\}$. The results are also shown as the granule network in Figure 7.

Fig. 7. An example of granule network

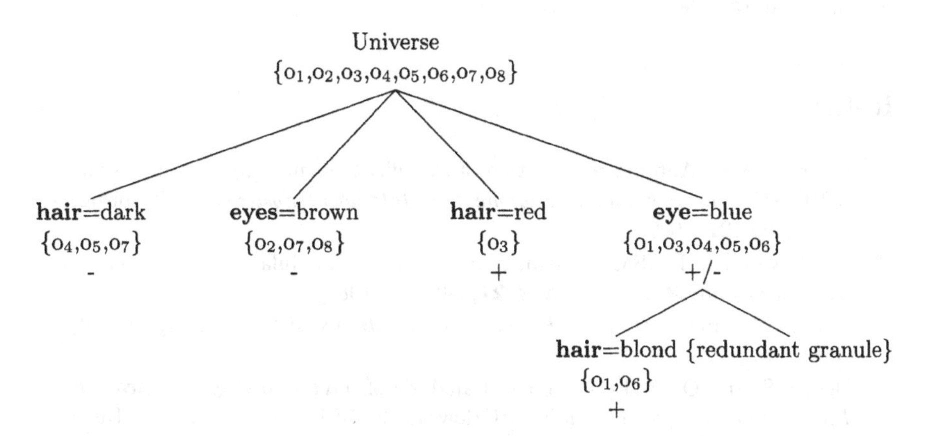

6 Conclusion

A consistent classification problem can be modelled as a search for a partition or a covering defined by a set of attribute values. In this chapter, we apply a granular computing model for solving classification problems. We explore the structures of classification of a universe. The consistent classification problems are expressed as the relationships between granules of the universe. Different classification lattices are introduced. Depending on the properties of classification rules, solutions to a consistent classification problem are definable granules in one of the lattices. Such a solution can be obtained by searching the lattice. The notion of a granule network is used to represent the classification knowledge. Our formulation is similar to the well established version space search method for machine learning [11].

The ID3, kLR and PRISM algorithms are examples of partition and covering search algorithms. As suggested by the No Free Lunch theorem [24], there is no algorithm which performs better than any other algorithms for all kinds of possible problems. It is useless to judge an algorithm irrespectively of the optimization problem. For some data sets, partition algorithm may be better than covering algorithm, for some other sets, the situation is vice versa. The new formulation enables us to precisely and concisely define many notions, and to present a more general framework for classification.

The granular computing classification approach discussed in this chapter provides more freedom of choice on heuristic and measures according to the user needs. The process is penetrated with the idea that the classification task can be more useful if it carries with the user preference and user interaction. In the future research, we will study various heuristics defined by the measures suggested in this chapter, and the evaluation of the proposed algorithm using real world data sets.

References

1. Bramer, M.A., Automatic induction of classification rules from examples using N-PRISM, *Research and Development in Intelligent Systems XVI*, Springer-Verlag, 99-121, 2000.
2. Cendrowska, J., PRISM: an algorithm for inducing modular rules, *International Journal of Man-Machine Studies*, **27**, 349-370, 1987.
3. Duda, R.O. and Hart, P.E., *Pattern Classification and Scene Analysis*, Wiley, New York, 1973.
4. Demri, S. and Orlowska, E., Logical analysis of indiscernibility, in: *Incomplete Information: Rough Set Analysis*, Orlowska, E. (Ed.), Physica-Verlag, Heidelberg, 347-380, 1998.
5. Fayyad, U.M. and Piatetsky-Shapiro, G. (Eds.) *Advances in Knowledge Discovery and Data Mining*, AAAI Press, 1996.
6. Ganascia, J.-G., TDIS: an algebraic formalization, *Proceedings of IJCAI'93*, 1008-1015, 1993.
7. Holte, R.C., Very simple classification rules perform well on most commonly used datasets, *Machine Learning*, **11**, 63-91, 1993.
8. Janikow, C., Fuzzy Decision Trees: Issues and Methods, *IEEE Transactions on Systems, Man, and Cybernetics - Part B: Cybernetics*, **28**(1), 1–14, 1998.
9. Mehta, M., Agrawal, R. and Rissanen, J., SLIQ: A fast scalable classifier for data mining, *Proceedings of International Conference on Extending Database Technology*, 1996.
10. Michalski, J.S., Carbonell, J.G., and Mirchell, T.M. (Eds.), *Machine Learning: An Artificial Intelligence Approach*, Morgan Kaufmann, Palo Alto, CA, 463-482, 1983.
11. Mitchell, T.M., Generalization as search, *Artificial Intelligence*, **18**, 203-226, 1982.
12. Mitchell, T.M., *Machine Learning*, McGraw-Hill, 1997.
13. Pawlak, Z., Rough sets, *International Journal of Computer and Information Sciences*, **11**(5), 341-356, 1982.
14. Pawlak, Z., *Rough Sets: Theoretical Aspects of Reasoning about Data*, Kluwer Academic Publishers, Dordrecht, 1991.
15. Quinlan, J.R., Learning efficient classification procedures and their application to chess end-games, in: *Machine Learning: An Artificial Intelligence Approach*,

Vol. 1, Michalski, J.S., Carbonell, J.G., and Mirchell, T.M. (Eds.), Morgan Kaufmann, Palo Alto, CA, 463-482, 1983.

16. Quinlan, J.R., Induction of decision trees, *Machine Learning*, **1**(1), 81–106, 1986.
17. Quinlan, J.R., *C4.5: Programs for Machine Learning*, Morgan Kaufmann, 1993.
18. Shafer, J.C., Agrawal, R. and Mehta, M., SPRINT: A Scalable Parallel Classifier for Data Mining, *Proceedings of VLDB'96*, 544-555, 1996.
19. Trochim, W., *The Research Methods Knowledge Base*, 2nd Edition. Atomic Dog Publishing, Cincinnati, OH., 2000.
20. Tsumoto, S., Modelling medical diagnostic rules based on rough sets, *Rough Sets and Current Trends in Computing, Lecture Notes in Artificial Intelligence, 1424*, Springer-Verlag, Berlin, 475-482, 1998.
21. Wang, G.Y., Algebra view and information view of rough sets theory, *Proceedings of SPIE'01*, **4384**: 200-207, 2001.
22. Wang, G.Y., Yu, H. and Yang, D.C., Decision table reduction based on conditional information entropy, *Chinese Journal of Computers*, **25**(7), 2002.
23. Wille, R., Concept lattices and conceptual knowledge systems, *Computers Mathematics with Applications*, **23**, 493-515, 1992.
24. Wolpert, D. H., and Macready, W.G., No free lunch theorems for optimization *IEEE Transactions on Evolutionary Computation*, **1**(1), 67-82, 1997.
25. Yao, J.T. and Yao, Y.Y., Induction of Classification Rules by Granular Computing,*Proceedings of the Third International Conference on Rough Sets and Current Trends in Computing, Lecture Notes in Artificial Intelligence*, 331-338, 2002.
26. Yao, J.T. and Yao, Y.Y., A granular computing approach to machine learning, *Proceedings of the First International Conference on Fuzzy Systems and Knowledge Discovery (FSKD'02)*, Singapore, pp732-736, 2002.
27. Yao, Y.Y., Granular computing: basic issues and possible solutions, *Proceedings of the Fifth Joint Conference on Information Sciences*, 186-189, 2000.
28. Yao, Y.Y., On Modeling data mining with granular computing, *Proceedings of COMPSAC'01*, 638-643, 2001.
29. Yao, Y.Y., Information granulation and rough set approximation, *International Journal of Intelligent Systems*, **16**, 87-104, 2001.
30. Yao, Y.Y. and Yao, J.T., Granular computing as a basis for consistent classification problems, *Communications of Institute of Information and Computing Machinery (Special Issue of PAKDD'02 Workshop on Toward the Foundation of Data Mining)*, **5**(2), 101-106, 2002.
31. Yao, Y.Y., Zhao, Y. and Yao, J.T., Level construction of decision trees in a partition-based framework for classification, *Proceedings of SEKE'04*, 199-205, 2004.
32. Yao, Y.Y. and Zhong, N., An analysis of quantitative measures associated with rules, *Proceedings of PAKDD'99*, 479-488, 1999.
33. Yao, Y.Y. and Zhong, N., Potential applications of granular computing in knowledge discovery and data mining, *Proceedings of World Multiconference on Systemics, Cybernetics and Informatics*, 573-580, 1999.
34. Zadeh, L.A., Towards a theory of fuzzy information granulation and its centrality in human reasoning and fuzzy logic, *Fuzzy Sets and Systems*, **19**, 111-127, 1997.

Data Mining as Generalization: A Formal Model

Ernestina Menasalvas[1] and Anita Wasilewska[2]

[1] Departamento de Lenguajes y Sistemas Informaticos Facultad de Informatica, U.P.M, Madrid, Spain ernes@fi.upm.es

[2] Department of Computer Science, State University of New York, Stony Brook, NY, USA anita@cs.sunysb.edu

Summary. The model we present here formalizes the definition of Data Mining as the process of information generalization. In the model the Data Mining algorithms are defined as generalization operators. We show that only three generalizations operators: classification operator, clustering operator, and association operator are needed to express all Data Mining algorithms for classification, clustering, and association, respectively. The framework of the model allows to describe formally the hybrid systems; combination of classifiers into multi-classifiers, and combination of clustering with classification.

We use our framework to show that classification, clustering and association analysis fall into three different generalization categories.

1 Introduction

During the past several years researches and practitioners in Data Mining and Machine Learning have created a number of complex systems that combine Data Mining or Machine Learning algorithms. The most common, successful and longstanding ([9]) is combination of classifiers. For example, a protein secondary structure prediction system **SSPro** ([1, 2]), combines 11 bidirectional recurrent neural networks. The final predictions are obtained averaging the network outputs for each residue. Another system, named **Prof** ([17]) combines (in multiple stages) different types of neural network classifiers with linear discrimination methods ([20]).

The most recent system for protein secondary structure prediction presented in [21] uses an approach of *stacked generalization*([?]). It builds layers of classifiers such that each layer is used to combine the predictions of the classifiers of its preceding layer. A single classifier at the top-most level outputs the ultimate prediction. Predictive accuracy of [21] system outperforms six of the best secondary structure predictors by about 2%.

Natural questions arise: which methods can be combined and which can not, and if one combines them how to interpret the results in a correct and consistent manner.

In attempt to address these questions we build a Formal Model in which basic Data Mining and Machine Learning algorithms and methods can be defined and discussed.

2 The General Model Framework

One of the main goals of Data Mining is to provide a comprehensible description of information we extract from the data bases. The description comes in different forms. In case of classification problems it might be a set of characteristic or discriminant rules, it might be a decision tree or a neural network with fixed set of weights. In case of association analysis it is a set of associations, or association rules (with accuracy parameters). In case of cluster analysis it is a set of clusters, each of which has its own description and a cluster name (class name) that can be written in a form of a set of discriminant rules. In case of approximate classification by the Rough Set analysis ([19],[22], [25], [26], [27]) it is usually a set of discriminant or characteristic rules (with or without accuracy parameters) or set of decision tables.

Of course any data table, or database can be always re-written as a set of rules, or some descriptive formulas by re-writing each record tuple (in attribute=value convention) as $a_1 = v_1 \cap a_2 = v_2 .... \cap a_n = v_n$ or as $a_1 = v_1 \cap a_2 = v_2 .... \cap a_n = v_n \Rightarrow c = c_k$, where a_i, v_i are attributes and their values defining the record and c is the classification (class) attribute with corresponding value c_i, if applicable. But by doing so we just choose another form of knowledge representation or syntactic form of the record (data base) and there is no generalization involved. The cardinality of the set of descriptions might be less then the number of records (some records might have the same descriptions once the key attribute is removed) but it is still far from our goal to have a short, comprehensible description of our information. In data mining methods we also look for shortening the length of descriptions. A database with for example 100 attributes would naturally "produce" descriptions of uncomprehensible length. The "goodness" of data mining algorithm is being judged, in between, by these two factors: number of descriptions and their length. If these two conditions are met we say that we generalized our knowledge from the initial database, that we "mined" a more comprehensive, more general information. If we for example would reduce a 100,000 descriptions (records of the database) of size 500 (number of attributes of the initial database) to 5 descriptions of size 4 we surely would say that we generalized our information, but the question would arise of the quality of such generalization and hence the validity of our method.

The formal model we present here addresses all problems described above. It also provides framework for a formal definitions of intuitive notion of generalization and generalization process. We build the model on two levels: syntactic and semantic. On syntactic level we define what does it mean for example that one set of rules is more general then another. This is being established in the model by a notion of descriptive generality relation.The quality of descriptive generalization (set of rules) is defined by its connection with semantic model. The generalization process is described in terms of generalization operators on both, syntactic and semantic levels.

The syntactic view of Data Mining as information generalization is modelled in our General Model by a Descriptive Generalization Model defined below. The definition of a Semantic Generalization Model follows.

Definition 1. *A* **Descriptive Generalization Model** *is a system*

$$\mathcal{D}\mathrm{M} = (\ \mathcal{L},\ \mathcal{DK},\ \mathcal{DG},\ \prec_{\mathcal{D}}\ )$$

where

$\mathcal{L} = (\ \mathcal{A},\ \mathcal{E}\ )$ *is called a* **descriptive language**.
$\mathcal{A}$ *is a countably infinite set called the* alphabet .
$\mathcal{E} \neq \emptyset$ *and* $\mathcal{E} \subseteq \mathcal{A}^*$ *is the set of* descriptive expressions *of* $\mathcal{L}$.
$\mathcal{DK} \neq \emptyset$ *and* $\mathcal{DK} \subseteq \mathcal{P}(\mathcal{E})$ *is a set of* **descriptions of knowledge states**.
$\prec_{\mathcal{D}} \subseteq \mathcal{DK} \times \mathcal{DK}$ *is called a* **descriptive generalization relation**.
We say that DS' is more general then DS when $\mathcal{DS} \prec_{\mathcal{D}} \mathcal{DS}'$.
$\mathcal{DG} \neq \emptyset$ *is a set of* **descriptive generalization operators** *and for each* $\mathcal{G}_d \in \mathcal{DG}, \mathcal{DS}, \mathcal{DS}' \in \mathcal{DK}$,

$$\mathcal{G}_d(\mathcal{DS}) = \mathcal{DS}' \quad iff \quad \mathcal{DS} \prec_{\mathcal{D}} \mathcal{DS}'$$

In above definition we assume that the generalization operators transform sets of descriptive expressions into a more general sets of descriptions, as defined by generalization relation. For a given Data Mining application or method one defines in detail all the components of the model (and how to do it will be a subject of a separate paper) but in the general model we only assume existence and a form of such definitions.

For example we can say that a set S1 of rules (descriptions of knowledge states) is more general then a set S2 (of descriptions of knowledge states) if it contains some rules that have shorter length (i.e. when the rules use less attributes) then any rule in the set S2, or if it has a smaller number of elements (rules) for example contains 40% less rules then the set S2, or both, or some other conditions. All this must happen while the sets S1 and S2 describe semantically the same set of objects (records). This relationship between syntactic (definition 1) and semantic (definition 2) models is being described in our General Model (definition4) by a satisfaction relation.

The generalization operators would then would transform one set of descriptions of knowledge states into the other, more general one and by composing them we can describe the generalization process, which always stops at the certain generalization level. The structure of levels of generalization depends of methods used and will be discussed in a separate paper.

We usually view Data Mining results and present them to the user in their syntactic form as it is the most natural form of communication. The algorithms process records finding similarities which are then presented in a corresponding descriptive i.e. syntactic form. But the Data Mining process is deeply semantical in its nature and in order to capture this fact we not only define a semantic model but also its relationship with the syntactic descriptions. It means that we have to assign a semantic meaning to descriptive generalization operators and this is done in the General Model by the satisfiability relation.

Definition 2. *A* **Semantic Generalization Model** *is a system*

$$\mathcal{S}\mathbf{M} = (U,\ \mathcal{K},\ \mathcal{G},\ \preceq)$$

where

$U \neq \emptyset$ *is the* **universe**,
$\mathcal{K} \neq \emptyset$ *is the set of* **knowledge states**,
$\prec\ \subseteq\ \mathcal{K} \times \mathcal{K}$ *is the* **weak information generalization relation**; $(K, K') \in \prec$ *iff* K' *is more or equally general then* K, *i.e. we assume that the relation* $\preceq$ *is reflexive.*
$\mathcal{G} \neq \emptyset$ *is the set of* **generalizations operators** *such that for every* $G \in \mathcal{G}$, *for every* $K, K' \in \mathcal{K}$,

$$G(K) = K' \ \ iff \ \ K \preceq K'.$$

Definition 3. *A* **Strong Semantic Generalization Model** *is the semantic generalization model (definition 2) in which the information generalization relation is not reflexive.*

We denote $\preceq = \prec$ *and call* $\prec$ *a* **strong information generalization relation**.

We write the resulting system as

$$\mathcal{S}\mathbf{SM} = (U,\ \mathcal{K},\ \mathcal{G},\ \prec).$$

Let DM, SSM be descriptive (definition 1) and strong semantic generalization models (definition 3). We define our General Model as follows.

Definition 4. *A* **General Model** *is a system*

$$\mathbf{M} \ = \ (\ \mathcal{D}\mathbf{M},\ \ \mathcal{S}\mathbf{SM},\ \models\ )$$

where

$\mathcal{D}$M *is a* **descriptive generalization model,**
$\mathcal{S}$SM *is the* **strong semantic generalization model,** *and*
$\models \subseteq \mathcal{E} \times U$ *is a* **satisfaction relation** *that establishes the relationship between the descriptive model* $\mathcal{D}$M, *and the strong semantic model* $\mathcal{S}$SM.

Let $\models$ be the satisfaction relation. For any $x \in U$, any $d \in \mathcal{E}$ such that $(d, x) \in \models$ we write $x \models d$ and say: *x satisfies d.* We denote, after [19].

$$e(d) = \{x \in U : \quad x \models d\}.$$

Let $\mathcal{DS} \subseteq \mathcal{E}$, we consider

$$(\mathcal{DS})^* = \{e(s) : \ s \in \mathcal{DS}\}.$$

By definition

$$\bigcup(\mathcal{DS})^* \subseteq U.$$

When $\bigcup(\mathcal{DS})^* = U$ we say that the descriptions from $\mathcal{DS}$ cover the universe, i.e. they are sufficient to describe all of the universe U. What is really the universe U?? Following the Rough Set theory we take as U the set of record identifiers for all records in our database (in the Rough Set community they are called objects) and treat them as records names. For a given record practically it is the key attribute, theoretically, it is just any name. When we perform the data mining procedures the first step in any of them is to drop the key attribute. This step let allows to examine similarities in the database as without it each record has a unique identification. For a record x (after dropping the key attribute) with attributes $a_1, ..a_n$ with corresponding values $v_1, ..v_n$, the formula s: $a_1 = v_1 \cap a_2 = v_2 .... \cap a_n = v_n$ would be an element of the set descriptors, i.e. $s \in \mathcal{E}$, and the name "x" of this record would be an element of our model universe, i.e. $x \in U$. The set $e(d) = \{x \in U : \quad x \models d\}$ contains all records (i.e. their identifiers) with the same description. The description does not need to utilize all attributes, as it is often the case, and one of ultimate goals of data mining is to find descriptions with as few attributes as possible.

In association analysis the description can be for example a frequent three itemset, i.e. s is $i_1 i_2 i_3$ and $e(s)$ represents all all transactions that contain items i_1, i_2, i_3.

Both components of our General Model contain sets of generalization operators. We define the relationship between the syntactic and semantic generalization operators in terms of correspondence relation $\models^* \subseteq \mathcal{G} \times \mathcal{G}_d$ and we assume that $\forall G \in \mathcal{G} \exists F \in \mathcal{G}_d (F \models^* G)$.
We use in our definition of the General Model the strong semantic generalization model as its component because our goal here is to build the full model (i.e. syntax and semantics) of Data Mining proper stage of data mining process.

The semantic generalization model is more general and we introduce it as it will allow us to describe in it the preprocessing part of data mining process as well. The preprocessing model will be a subject of a separate paper.

The details of the semantical model are presented in the next section. The details of descriptive generalization model and their relationship within the general model will be published in the next paper.

3 Semantic Generalization Model

In this section we define in detail and discuss the component of the semantic model $\mathcal{S}\mathbf{M}$. The model was defined (definition 2) as

$$\mathcal{S}\mathbf{M} = (U,\ \mathcal{K},\ \mathcal{G},\ \preceq).$$

The universe U and the set $\mathcal{K}$ are defined in terms of *Knowledge Generalization Systems* which are discussed in next subsection 3.1.

The the set $\mathcal{K}$ of knowledge states, the generalization relation $\prec$ are discussed the subsection **??** and the set $\mathcal{G}$ of generalization operators id defined in subsection **??**).

The model presented here follows and generalizes ideas developed during years of investigations; first in the development of Rough Sets Theory (to include only few recent publications [25], [11], [?], [13], [27], [12]); then in building Rough Sets inspired foundations of information generalization ([7], [15], [16],[14]).

3.1 Knowledge Generalization System

The *knowledge generalization system* is a generalization of the notion of an information system. The information system was introduced in [18] as a database model. The information system represents the relational table with key attribute acting as object attribute and is defined as follows.

Definition 5. *Pawlak's* **Information System** *is a system*

$$I = (U, A, V_A, f),$$

where

$U \neq \emptyset$ *is called a set of* **objects**
$A \neq \emptyset$, $V_A \neq \emptyset$ *are called the set of* **attributes** *and* **values of of attributes**, *respectively,*
f *is called an* **information** *function and*

$$f : U \times A \longrightarrow V_A$$

In the data analysis, preprocessing and data mining although we start the process with the information table (i.e. we define the lowest level of information generalization as the relational table) the meaning of the intermediate and final results are considered to be of a higher level of generalization. We represent those levels of generalization by a sets of objects of the given (data mining) universe U, as in [7], [15].

This approach follows the granular view of the data mining and is formalized within a notion of knowledge generalization system, defined as follows.

Definition 6. *A* **knowledge generalization system** *based on the information system* $I = (U, A, V_A, f)$ *is a system*

$$K_I = (\mathcal{P}(U), A, E, V_A, V_E, g)$$

where

E *is a finite set of* **knowledge attributes** *(k-attributes) such that* $A \cap E = \emptyset$.
V_E *is a finite set of* **values of k- attributes.**
g *is a partial function called* **knowledge information function** *(k-function)*

$$g : \mathcal{P}(U) \times (A \cup E) \longrightarrow (V_A \cup V_E)$$

such that
(i) $g \mid (\bigcup_{x \in U}\{x\} \times A) = f$
(ii) $\forall_{S \in \mathcal{P}(U)} \forall_{a \in A}((S, a) \in dom(g) \Rightarrow g(S, a) \in V_A)$
(iii) $\forall_{S \in \mathcal{P}(U)} \forall_{e \in E}((S, a) \in dom(g) \Rightarrow g(S, e) \in V_E)$

Any set $S \in \mathcal{P}(U)$ i.e. $S \subseteq U$ is often called **a granule** or **a group** of objects.

Definition 7. *The set*

$$G_K = \{S \in \mathcal{P}(U) : \exists b \in (E \cup A)((S, b) \in dom(g))\}$$

is called **a granule universe** *of* K_I.

Observe that g is a total function on G_K.

The condition **(i)** of definition 6 says that when $E = \emptyset$ the k-function g is total on the set $\{\{x\} : x \in U\} \times A$ and $\forall x \in U \forall a \in A(g(\{x\}, a) = f(x, a))$. Denote

$$\mathcal{P}^{one}(U) = \{\{x\} : x \in U\}.$$

Lemma 1. *For any information system*

$$I = (U, A, V_A, f),$$

the knowledge generalization system

$$K_I^{one} = (\mathcal{P}^{one}(U), A, \emptyset, V_A, \emptyset, g) = (\mathcal{P}^{one}(U), A, V_A, g)$$

is isomorphic with I. We denote it by

$$I \simeq K_I^{one}.$$

The function

$$F : U \longrightarrow \mathcal{P}^{one}(U), \quad F(x) = \{x\}$$

establishes (by condition **(i)** of definition 6) the required isomorphism of K_I^{one} and I.

Given a knowledge generalization system $K_I = (\mathcal{P}(U), A, E, V_A, V_E, g)$ based on $I = (U, A, V_A, f)$.

Definition 8. *We call $K_I^{one} \simeq I$ the* **object knowledge generalization system,** *and the system $K = (G_K, E, V_E, g)$ is called a* **granule knowledge generalization system.**

The following examples are adapted from examples on pages 260 and 261 of [6].

Example 1. The input Information System in the paper contains 6650 medical records of patients who were submitted to three medical tests t_1, t_2, t_3 and were diagnosed, or not with the disease d. We represent it as

$$I = (U, A, V_A, f)$$

for $|U| = 6650, A = \{t_1, t_2, t_3, d\}, \ V_A = \{+, -\}$, where $+, -$ denotes the presence or absence of the test/disease, respectively. The function f can be reconstructed from the table for g in the next example.

The first step of the data mining process described in the paper unifies all records that have the same values for all attributes.

Example 2. As the result of the first step we get a knowledge generalization system:

$$K1 = (\mathcal{P}(U), A, E, V_A, V_E, g)$$

with $E = \emptyset = V_E$ and g given by the table (Gn:k denotes the fact that the granule (group) n contains k objects; we don't list them explicitly):

G_K	t_1	t_2	t_3	d
G1:300	-	-	-	-
G2:20	-	-	-	+
G3:40	+	-	-	-
G4:850	+	-	-	+
G5:10	-	+	-	-
G6:350	-	+	-	+
G7:500	+	+	-	-
G8:250	+	+	-	+
G9:50	-	-	+	-
G10:150	-	-	+	+
G11:150	+	-	+	-
G12:500	+	-	+	+
G13:150	-	+	+	-
G14:300	-	+	+	+
G15:30	+	+	+	-
G16:3000	+	+	+	+

Data Mining process presented in the paper ([6]) is a classification process with extra measures: confidence, cooperative game theory Shapley and Banzhaf values, and interaction fuzzy measures incorporated in *Möbius* representation.

Example 3. The final results of [6] can be presented as the following knowledge generalization system.

$$K2 = (\mathcal{P}(U), A, E, V_A, V_E, g)$$

where $A = \{t_1, t_2, t_3, d\}$, $V_A = \{+, -\}$, $E = \{Conf, Möbius, Shapley, Banzhaf\}, V_E \in [-1, 1]$ and g is given by the table below.(G_k denotes a granule (group) k; we don't list them explicitly.)

G_K	t_1	t_2	t_3	d	Conf	*Möbius*	Shapley	Banzhaf
G1	-	-	-	+	.815	0	0	0
G2	+	-	-	+	.865	.050	.039	.058
G3	-	+	-	+	.850	.035	.030	.049
G4	+	+	-	+	.860	-.039	-.018	-.018
G5	-	-	+	+	.912	.097	.107	.103
G6	+	-	+	+	.951	-.018	.010	.011
G7	+	+	-	+	.860	-.039	-.018	-.018

3.2 Model Components: Universe and Knowledge States

Any Data Mining process starts with a certain initial set of data. The model of such a process depends on representation of this data, i.e. it starts with an initial information system

$$I_0 = (U_0, A_0, V_{A_0}, f_0)$$

and we adopt the **universe U_0 as the universe of the model**, i.e.

$$\mathcal{S}\mathrm{M} = (U_0,\ \mathcal{K},\ \mathcal{G},\ \preceq).$$

In preprocessing stage of data mining process we might perform the following standard operations:

1. eliminate some records, obtaining as result a new information system with an universe $U \subseteq U_0$, or

2. eliminate some attributes, obtaining as result the information system I with the set of attributes $A \subset A_0$, or

3. perform some operations on values of attributes: normalization, clustering, application of concept hierarchy on, etc..., obtaining some set V_A of values of attributes that is similar, or equivalent to V_0. We denote it by

$$V_A \sim V_0.$$

Given an attribute value $v_a \in V_A$ and a corresponding attribute $v_a^0 \in V_0$ (for example v_a being a normalized form of v_a^0 or v_a being a more general form as defined by concept hierarchy of v_a^0) we denote this correspondence by

$$v_a \sim v_a^0.$$

We call any information system I obtained by any of the above operation **a subsystem of I_0**. We put it formally in the following definition.

Definition 9. *Given two information systems $I_0 = (U_0, A_0, V_{A_0}, f_0)$ and $I = (U, A, V_A, f)$, we say that I* **is a subsystem of I_0** *and denote it as*

$$I \subseteq I_0$$

if and only if the following conditions are satisfied

$U \subseteq U_0$,
$A \subseteq A_0$, $V_A \sim V_0$, and
the information functions f and f_0 are such that

$$\forall x \in U \forall a \in A (f(x,a) = v_a \ iff \ \exists v_a^0 \in V_0 (f_0(x,a) = v_a^0 \cap v_a^0 \sim v_a)).$$

Given initial information system $I_0 = (U_0, A_0, V_{A_0}, f_0)$, the object knowledge generalization system (definition 8)

$$K_{I_0}^{one} = (\mathcal{P}^{one}(U_0), A_0, \emptyset, V_{A_0}, \emptyset, g) = (\mathcal{P}^{one}(U_0), A, V_A, g)$$

isomorphic with I_0 i.e.

$$K_{I_0}^{one} \simeq I_0$$

is also called **the initial knowledge generalization system.**

Data Mining process in the preprocessing stage consists of transformations the initial I_0 into some of $I \subseteq I_0$ and subsequently, in the data mining proper stage, of transformations of knowledge generalizations systems K_I based on $I \subseteq I_0$. The transformations in practice are defined by different Data Mining algorithms, and in our model by appropriate generalization operators. Any data mining transformation starts, for unification purposes with corresponding initial knowledge generalization systems $K_I^{one} \simeq I$. We hence adopt the following definition of the set $\mathcal{K}$ of knowledge states.

Definition 10. *We adopt the set*

$$\mathcal{K} = \{K_I : \quad I \subseteq I_0\}$$

of all knowledge generalization systems based on the initial information system (input data) I_0 *as the set of* **knowledge states of** $\mathcal{SM}$.

The set $\mathcal{K}^{prep} \subseteq \mathcal{K}$ *such that*

$$\mathcal{K}^{prep} = \{K_I^{one} : \quad K_I^{one} \simeq I \text{ and } I \subseteq I_0\}$$

is called **a set of preprocessing knowledge states,** *or preprocessing knowledge systems of* $\mathcal{SM}$.

3.3 Model Components: Information Generalization Relations

Data Mining, as it is commonly said, is a process of generalization. In order to model this process we have first b define what does it mean that one stage (knowledge system) is more general then the other.The idea behind is very simple. It is the same as saying that $(a+b)^2 = a^2 + 2ab + b^2$ is a more general formula then the formula $(2+3)^2 = 2^2 + 2 \cdot 2 \cdot 3 + 3^2$. This means that one description (formula) is more general then the other if it describes more objects. From semantical point of view it means that generalization consists in putting objects (records) in sets of objects. From syntactical point of view generalization consists of building descriptions (in terms of attribute, values of attributes pairs) of these sets of objects, with some extra parameters (group attributes), if needed. Hence the definition of our knowledge generalization system. The generalization process starts with the input data I_0 i.e. with the initial knowledge generalization system$K^1_{I_0} \simeq I_0$ and produces systems which we call more general, with universes containing $S \in \mathcal{P}^1(U_0)$ with more then one element i.e. such that $|S| > 1$, still based on I_0 or any $I \subseteq I_0$. Formally we describe this in terms of generalization relation.

Definition 11. *Given set $\mathcal{K}$ of knowledge states (definition 10) based on the input data I_0 and $K, K' \in \mathcal{K}$ i.e.*

$$K = (\mathcal{P}(U_0), A, E, V_A, V_E, g), \quad K' = (\mathcal{P}(U_0), A', E', V_{A'}, V_{E'}, g').$$

Let $G_K, G_{K'}$ be granule universes (definition 7) of K, K' respectively. We define **a weak generalization relation**

$$\preceq \subseteq \mathcal{K} \times \mathcal{K}$$

as follows:

$$K \preceq K' \quad iff$$

i $|G_{K'}| \leq |G_K|$,
ii $A' \subseteq A$.

If $K \preceq K'$ we say that the system K' **is more or equally general as** K.

Observe that the relation $\preceq$ is reflexive and transitive, but is not antisymmetric, as systems K and K' such that $K \preceq K'$ may have different sets of knowledge attributes and knowledge functions.

Definition 12. *Let $\preceq \subseteq \mathcal{K} \times \mathcal{K}$ be relation defined in the definition 11. A relation*

$$\prec_{dm} \subseteq \preceq$$

such that it satisfies additional conditions:

iii $|G_{K'}| < |G_K|$,
iv $\exists S \in G_{K'}(|S| > 1)$

is called **a data mining generalization relation.**

Lemma 2. *The relation $\prec_{dm}$ is not reflexive, and the following properties hold.*

(1) The weak generalization relation of definition 11 is the weak information generalization relation of the semantic generalization model (definition 2),

(2) $\prec_{dm} \subset \preceq$,
(3) $\prec_{dm}$ *is a strong information generalization of the definition 3 and if* $K \prec_{dm} K'$ *we say that the system* K' **is more general then** K.

Consider the system $K2_I$ from example 3. $K2_I$ is more general then the system $K1_I$ from example 2 as $G_{K1} = \{G1, G2, ..., G16\}$ and $G_{K2} = \{G1, G2, ..., G7\}$ what proves condition **ii** of definition 11 and condition **iii** of definition 12. In this case the sets of attributes of $K1$ and $K2$ are the same. Moreover, any $Gi \in G_{K2_I}$ has more then one element, so condition **iv** of definition 12 is also satisfied and $\preceq$ is a data mining generalization relation.

Similarly we get that the systems $K1, K2$ are more general with respect to $\prec_{dm}$ then a system $K0 \simeq I$ for I the initial system from example 1. I.e we have the following properties.

Lemma 3. *Let system* $K0 \simeq I$ *for* I *the initial system from example 1,* $K1, K2$ *be systems from examples 2and 3, respectively. The following relationships hold.*

$$K1 \preceq K2,$$

$$K1 \prec_{dm} K2,$$

$$KO \prec_{dm} K1 \prec_{dm} K2.$$

The preprocessing of data is the initial (an crucial) step of the data mining process. We show now that we can talk about preprocessing operations within our generalization model. The detailed analysis of preprocessing methods and techniques within it will be a subject of separate paper.

Definition 13. *Let* $\mathcal{K}^{prep} \subseteq \mathcal{K}$ *be a the set of preprocessing states (definition 10). A relation* $\preceq_{prep} \subseteq \preceq$ *defined as follows:*

$$\preceq_{prep} = \{(K, K') \in \preceq : \quad K, K' \in \mathcal{K}^{prep}\}$$

is called **a preprocessing generalization relation.**

Lemma 4. *The preprocessing generalization relation is a weak generalization relation and is not a data mining generalization relation i.e.*

$$\preceq_{prep} \cap \prec_{dm} \; = \; \emptyset.$$

Within our framework the systems K, K' such that $K \preceq_{prep} K'$ are, in fact, equally general. So why do we call some preprocessing operations a "generalization"? There are two reasons. One is that traditionally some preprocessing operations have been always called by this name. For example we usually state that we "generalize" attributes by clustering, by introducing attributes hierarchy, by aggregation, etc. as stated on page 114 of the most comprehensive, as far, Data Mining book ([8]).

> ...."Data transformation (preprocessing stage) can involve the following .....
> **Generalization** of the data , where low-level or "primitive" (raw) data are replaced by higher -level concepts through the use of concept hierarchies. For example, categorical attributes ..... can be generalized to higher level concepts. Similarly, values for numeric attributes,like ... may be mapped to higher level concepts." ....

The second, more important reason to call some preprocessing operations a (weak) generalization is that they lead to the "strong" information generalization in the next, data mining proper stage and we perform them in order to improve the quality (granularity) of the generalization.

We present the case of attributes values generalization is presented below in the examples 4, 5 and the data ming stage based on this preprocessing in the example 6.

Example 4. Let the input data be an information system

$$I1 = (U1, A1, V_{A1}, f_1),$$

where $A1 = \{a1 = size, a2 = lenght, a3 = hight, a4 = color\}$, with $V_{a1} = \{small, medium, big\}, V_{a2} = \{1.0, 1.5, 2.2, 2.4, 3.1, 3.5, 3.6\}$,
$V_{a3} = \{4.1, 4.6, 5.6, 5.7, 6.0, 6.1\}, V_{a4} = \{blue\}$. The set $V_{A1} = \bigcup_{i=1}^{i=4} V_{ai}$. The function f_1 is such that

$$\forall x \in U1 \forall a \in A1 (f_1(x, a) = g_1(\{x\}, a),$$

where the function g_1 is defined by the table below.

$\mathcal{P}^1(U1)$	a1	a2	a3	a4
{x1}	small	1.5	5.6	blue
{x2}	medium	2.2	5.7	blue
{x3}	small	1.0	5.7	blue
{x4}	big	2.2	4.6	blue
{x5}	medium	3.1	6.0	blue
{x6}	small	2.4	5.7	blue
{x7}	big	1.5	4.1	blue
{x8}	medium	3.6	6.1	blue
{x9}	small	2.2	5.7	blue
{x10}	big	1.0	4.6	blue
{x11}	medium	3.1	4.1	blue
{x12}	small	1.0	5.7	blue
{x13}	big	1.5	6.1	blue
{x14}	medium	2.4	4.6	blue
{x15}	medium	2.2	4.1	blue
{x16}	small	1.0	5.6	blue

Observe that by Fact 1 the function g_1 defines a knowledge system

$$K^1_{I1} == (\mathcal{P}^1(U1), A1, V_{A1}, g_1)$$

isomorphic with $I1$ i.e.

$$K^1_{I1} \simeq I1.$$

Assume now that in the preprocessing stage we generalize our data by applying the **concept hierarchy** on values of attributes $a2$ and $a3$ as follows. For $V_{a2} = \{1.0 = 1.5 = small, 2.2 = 2.4 = medium, 3.1 = 3.5 = 3.6 = big\}, V_{a3} = \{4.1 = 4.6 = small, 5.6 = 5.7 = medium, 6.0 = 6.1 = big\}$. We also dismiss the attribute $a4$, as it has the same value for all records considered.The resulting system is called $I2$. Obviously, $I2 \subseteq I1$ and the system

$$K^1_{I2} == (\mathcal{P}^1(U2), A2, V_{A2}, g_2)$$

presented below is isomorphic with $I2$ i.e.

$$K^1_{I2} \simeq I2.$$

Example 5. Let $K^1_{I2} == (\mathcal{P}^1(U2), A2, V_{A2}, g_2)$, where $A2 = \{a1 = size, a2 = lenght, a3 = hight\}$, $V_{A2} = \{small, medium, big\}$, $U2 = U1$ and the function g_1 is defined by the table below.

$U(K^1_{I2})$	a1	a2	a3
{x1}	small	small	medium
{x2}	medium	small	medium
{x3}	small	small	medium
{x4}	big	small	small
{x5}	medium	medium	big
{x6}	small	small	medium
{x7}	big	small	small
{x8}	medium	medium	big
{x9}	small	small	medium
{x10}	big	small	medium
{x11}	medium	medium	small
{x12}	small	small	medium
{x13}	big	small	big
{x14}	medium	medium	small
{x15}	medium	small	small
{x16}	small	small	medium

Let $K2 = K^1_{I2}$ be the system defined in the example 5. Assume now that in the first step of our Data Mining process we unify all records that have the same values of attributes for all attributes involved. The result is a knowledge generalization system $K3$ defined below.

Example 6. The system $K3$ function g given by the following table.

G_{K3}	a1	a2	a3
{x1,x3,x6,x9,x12,x16}	small	small	medium
{x2}	medium	small	medium
{x4,x7}	big	small	small
{x5,x8}	medium	medium	big
{x10}	big	small	medium
{x11, x14}	medium	medium	small
{x13}	big	small	big
{x15}	medium	small	small

Lemma 5. *Let $K1 = K^1_{I1}, K2 = K^1_{I2}$ and $K3$ be systems defined in examples 4, 5, 5, respectively. The following relationships hold.*

(1) $K1 \preceq K2$, $K1 \not\prec_{dm} K2$, *but the weak generalization relation* $\preceq$ *on* $\{K1, K2\}$ *is a preprocessing generalization and*

(2) $K1 \prec_{prep} K2$,

(3) $K1 \preceq K2 \preceq K3$,

(4) $K1 \preceq_{prep} K2 \prec_{dm} K3$, *but* $K1 \not\prec_{dm} K3$.

The above examples show that preprocessing stage and data mining stage "generalize" data in a different way and that in fact, the generalization proper (i.e. strong generalization in our model) occurs only at the data mining stage.

3.4 Data Preprocessing, Data Mining, and Data Mining Process Semantic Models

It is natural that when building a model of the data mining process one has to include data preprocessing methods and algorithms, i.e. one has to model within it preprocessing stage as well as the data mining proper stage. In order to achieve this task we choose the notion of weak information generalization relation as a component of our (the most general) notion of the semantic generalization model (definition 2). We have then introduced the preprocessing and the data mining generalization relations (definitions 13, 12, respectively) and proved (lemma 4) that the preprocessing relation is a special case of the weak information generalization relation and it is disjoint with our data mining generalization relation. This means that within the framework of our general model we were able to distinguish (as we should have) the preprocessing generalization from the data mining proper stage generalization.

Consequently we define here the semantic models of data preprocessing, data mining, and data mining precess. They are all particular cases of our semantic generalization model (definition 2).

Definition 14. *When we adopt the preprocessing generalization relation* $\preceq_{prep}$ *(definition 13) as the information generalization relation of the semantic generalization model (definition 2) we call the model thus obtained a* **Semantic Preprocessing Model** *and denote it* $\mathcal{S}$**PM**, *i.e.*

$$\mathcal{S}\mathbf{PM} = (U,\ \mathcal{K}^{prep},\ \mathcal{G}_{prep},\ \prec_{prep})$$

where

$\mathcal{K}^{prep}$ *is the set of preprocessing knowledge states (definition 10),*
$\mathcal{G}_{prep} \subseteq \mathcal{G}$ *called a set of preprocessing generalization operators (to be defined separately).*

The data mining proper stage is determined by the data mining generalization relation and is defied formally as follows.

Definition 15. *Let* $\prec_{dm}$ *be the data mining generalization relation (definition 12). A* **Semantic Data Mining Model** *is a system*

$$\mathcal{S}\mathbf{DM} = (U,\ \mathcal{K},\ \mathcal{G}_{dm},\ \prec_{dm})$$

where

$$\mathcal{G}_{dm} \subseteq \mathcal{G}$$

for $\mathcal{G}_{dm} \neq \emptyset$ *being a set of data mining generalization operators defined in the next section.*

Now, we express the whole data mining process within our semantic generalization model as follows.

Definition 16. *A* **Semantic Data Mining Process Model** *is a system*

$$\mathcal{S}\mathbf{DMP} = (U,\ \mathcal{K},\ \mathcal{G}_p,\ \preceq_p)$$

where

$\preceq_p = \preceq_{prep} \cup \prec_{dm}$,
$\mathcal{G}_p = \mathcal{G}_{prep} \cup \mathcal{G}_{dm}$.

We have already defined all components of our respective semantic models except for the most important one: the set of generalization operators. This subject is introduced and discussed in the next section.

4 Generalization Operators

The main idea behind the concept of generalization operator is to capture not only the fact that data mining techniques generalize the data but also to categorize the possible existing methods. We want to do it in as exlusive/inclusive sense as possible. It means that we want to make sure that our categorization will to distinguish as it should, for example clustering from classification making at the same time all classification methods fell into one category, called classification operator while all clustering methods would fall into the category called clustering operator. The third category is the association analysis described in our framework by association operator. We don't include in our analysis purely statistical methods like regression, etc... This gives us only three data mining generalization operators to consider: classification, clustering, and association.

As we have already observed the preprocessing and data mining are also disjoint , inclusive/exlusive categories and can be

described as such in our framework. Hence if we use for example clustering in the preprocessing stage it should (and will be) described by a different clustering operator then the data mining clustering operator. The preprocessing is an integral and most important stage of the data mining process and needs as careful analysis and the data mining itself. We have already included some of basic definitions and the language in which we plan to further develop it analysis, but its full analysis will be a subject of future investigations.

4.1 Classification Operator

In the classification process we are given a data set (set of records) with a special attribute C, called a class attribute. The values $c_1, c_2, ...c_n$ of the class attribute C are called class labels. We call such data set a classification data. An algorithm (method) is called a classification algorithm if it uses the classification data to build a set of patters: discriminant and/or characterization rules or other pattern descriptions that then can be used to classify unknown sets of records (objects). For that reason, and because of the goal, the classification algorithm is often called shortly a classifier. The name classifier implies more then just the classification algorithm. The classifier is in fact a final product of the classification data set and a set of patterns created by a classification algorithm. Building a classifier (as a product ready for future use) involves two major phases called training and testing. In both phases we use our given classification data set in which the class labels for all records are known. We split it into two or more subsets called training and testing sets, respectively. We use the training data set and the classification algorithm to build our sets of patterns (rules, trees, Neural or Bayesian Network). We use the test data to evaluate created patterns. The measure of goodness of the classification process is called a predictive accuracy. The classifier is build i.e. wc terminate the process if it has been trained and tested and predictive accuracy is on an acceptable level.

As we can see the classification process is both semantical (grouping objects in sets that would fit the classes) and syntactical (finding the descriptions of those sets in order to use them for testing and future classification). In fact all data mining techniques share the same characteristics of semantical/syntactical duality. We have expressed this duality (and its importance) in our notion of General Model (definition 4). We also included it in the notion of Knowledge Generalization System (definition 6). The duality in the General Model is explicit, the duality included in the Knowledge Generalization System is more implicit. The knowledge generalization system is a basic component of the Semantic Generalization Model (definition 2. We illustrate these points in the following simple examples.

Example 7. Let $K^1 == (\mathcal{P}^1(\{x1, ..., x16\}), A = \{a1, a2\} \cup \{\mathbf{C}\}, V_A, g)$, where $a1 = size, a2 = lenght, \mathbf{C} = hight, V_A = \{small, medium, big\}$. The attribute $\mathbf{C}$ is the classification attribute. We denote the values $c1, c2, c3$ of the classification attribute $\mathbf{C}$ by $c1 = \mathbf{s}$ for *small*, $c2 = \mathbf{m}$ for *medium* and $c3 = \mathbf{b}$ for *big*. The function g given by the table below.

$U(K^1)$	a1	a2	C
{x1}	small	small	**m**
{x2}	medium	small	**m**
{x3}	small	small	**m**
{x4}	big	small	**s**
{x5}	medium	medium	**b**
{x6}	small	small	**m**
{x7}	big	small	**s**
{x8}	medium	medium	**b**
{x9}	small	small	**m**
{x10}	big	small	**m**
{x11}	medium	medium	**s**
{x12}	small	small	**m**
{x13}	big	small	**b**
{x14}	medium	medium	**s**
{x15}	medium	small	**s**
{x16}	small	small	**m**

The classification attribute **C** classifiers the records (objects) in the following classes $C1, C2, C3$ (sets, groups, granules):
$C1 = \{x4, x7, x11, x14, x15\}$ with description **C** = **s**,
$C2 = \{x1, x2, x3, x6, x9, x10, x12, x16\}$ with description **C** = **m**, and
$C3 = \{x5, x8, x13\}$ with description **C** = **b**.
We also say that the semantics (meaning) of the description (syntax) **C** = **c** with respect to the given the data set K^1 is the class (set) $C1 = \{x4, x7, x11, x14, x15\}$. The corresponding semantics of descriptions **C** = **m**, **C** = **b** are the sets $C2, C3$, respectively. The above is also an example and motivation for the need of the descriptive and semantic components together with the satisfaction relation in out General Model (definition 4.

Assume now that we run a classification algorithm and we get the following system as result.

Example 8. Let K' be a system obtained from K. The function g of K' is given by the following table.

$G_{K'}$	a1	a2	C
G1 = {x1,x3,x6,x9,x12,x16}	small	small	**m**
G2 = {x2}	medium	small	**m**
G3 = {x4,x7}	big	small	**s**
G4 = {x5,x8}	medium	medium	**b**
G5 = {x10}	big	small	**m**
G6 = {x11, x14}	medium	medium	**s**
G7 = {x13}	big	small	**b**
G8 = {x15}	medium	small	**s**

Of course, it is obvious that

$$K \prec_{dm} K'.$$

The correctness of the final result of our algorithm is **semantically** verified by the fact that the following is true.

Lemma 6. *Let K, K', $C1, C2, C3$ be defined in and by the above examples 7, 8. The following set inclusions hold.*

(1) $G3 \subseteq C1$, $G6 \subseteq C1$, $G8 \subseteq C1$.
(2) $G1 \subseteq C1$, $G2 \subseteq C1$, $G5 \subseteq C1$. C
(3) $G4 \subseteq C1$, $G7 \subseteq C1$.

Observe that the sets $G3, G6, G8$ provide semantics for the following descriptions which we read from the system K'. $(a1 = big \cap a2 = small)$, $(a1 = medium \cap a2 = medium)$, and $(a1 = medium \cap a2 = small)$, for $G3, G6, G8$, respectively.
The sets $G1, G2, G5$ are semantics for $(a1 = small \cap a2 = small), (a1 = medium \cap a2 = small)$, and $(a1 = big \cap a2 = small)$, respectively.
And the sets $G4, G7$ are semantics for $(a1 = medium \cap a2 = medium)$ and $(a1 = big \cap a2 = small)$, respectively.

The set inclusions of the lemma 6 prove the correctness of the following classification discriminant rules, and hence the correctness of the syntactic output of the classification algorithm used to obtain the system K' from the input K.

(1) $((a1 = small \cap a2 = small) \Rightarrow C = s)$, $((a1 = medium \cap a2 = small) \Rightarrow C = s)$, and $((a1 = big \cap a2 = small) \Rightarrow C = s)$.
(2) $((a1 = small \cap a2 = small) \Rightarrow C = m)$, $((a1 = medium \cap a2 = small) \Rightarrow C = m)$, and $(a1 = big \cap a2 = small) \Rightarrow C = m)$.
(3) $((a1 = medium \cap a2 = medium) \Rightarrow C = b)$ and $(a1 = big \cap a2 = small) \Rightarrow C = b)$.

We call our model a semantic model even if it does carry all syntactic information we need, because we consider in it, on this stage, only a semantic notion of information generalization. The similar investigations into the descriptive part will be a subject of next paper.

The formal definitions of all ideas presented intuitively above follow.

Definition 17. *Any information system $I = (U, A, V_A, f)$ with a distinguished class attribute is called a* **classification information system**, *or* **classification system** *for short. We represent A in this case as $A = A' \cup \{C\}$, $V_A = V_{A-\{C\}} \cup V_{\{C\}}$. We call C the* **class attribute** *and assume that the set of values of the class attribute C has at least two element (represents at least two distinguished classes). We write the resulting system as*

$$CI = (U, A' \cup \{C\}, V_{A-\{C\}} \cup V_{\{C\}}, f).$$

The classification information system is called in the Rough Set community and literature ([26], [19], [22], [11], [27]) a decision information system with the **decision attribute** C. We adopted our definition from the Rough Set literature for our more general framework.

We assume here, as it is the case in usual classification problems, that we have only one classification attribute. It is possible, as the Rough Set community does, to consider the decision information systems with any non empty set $C \subset A$ of decision attributes.

Definition 18. *The corresponding set $\mathcal{K}$ of knowledge systems based on the initial classification information system CI_0 as in definition 10 is called the set of* **classification knowledge systems** *and denoted by $\mathcal{CK}$.*

Let $\mathcal{S}$**DM** $= (U,\ \mathcal{K},\ \mathcal{G}_{dm},\ \prec_{dm})$ be the **Semantic Data Mining Model** (definition 15).

Definition 19. *An operator $G \in \mathcal{G}_{dm}$ is called a* **classification operator** *iff G is a partial function that maps $\mathcal{CK}$ into $\mathcal{CK}$, i.e.*

$$G:\ \mathcal{CK} \longrightarrow \mathcal{CK}.$$

We denote the set of classification operators by $\mathcal{G}_{class}$.

Example 9. Let G be a function with the domain $\{K\}$, such that $G(K) = K'$, where K is the system defined in example 7 and K' is the system from example 8. Obviously G is a classification operator.

Lemma 7. *By definition*

$$\mathcal{G}_{clas} \subseteq \mathcal{G}_{dm}$$

and for any $G \in \mathcal{G}_{class}$, for any $K \in \mathcal{CK}$ such that $K \in domG$,

$$G(K) = K' \quad iff \quad K \prec_{dm} K'.$$

Observe that our definition of the classification operators gives us freedom of choosing the level of generality (granularity) with which we want to describe our data mining technique, in this case the classification process. We illustrate this in the following example.

Example 10. Let K be a classification generalization system isomorphic with the classification information system (decision system) I, i.e. $K \simeq I$, and I is defined in the example 1. The classification (decision) attribute is the attribute d.

Let $K1, K2$ be classification systems defined in examples 2, and 3, respectively.

Consider $G, G' \in \mathcal{G}_{clas}$ such that $K, K1 \in domG$ and $G(K) = K1, G(K1) = K2$ and $K \in domG'$ but $K1 \notin domG'$ and $G'(K) = K2$.

Obviously the classification operator G describes steps in obtaining the final result $K2$ from the input data K with more detail then the operator G'.

4.2 Clustering Operator

In intuitive sense the term *clustering* refers to the process of grouping physical or abstract objets into classes of similar objects. It is also called *unsupervised* learning or unsupervised classification. *A cluster* is a collection of data objects that are similar to one another within the collection and are dissimilar to the objects in other clusters ([3]). Clustering analysis constructs hence meaningful partitioning of large sets of objects into smaller components. Our approach to the definition of clustering operator is semantical and we are not expressing explicitly in our semantical model the second, syntactical, property of the clusters: their measures of similarity and dissimilarity. Nevertheless these measures must be present in our definition of the knowledge generalization system applied to clustering analysis. We define hence a notion of clustering knowledge system as follows.

Definition 20. *A knowledge generalization system $K \in \mathcal{K}$,*

$$K = (\mathcal{P}(U), A, E, V_A, V_E, g)$$

is called a **clustering knowledge system** *iff $E \neq \emptyset$ and there are two knowledge attributes $s, ds \in E$ such that for any $S \in G_K$ (definition 7)*

$g(S,s)$ is a measure of similarity of objects in S,
$g(S,ds)$ is a measure of similarity of objects in S.

We put

$$\mathcal{K}^{cluster} = \{K \in \mathcal{K} : K \textit{ is a clustering system}\}.$$

Definition 21. *We denote $\neg\mathcal{CK}^{one}$ the set of all* **object** *knowledge generalization systems that are not classification systems, i.e.*

$$\neg\mathcal{CK}^{one} = \mathcal{K}^{one} \cap \neg\mathcal{CK}.$$

Definition 22. *An operator* $G \in \mathcal{G}_{dm}$ *is called a* **clustering operator** *iff* G *is a partial function that maps* $\neg \mathcal{CK}^{one}$ *into* $\mathcal{K}^{cluster}$, *i.e.*

$$G: \neg \mathcal{CK}^{one} \longrightarrow \mathcal{K}^{cluster}$$

and for any $K \in domG$ *such that*

$$G(K) = K'$$

the granule universe (definition 7) of K' *is a* **partition** *of* U.

Elements of the granule universe $G_{K'}$ *of* K' *are called* **clusters** *defined (generated) by the operator* G.
We denote the set of all clustering operators by $\mathcal{G}_{cluster}$.

To include clustering method that return not always disjoint clusters (with for example some measure of overlapping) we introduce a cover cluster operator.

Definition 23. *An operator* $G \in \mathcal{G}_{dm}$ *is called a* **cover clustering operator** *iff* G *is a partial function that maps* $\neg \mathcal{CK}^{one}$ *into* $\mathcal{K}^{cluster}$, *i.e.*

$$G: \neg \mathcal{CK}^{one} \longrightarrow \mathcal{K}^{cluster}$$

and for any $K \in domG$ *such that*

$$G(K) = K'$$

the granule universe of K' *is a* **cover** *of* U *and* K' *contains a knowledge attribute describing the measure of overlapping of not disjoint elements of the granule universe.*
We denote the set of all cover clustering operators by $\mathcal{G}_{ccluster}$.

Observe that we allow the cluster knowledge system be a classification system. It allows the cluster operator to return not only clusters it has generated, their descriptions (with similarity measures) but their names, if needed.

In order to incorporate the notion of classification by clustering (for example k-nearest neighbor algorithm) we need only to change the domain of the cluster operator to allow the use of training examples.

Definition 24. *The clustering operator* G *(definition 24) is a* **classification clustering operator** *if the domain of* G *is* $\mathcal{CK}^{one}$ *instead of* $\neg \mathcal{CK}^{one}$.
We denote the set of all classification clustering operators by $\mathcal{G}_{clcluster}$.

The most general clustering operator, from which all other clustering operators can be obtained as particular cases is the following.

Definition 25. *An operator* $G \in \mathcal{G}_{dm}$ *is called a* **general clustering operator** *iff* G *is a partial function that maps* $\mathcal{K}^{one}$ *into* $\mathcal{K}^{cluster}$*, i.e.*

$$G: \mathcal{K}^{one} \longrightarrow \mathcal{K}^{cluster}$$

and for any $K \in domG$ *such that*

$$G(K) = K'$$

the granule universe of K' *is a* **cover** *of* U *and* K' *contains a knowledge attribute describing the measure of overlapping of not disjoint elements of the granule universe.*
We denote the set of all cover clustering operators by $\mathcal{G}_{gcluster}$.

Observe that the domain of the general cluster operator is $\mathcal{K}^{one}$ and G is a partial function, so if we G is defined (partially) only on $\neg\mathcal{C}\mathcal{K}^{one}$ we get the domain of the clustering operator. If G is defined (partially) only on $\mathcal{C}\mathcal{K}^{one}$ we get the domain of the classification clustering operator. The partition is a particular case of the cover and when the measure of overlapping sets is 0 for all sets involved. We have hence proved the following.

Lemma 8. *Let* $\mathcal{G}_{gcluster}, \mathcal{G}_{clcluster}, \mathcal{G}_{ccluster}$ *and* $\mathcal{G}_{cluster}$ *be the sets of general clustering, classification clustering, cover clustering, and clustering operators. The following relationships hold.*

(1) $\mathcal{G}_{clcluster} \subseteq \mathcal{G}_{gcluster}$, $\mathcal{G}_{ccluster} \subseteq \mathcal{G}_{gcluster}$ *and* $\mathcal{G}_{cluster} \subseteq \mathcal{G}_{gcluster}$.
(2) $\mathcal{G}_{ccluster} \subseteq \mathcal{G}_{cluster}$.
(3) $\mathcal{G}_{clcluster} \cap \mathcal{G}_{ccluster} = \emptyset$.
(4) $\mathcal{G}_{clcluster} \cap \mathcal{G}_{cluster} = \emptyset$.

4.3 Association Operator

The association analysis is yet another important subject and will be treated in a separate paper. We present here a short overview of main ideas and describe how they fit into our framework. We start with few observations.

1. In all of the association analysis computation of association rules follow directly from the computed associations and it is done after the set of associations (frequent itemsets in basket analysis) had been found. We do not concern us here with rule computation part of the process which is syntactic in its nature and will be treated in the descriptive model. Our association operator computes the associations only.

2. Transactional "buy items" database can be represented as a relational database with "buy item i" (or just "item i") as the attributes with values 1 (buy=yes) and 0 (buy=no). The rules thus generated are called single-dimensional ([8]).
3. A multidimensional association rules extend the original single-dimensional analysis. For example they allow to produce rules that say: *Rich and young people buy fancy cars.* This statement had been re-written from the following attribute-value description $(income = rich \cap age = young) \Rightarrow buy = fancycar$. This rule had been produced from a frequent $3 - association$ (of attribute-values pairs): $income = rich, age = young, buy = fancycar$. Of course, single-dimensional rule is a particular case of the multidimensional and an one dimensional frequent $1 - association$: $buy = i_1, buy = i_2, ..., buy = i_k$ represents a frequent k-itemset $i_1, i_2, ..., i_k$.

We define a special knowledge generalization system system AK, called an **association system**. All frequent $k-associations$ are represented in it (with the support count), as well as all information needed to compute association rules that follow from them.

We put

$$\mathcal{K}^{assoc} = \{K \in \mathcal{K} : K \text{ is an association system}\}.$$

Definition 26. *An operator $G \in \mathcal{G}_{dm}$ is called an* **association operator** *iff G is a partial function that maps $\mathcal{K}^{one}$ into $\mathcal{K}^{assoc}$, i.e.*

$$G : \mathcal{K}^{one} \longrightarrow \mathcal{K}^{assoc}$$

References

1. P. Baldi, S. Brunak, P. Frasconi, G. Pollastri, G. Soda *Bidirectional dynamics for protein secondary structure prediction* Proceedings of Sixth International Joint Conference on Artificial Intelligence (IJCAI99). Stockholm, Sweden.
2. P. Baldi, S. Brunak, P. Frasconi, G. Pollastri, G. Soda *Exploiting the past and the future in protein secondary structure prediction* Bioinformatics, 15:937- 946, 1999.
3. Z. Chen. *Data Mining and Uncertain Reasoning. An Integrated Approach* J. Wiley and Sons, Inc. 2001.
4. J. Garnier, D.J. Osguthorpe, B. Robson *Analysis of the accuracy and implications of simple methods for protein secondary structure prediction* Journal of Molecular Biology, 120:97-120. 1978.
5. J.F. Gibrat, J. Garnier, B. Robson *Further developements of protein secondary structure prediction using information theory* Journal of Molecular Biology, 198:425-443. 1987.
6. Salvatore Greco, Benedetto Matarazzo, Roman Slowinski, Jerzy Stefanowski. *Importance and Interaction of Conditions in Decision Rules* Proceedings of Third International RSCTC'02 Conference, Malvern, PA, USA, October 2002, pp. 255-262. Springer Lecture Notes in Artificial Intelligence.

7. M. Hadjimichael, A. Wasilewska. *A Hierarchical Model for Information Generalization.* Proceedings of the 4th Joint Conference on Information Sciences, Rough Sets, Data Mining and Granual Computing (RSDMGrC'98), North Carolona, USA, vol.II, 306–309.
8. J. Han, M. Kamber. *Data Mining: Concepts and Techniques* Morgan, Kauffman, 2000
9. T.K. Ho, S.N. Srihati. *Decisions combination in multiple classifier system* IEEE Transactions on Pattern Analysis and Machine Learning, 11:63-90, 1994.
10. Johnson T., Lakshmanan L., Ng R. *The 3W Model and Algebra for Unified Data Mining* Proceedings of the 26th VLDB Conference, Cairo, Egypt, 2000
11. M. Inuiguchi, T. Tanino. *Classification versus Approximation oriented Generalization of Rough Sets* Bulletin of International Rough Set Society, Volume 7, No. 1/2.2003
12. J. Komorowski. *Modelling Biological Phenomena with Rough Sets Proceedings of Third International Conference RSCTC'02*, Malvern, PA, USA, October 2002, p13. Springer Lecture Notes in Artificial Intelligence.
13. T.Y. Lin. *Database Mining on Derived Attributes Proceedings of Third International Conference RSCTC'02*, Malvern, PA, USA, October 2002, pp. 14 - 32. Springer Lecture Notes in Artificial Intelligence.
14. Juan F.Martinez, Ernestina Menasalvas, Anita Wasilewska, Covadonga Fernández, M. Hadjimichael. *Extension of Relational Management System with Data Mining Capabilities* Proceedings of Third International Conference RSCTC'02, Malvern, PA, USA, October 2002, pp. 421- 428. Springer Lecture Notes in Artificial Intelligence.
15. Ernestina Menasalvas, Anita Wasilewska, Covadonga Fernández *The lattice structure of the KDD process: Mathematical expression of the model and its operators* International Journal of Information Systems) and FI (Fundamenta Informaticae; special isue2001, pp. 48 - 62.
16. Ernestina Menasalvas, Anita Wasilewska, Covadonga Fernández, Juan F. Martinez. *Data Mining- A Semantical Model* Proceedings of 2002 World Congres on Computational Intelligence, Honolulu, Hawai, USA, May 11- 17, 2002, pp. 435 - 441.
17. M. Ouali, R.D. King. *Cascaded multiple classifiers for protein secondary structure prediction* Protein Science, 9:1162- 1176, 2000.
18. Pawlak, Z. *Information systems - theoretical foundations* Information systems, 6 (1981), pp. 205-218
19. Pawlak, Z. *Rough Sets- theoretical Aspects Reasoning About Data* Kluwer Academic Publishers 1991
20. B. Rost, C. Sander, R. Schneider. *PHD - an automatic mail server for protein secondary structure prediction* Journal of Molecular Biology, 235:13-26, 1994.
21. Victor Robles, Pedro Larranaga, J. M. Pena, Ernestina Menasalvas, Maria S. Perez, Vanessa Herves, Anita Wasilewska. *Bayesian Network Multi-classifiers for Protein Secondary Structure Prediction* Artificial Intelligence in Medicine -special volume; to appear.
22. Skowron, A. *Data Filtration: A Rough Set Approach* Proceedings de Rough Sets, Fuzzy Sets and Knowledge Discovery. (1993). Pag. 108-118
23. A. Wasilewska, Ernestina Menasalvas Ruiz, María C. Fernández-Baizan. *Modelization of rough set functions in the KDD frame* 1st International Conference on Rough Sets and Current Trends in Computing (RSCTC'98) June 22 - 26 1998, Warsaw, Poland.

24. D.H. Wolpert. *Stacked Generalization* Neural Networks, 5:241-259, 1992.
25. Wojciech Ziarko, Xue Fei. *VPRSM Approach to WEB Searching* Proceedings of Third International RSCTC'02 Conference, Malvern, PA, USA, October 2002, pp. 514- 522. Springer Lecture Notes in Artificial Intelligence.
26. Wojciech Ziarko. *Variable Precision Rough Set Model* Journal of Computer and Systen Sciences, Vol.46. No.1, pp. 39-59, 1993.
27. J.T. Yao, Y.Y. Yao. *Induction of Classification Rules by Granular Computing* Proceedings of Third International RSCTC'02 Conference, Malvern, PA, USA, October 2002, pp. 331-338. Springer Lecture Notes in Artificial Intelligence.

Part II

Novel Approaches

SVM-OD: SVM Method to Detect Outliers[1]

Jiaqi Wang[1], Chengqi Zhang[1], Xindong Wu[2], Hongwei Qi[3], Jue Wang[3]

[1] Faculty of Information Technology, University of Technology, Sydney, Australia

[2] Department of Computer Science, University of Vermont, USA

[3] Laboratory of Complex Systems and Intelligence Science, Institute of Automation, Chinese Academy of Sciences, China

ABSTRACT

Outlier detection is an important task in data mining because outliers can be either useful knowledge or noise. Many statistical methods have been applied to detect outliers, but they usually assume a given distribution of data and it is difficult to deal with high dimensional data. The Statistical Learning Theory (SLT) established by Vapnik et al. provides a new way to overcome these drawbacks. According to SLT Schölkopf et al. proposed a ν-Support Vector Machine (ν-SVM) and applied it to detect outliers. However, it is still difficult for data mining users to decide one key parameter in ν-SVM. This paper proposes a new SVM method to detect outliers, SVM-OD, which can avoid this parameter. We provide the theoretical analysis based on SLT as well as experiments to verify the effectiveness of our method. Moreover, an experiment on synthetic data shows that SVM-OD can detect some local outliers near the cluster with some distribution while ν-SVM cannot do that.

[1] This research is partly supported by the National Key Project for Basic Research in China (G1998030508).

1 Introduction

Outliers are abnormal observations from the main group, and are either noise or new knowledge hidden in the data. Researchers always wish to remove noise in the data during the pre-processing of data mining because noise may prevent many data mining tasks. In addition, data mining users are interested in new knowledge or unusual behaviors hidden behind data such as the fraud behavior of credit cards. Therefore, outlier detection is one of the important data mining tasks.

Statistics has been an important tool for outlier detection [1], and many researchers have tried to define outliers using statistical terms. Ferguson pointed out [4] that, "In a sample of moderate size taken from a certain population it appears that one or two values are surprisingly far away from the main group." Barnett et al. gave another definition [1], "An outlier in a set of data is an observation (or a subset of observations) which appears to be inconsistent with the remainder of that set of data." Hawkins characterized an outlier in a quite intuitive way [7], "An outlier is an observation that deviates so much from other observations as to arouse suspicion that it was generated by a different mechanism."

All these definitions imply that outliers in a given data set are events with a very low probability or even those generated by the different distribution from most data. Although statistical methods have been applied to detect outliers, usually they need to assume some distribution of data. It is also difficult for statistical methods to deal with high dimensional data [3, 6].

To some extent, the Statistical Learning Theory (SLT) established by Vapnik et al. and the corresponding algorithms can overcome these drawbacks [12]. According to this theory Schölkopf et al. proposed a ν-Support Vector Machine (ν-SVM) to estimate the support of a high dimensional distribution of data and applied it to detect outliers [11]. As pointed out by Schölkopf et al., a practical method has not been provided to decide the key parameter ν in ν-SVM though Theorem 7 in [11] gives the confidence that ν is a proper parameter to adjust. Therefore, it is still difficult for data mining users to decide this parameter.

In fact, we find in some experiments that this parameter can be avoided if another strategy is adopted. This strategy consists of two components: (1) a geometric method is applied to solve ν-SVM without the penalty term and (2) the support vector with the maximal coefficient is selected as the outlier. This method is called SVM-OD in this paper.

Vapnik et al. originally provided a standard SVM without the penalty term to solve the classification problem and then added the penalty term to deal with noise and nonlinear separability in the feature space [12]. In this paper, SVM-OD tries to detect outliers by solving ν-SVM without the penalty term.

Although removing the penalty term from ν-SVM may drastically change the classification model, we can theoretically analyze why the strategy adopted in SVM-OD is reasonable for outlier detection based on SLT and the popular definition of outliers.

Some experiments on toy and real-world data show the effectiveness of SVM-OD. The experiment on a synthetic data set shows that when the kernel parameter is given, SVM-OD can detect some local outliers near the cluster with some distribution (e.g. *O1* and *O2* in Fig. 1) while ν-SVM cannot do that. The details about local outliers can be found in [3]. Another interesting phenomenon is found in the experiment on stock data that SVM-OD is insensitive for some values of the kernel parameter compared with ν-SVM though this still needs to be verified by the theoretical analysis and more experiments.

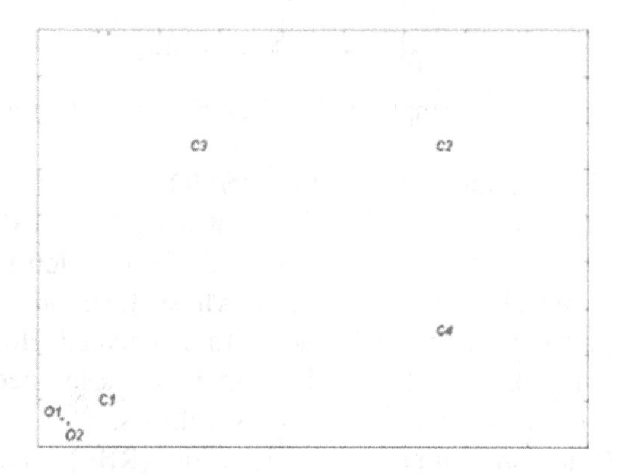

Fig. 1. Four clusters with different distributions and two local outliers (bigger points *O1*, *O2*)

The other sections in this paper are organized as follows. SVM-OD and ν-SVM are introduced in Section 2 and the theoretical analysis of SVM-OD is given in Section 3. Experiments are provided to illustrate the effectiveness of SVM-OD in Section 4. A discussion and conclusions are given in Sections 5 and 6. The notations used in this paper are shown in Table 1.

Table 1. Notations and their meanings

Notations	Meaning	Notations	Meaning
χ	Sample space	$\varphi(x)$	Points in kernel space
X	Sample set of size l	$(\cdot)$	Inner product
x_o	Outlier	Ω	Index set of sample
$K(x_i, x_j)$	Kernel function	Λ	Index set of SV

2 SVM-OD for Outlier Detection

In 2001, Schölkopf et al. presented ν-SVM to estimate the support of a high dimensional distribution of data and applied it to detect outliers [11]. ν-SVM solves the following optimization problem in the kernel space:

$$min\ 0.5\times\|w\|^2-\rho+\sum_i\xi_i/\nu l \quad s.t.\ (w\bullet\varphi(x_i))\geq\rho-\xi_i,\ \xi_i\geq 0,\ i=1...l. \tag{1}$$

The geometric description of ν-SVM is shown in Fig. 2:

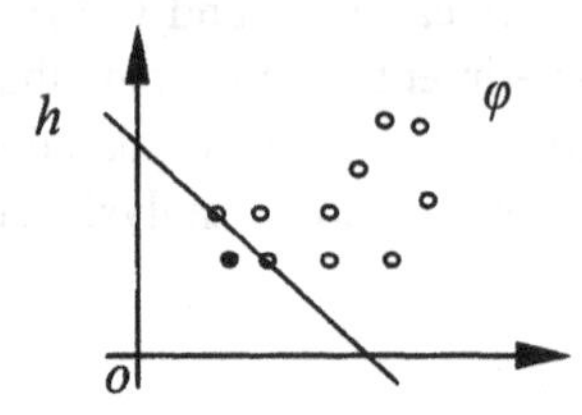

Fig. 2. o is the origin, φ is the sample set, h is the hyper-plane. h separates φ and o

Sequential Minimization Optimization (SMO), an optimization method proposed by Platt to solve classification problems [10], is extended to solve the optimization problem in (1). After the decision function f is obtained by solving (1), ν-SVM selects the samples x_i whose function value $f(x_i)<0$ as outliers. A key parameter ν in ν-SVM needs to be decided. However, it is not easy for data mining users to decide this parameter as implied in [11]. Thus SVM-OD, which can avoid ν, is introduced as follows.

Firstly, the Gaussian Radius Basis Function (RBF) $K(x,y) = exp\{-\|x-y\|^2/2\sigma^2\}$ is used as the kernel function in SVM-OD. The three properties of the RBF kernel, which are implied in [11] and will lead to the theorems in this paper, are discussed here.

Property 1 $\forall\ x\in X,\ K(x,x)=1.$

Property 2 For any $x,y\in X$ and $x\neq y$, $0<K(x,y)<1.$

Property 3 In the kernel space spanned by RBF, the origin o and the mapped point set φ are linearly separable.

Proof According to Properties 1 and 2,

$$\exists\ x_i\in X,\ w=\varphi(x_i),\forall\ x_j\in X,\ (w\bullet\varphi(x_j))>0\,.$$

Let $0<\varepsilon<min(w\bullet\varphi(x_j))\ (x_j\in X)$, then $\forall\ x_j\in X,\ (w\bullet\varphi(x_j)) - \varepsilon > 0$ and $(w\bullet o)-\varepsilon<0$, where o is the origin in the kernel space. Therefore the hyperplane $(w\bullet\varphi(x))-\varepsilon>0$ can separate φ and the origin o in the kernel space. ■

Secondly, SVM-OD solves the following optimization problem in the kernel space:

$$min\ 0.5\times\|w\|^2-\rho \quad s.t.\ (w\bullet\varphi(x_i))\geq\rho,\ i=1...l. \tag{2}$$

Although the optimization problem in (2) removes the penalty term in (1), the solution of (2) always exists for any given data because Property 3 guarantees that the mapped point set φ and the origin o in the kernel space are linearly separable. It can be proved that the optimization problem in (2) is equivalent to finding the shortest distance from the origin o to the convex hull C (see Fig. 3). This is the special case of the nearest point problem of two convex hulls and the proof can be found in [8].

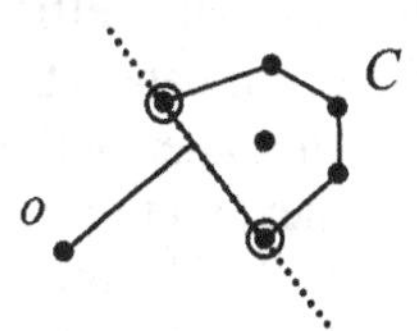

Fig. 3. Minimal normal problem (the shortest distance from the origin o to the convex hull C)

The corresponding optimization problem is as follows:

$$min\ \|\textstyle\sum_{i=1}^{l}\beta_i x_i\|^2 \ \ s.t.\ \sum_i\beta_i = 1,\ \beta_i\geq 0,\ i=1...l. \tag{3}$$

Many geometric methods have been designed to solve (3), e.g. the Gilbert algorithm [5], and the MDM algorithm [9]. Since program developers can implement these geometric algorithms very easily, we combine them with the kernel function to solve the optimization problems in (2) or (3) in the kernel space. In this paper, the Gilbert algorithm is used to solve the problems in (2) or (3).

Finally, the function obtained by solving the optimization problem in (3) in the kernel space is as follows:

$$f(x)=\textstyle\sum_i \alpha_i K(x,x_i)-\rho,\ i\in\Lambda. \tag{4}$$

where $\rho=\sum_{i,j}\alpha_i K(x_j,x_i)$ and $i,j\in\Lambda$.

According to the function in (4), we define a decision function class as follows:

$$\{f_k|\ f_k(x)=\textstyle\sum_i\alpha_i K(x,x_i)-\rho+\alpha_k-\varepsilon,\ i\in\Lambda\text{-}\{k\},\ k\in\Lambda\}. \tag{5}$$

where $\varepsilon=min_i\alpha_i(1\text{-}K),\ i\in\Lambda,\ K=max_{i,j}K(x_i,x_j),\ i,j\in\Omega,\ i\neq j$.

The decision region of a decision function $f_k(x)$ in the decision function class in (5) is $R_{k,O}=\{x: f_k(x)\geq 0\}$. The following property holds true for the decision function class in (5).

Property 4 For each decision function in the decision function class in (5) $f_k(x)$,

$$f_k(x_k)<0 \text{ and } \forall\ i\in\Omega\text{-}\{k\},\ f_k(x_i)\geq 0.$$

Proof $f_k(x)=\sum_i\alpha_i K(x,x_i)\text{-}\rho+\alpha_k(1\text{-}K(x,x_k))-\varepsilon\ (i\in\Lambda)$ and according to Property 2, $0<\varepsilon<1,\ 0<K<1$.

(1) $k\in\Lambda$, that is, x_k is a support vector, therefore $\sum_i\alpha_i K(x_k,x_i)\text{-}\rho=0\ (i\in\Lambda)$. According to Property 1, $f_k(x_k)=\alpha_k(1\text{-}K(x_k,x_k))-\varepsilon=\text{-}\varepsilon<0$.

(2) $\forall j \in \Omega - \{k\}$, $\sum_i \alpha_i K(x_j, x_i) - \rho \geq 0$ $(i \in \Lambda)$. According to the definition of ε,

$$f_k(x_j) \geq \alpha_k(1 - K(x_j, x_k)) - \varepsilon \geq 0.$$

Therefore Property 4 holds true. ■

Then SVM-OD selects a function $f_o(x)$ in the decision function class in (5), where o is the index of α_o and $\alpha_o = max_i \alpha_i$ $(i \in \Lambda)$, as the decision function and the support vector x_o with the maximal coefficient α_o as the outlier. After removing the outlier x_o from the given data set, according to the same strategy we re-train the data and select the other outlier. The theoretical analysis for this strategy will be given in the next section.

The steps of SVM-OD are described below:

Step 1 Use the Gilbert algorithm to solve the optimization problem in (2) or (3) in the kernel space.

Step 2 Select the support vector with the maximal coefficient as an outlier.

Step 3 Remove this outlier from the given data set and go to Step 1.

3 Theoretical Analysis

According to statistics, a sample x_o will be regarded as an outlier if it falls in the region with a very low probability compared with other samples in the given set. The statistical methods to detect outliers usually assume some given distribution of data and then detect outliers according to the estimated probability. However, the real distribution of data is often unknown. When analyzing ν-SVM according to SLT, Schölkopf et al. provided the bound of the probability of the non-decision region, where outliers fall.

Definition 1 (Definition 6 in [11]) Suppose that f is a real-valued function on χ. Fix $\theta \in R$. For $x \in \chi$, let $d(X, f, \theta) = max\{0, \theta - f(x)\}$. And for a given data set X, we define $D(X, f, \theta) = \sum_x d(X, f, \theta)$ $(x \in X)$.

Then the following two theorems can be proved according to the results in [11], Definition 1 and four properties discussed in Section 2.

Theorem 2 Suppose that an independent identically distributed sample set X of size l is generated from an unknown distribution P that does not contain discrete components. For a decision function $f_k(x)$ in the decision function class in (5), the corresponding decision region is $R_{k,0} = \{x: f_k(x) \geq 0\}$, $k \in \Lambda$. Then with probability $1-\delta$ over randomly drawn training sequences X of size l, for all $\gamma > 0$ and any $k \in \Lambda$,

$$P\{x: x \notin R_{k,-\gamma}\} \leq 2 \times (k + log(l^2/2\delta))/l$$

where $c_1 = 4c^2$, $c_2 = ln(2)/c^2$, $c = 103$, $\hat{\gamma} = \gamma/\|w_k\|$, and $\|w_k\| = (\sum_{i1,i2} \alpha_{i1} \alpha_{i2} K(x_{i1}, x_{i2}))^{1/2}$ $(i1, i2 \in \Lambda - \{k\})$, $k = (c_1 / \hat{\gamma}^2) log(c_2 l \hat{\gamma}^2) + (\varepsilon/\hat{\gamma}) log(e((2l-1)\hat{\gamma}/\varepsilon + 1)) + 2$.

Proof According to Property 4, $f_k(x_k)=-\varepsilon$ and $\forall\ i \in \Omega\text{-}\{k\},\ f_k(x_i)\geq 0$. So $D=D(X,f_k,0)=\varepsilon$, where D is defined in Definition 1. Theorem 2 holds true according to Theorem 17 in [11]. ■

Theorem 3 For any decision function in the decision function class in (5) $f_k(x),\ (k\in\Lambda)$, $\|w_k\|^2=\sum_{i,j}\alpha_i\alpha_j K(x_i,x_j),\ (i,j\in\Lambda\text{-}\{k\},\ k\in\Lambda)$. Then

$$\|w_o\|^2=min_k\|w_k\|^2 \text{ iff } \alpha_o=max_k\alpha_k\ (k\in\Lambda).$$

Proof For $o\in\Lambda$ and any $k\in\Lambda$ $k\neq o$, $\alpha_k+\sum_i\alpha_i K(x_k,x_i)=\alpha_o+\sum_j\alpha_j K(x_o,x_j)=\rho$ where $i\in\Lambda$ -$\{k\}$, $j\in\Lambda\text{-}\{o\}$ and ρ is defined in the function in (4). So $\sum_i\alpha_i K(x_k,x_i)=\alpha_o\text{-}\alpha_k+\sum_j\alpha_j K(x_o,x_j)\ (i\in\Lambda\text{-}\{k\}, j\in\Lambda\text{-}\{o\})$.

$\|w_o\|^2=\sum_{i,j}\alpha_i\alpha_j K(x_i,x_j)(i,j\in\Lambda\text{-}\{o\})=\sum_{i,j}\alpha_i\alpha_j K(x_i,x_j)(i,j\in\Lambda)\text{-}2\alpha_o\sum_i\alpha_i K(x_o,x_i)(i\in\Lambda\text{-}\{o\})\text{-}\alpha_o^2$

$\|w_k\|^2=\sum_{i,j}\alpha_i\alpha_j K(x_i,x_j)(i,j\in\Lambda\text{-}\{k\})=\sum_{i,j}\alpha_i\alpha_j K(x_i,x_j)(i,j\in\Lambda)\text{-}2\alpha_k\sum_i\alpha_i K(x_k,x_i)(i\in\Lambda\text{-}\{k\})\text{-}\alpha_k^2$

So $\|w_o\|^2\text{-}\|w_k\|^2=2\alpha_k\sum_i\alpha_i K(x_k,x_i)\ (i\in\Lambda\text{-}\{k\})$ - $2\alpha_o\sum_i\alpha_i K(x_o,x_i)\ (i\in\Lambda\text{-}\{o\})$ + α_k^2 – α_o^2

$$
\begin{aligned}
&=2\alpha_k(\alpha_o\text{-}\alpha_k+\textstyle\sum_i\alpha_i K(x_o,x_i))\text{-}2\alpha_o\sum_i\alpha_i K(x_o,x_i))+\alpha_k^2-\alpha_o^2\ (i\in\Lambda\text{-}\{o\})\\
&=2\alpha_k\alpha_o-\alpha_k^2\text{-}\alpha_o^2+2(\alpha_k\text{-}\alpha_o)\textstyle\sum_i\alpha_i K(x_o,x_i)\ (i\in\Lambda\text{-}\{o\})\\
&= \text{-}(\alpha_o\text{-}\alpha_k)^2-2(\alpha_o\text{-}\alpha_k)\textstyle\sum_i\alpha_i K(x_o,x_i)\ (i\in\Lambda\text{-}\{o\})\\
&= \text{-}(\alpha_o\text{-}\alpha_k)(\alpha_o\text{-}\alpha_k+2\textstyle\sum_i\alpha_i K(x_o,x_i))\ (i\in\Lambda\text{-}\{o\})\\
&= \text{-}(\alpha_o\text{-}\alpha_k)(\textstyle\sum_i\alpha_i K(x_o,x_i)+\sum_j\alpha_j K(x_k,x_j))\ (i\in\Lambda\text{-}\{o\}, j\in\Lambda\text{-}\{k\}).
\end{aligned}
$$

According to Properties 1 and 2 of RBF, the following inequality always holds true:

$$\forall\ i,j\in\Omega,\ K(x_i,x_j)>0.$$

And $\forall\ i\in\Lambda,\ \alpha_i>0$, therefore $\|w_o\|^2=min_k\|w_i\|^2$ iff $\alpha_o=max_k\alpha_k\ (k\in\Lambda)$. ■

Note that ε in Theorem 2 is a constant for any decision function in the decision function class in (5). Therefore Theorem 2 shows that the smaller value of $\|w\|^2$, the lower bound of probability of the non-decision region decided by the decision function. Furthermore, Theorem 3 shows that we can obtain the smaller value of $\|w\|^2$ if the function $f_o(x)$ in the decision function class in (5), where o is the index of α_o, is chosen as the decision function. This means that the probability of the non-decision region decided by $f_o(x)$ is low compared with others. In addition, according to Property 4, the following inequalities hold true: $\forall\ i\in\Omega\text{-}\{o\},\ f_o(x_i)\geq 0$ and $f_o(x_o)<0$. This means that the support vector x_o with the maximal coefficient falls in the non-decision region decided by $f_o(x)$. Thus we select x_o as an outlier.

4 Experiments

Experiment 1. This experiment is performed on a synthetic data set including local outliers. There are respectively 400 and 100 points in the clusters *C1* and *C2* with the uniform distribution and respectively 300 and 200 points in the clusters *C3* and *C4* with the Gaussian distribution. In addition, there are two local outliers *O1* and *O2* near the cluster *C1* (see Fig. 1). The details about local outliers can be found in [3]. The goal of this experiment is to test whether ν-SVM and SVM-OD only detect these two local outliers since other points are regarded as the normal data from some distributions. $\sigma=5$ is set as the kernel parameter value. In Fig. 1, bigger points are two local outliers. In Fig. 4 and Fig. 5, bigger points are some "outliers" detected by ν-SVM and SVM-OD. Comparing SVM-OD and ν-SVM, we find that SVM-OD can detect *O1* and *O2* after two loops. However, ν-SVM either does not detect both *O1* and *O2* or detect other normal data other than these two local outliers (see Fig. 4 and Fig. 5) when the different values of ν are tested. So in this experiment SVM-OD is more effective for detecting some kinds of local outliers than ν-SVM.

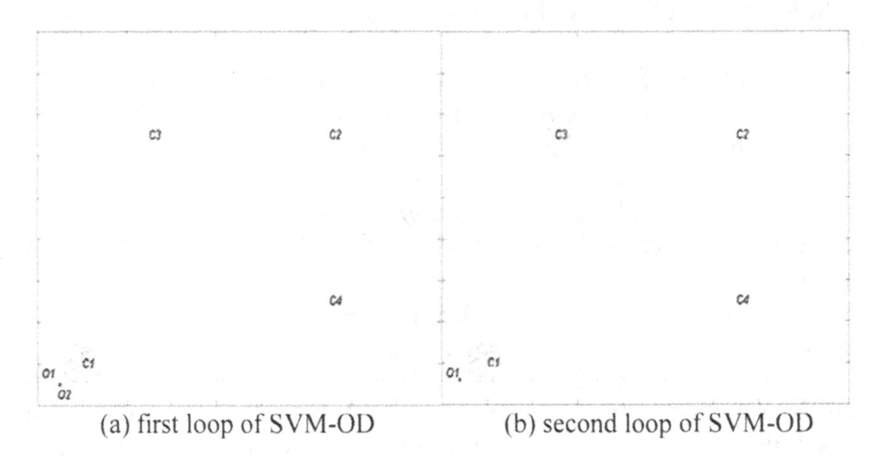

(a) first loop of SVM-OD (b) second loop of SVM-OD

Fig. 4. (a) only *O2* is detected as "outlier", (b) only *O1* is detected as "outlier"

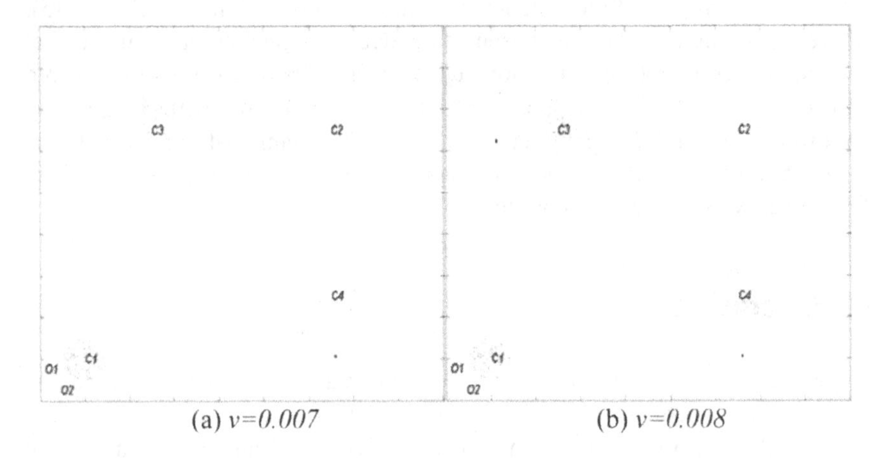

(a) $\nu=0.007$ (b) $\nu=0.008$

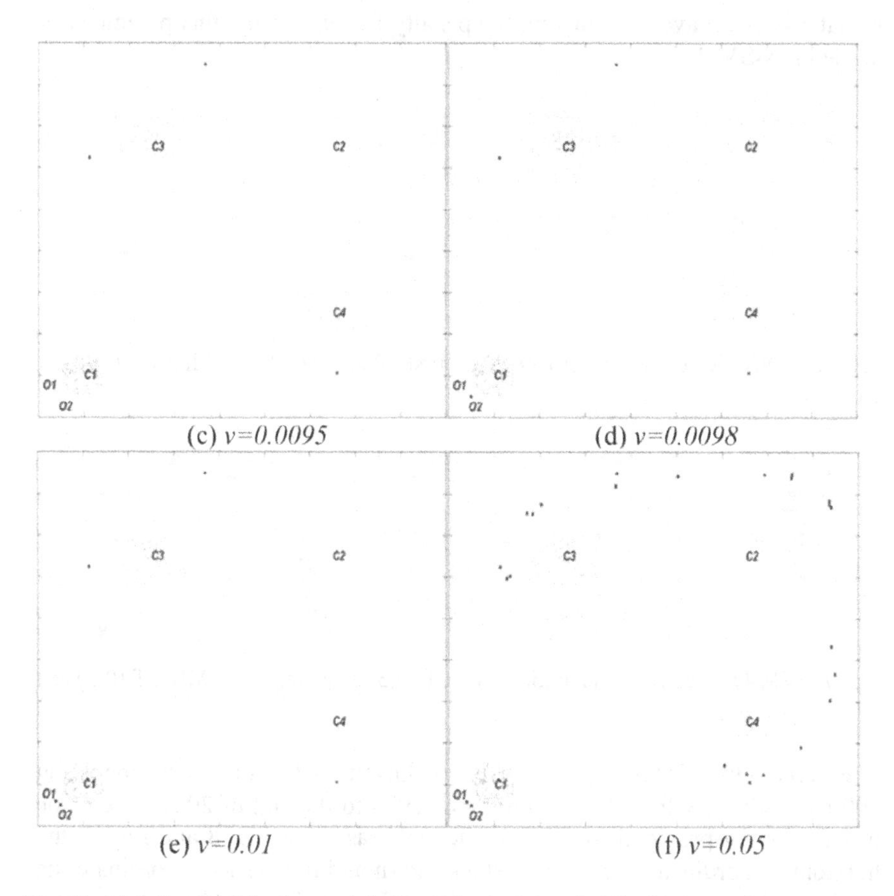

Fig. 5. (a) 1 "outlier" is detected (without both *O1* and *O2*), (b) 2 "outliers" are detected (without both *O1* and *O2*), (c) 3 "outliers" are detected (without both *O1* and *O2*), (d) 4 "outliers" are detected (without *O1* and with *O2*), (e) 5 "outliers" are detected (with both *O1*, *O2* and others), (f) many "outliers" are detected (with both *O1*, *O2* and others)

Experiment 2. This experiment is performed on the first 5000 samples of the MNIST test set because there are fewer outliers in these samples than the last 5000. Data can be available from the website "yann.lecun.com/exdb/mnist/". In both SVM-OD and ν-SVM, $\sigma=8\times256$ is selected as the kernel parameter value. For ten hand-digits (0-9), the penalty factor ν in ν-SVM is *5.57%, 5.47%, 5.9%, 5.71%, 5.88%, 7.3%, 5.41%, 6.51%, 5.84%,* and *4.81%*, respectively. Classes (0-9) are labeled on the left side of Fig. 6 and Fig. 7. From both Fig. 6 and Fig. 7, a number of samples detected are either abnormal or mislabeled. The results of this experiment show that SVM-OD is effective for outlier detection. What is more important is that SVM-OD avoids adjusting the penalty factor ν, while this parameter is needed in ν-SVM.

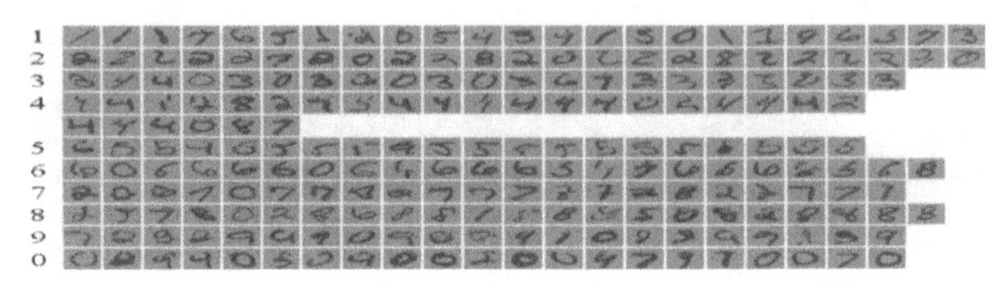

Fig. 6. ν-SVM detects some outliers in the first 5000 examples of MNIST (0-9) test set

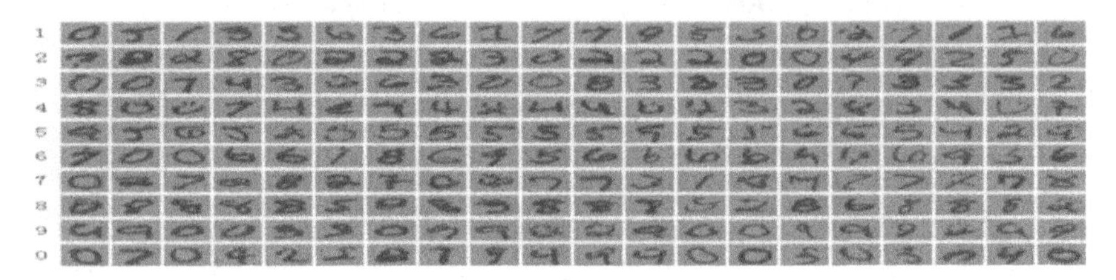

Fig. 7. SVM-OD detects some outliers in the first 5000 examples of MNIST (0-9) test set

Experiment 3. This experiment is conducted on the stock data including 909 daily samples from the beginning of 1998 to the end of 2001. The name of stock is not provided for the commercial reason. In Fig. 8 and Fig. 9, the horizontal coordinate refers to the stock return and the vertical coordinate the trading volume. The penalty factor ν in ν-SVM is *5.5%* and the kernel parameter σ in SVM-OD and ν-SVM are shown in Fig. 8 and Fig. 9. Bigger points in these two figures are outliers detected by SVM-OD and ν-SVM respectively. The goal of this experiment is to show that SVM-OD can also detect those points far away from the main group, though it is still necessary to verify whether or not outliers detected by SVM-OD and ν-SVM are unusual behaviors in the stock market. An interesting phenomenon is also found that SVM-OD is more insensitive for some values of the kernel parameter than ν-SVM though more experiments and the theoretical analysis are needed (see Fig. 8 and Fig. 9).

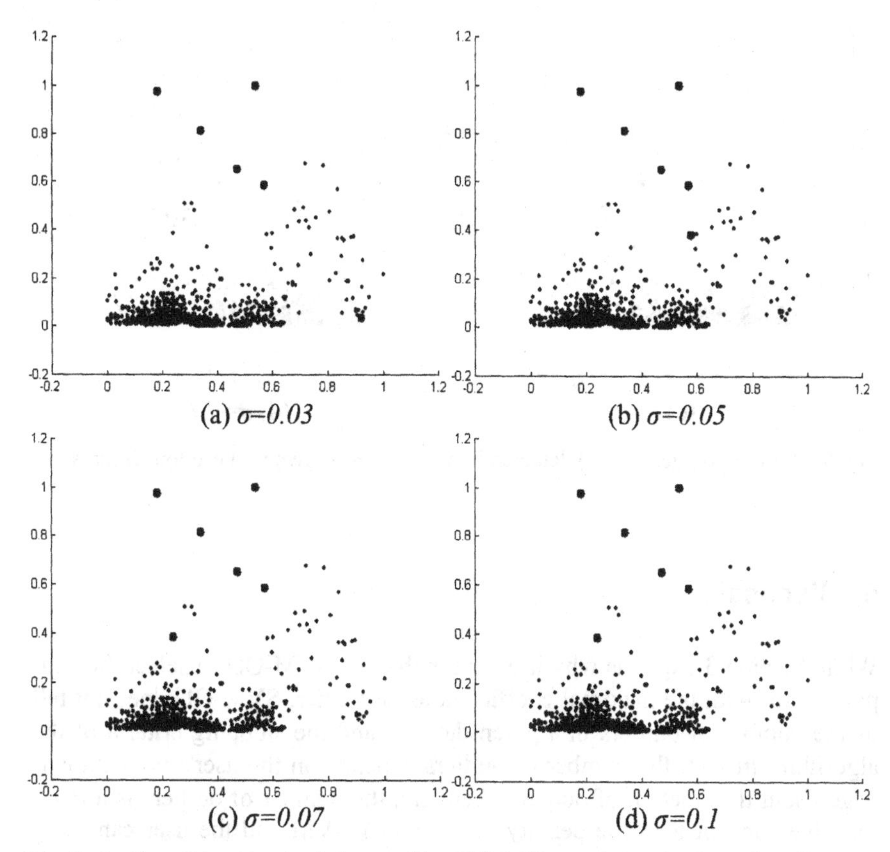

(a) σ=0.03 (b) σ=0.05

(c) σ=0.07 (d) σ=0.1

Fig. 8. Outliers (bigger points) detected by SVM-OD are shown in these four figures

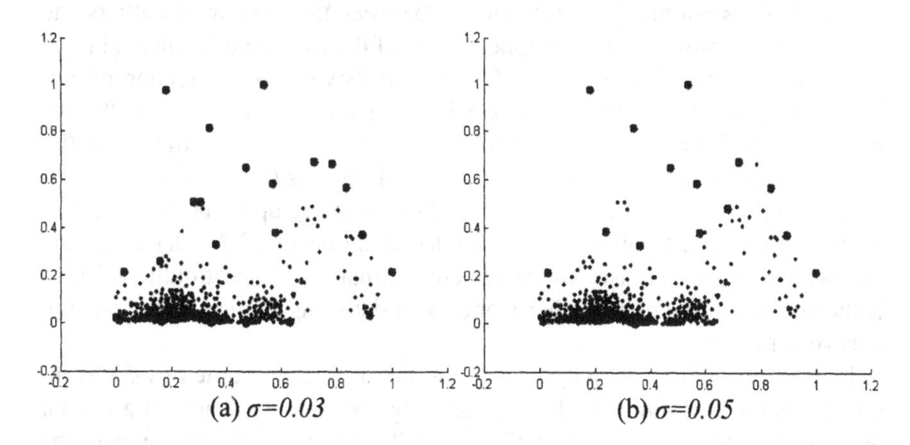

(a) σ=0.03 (b) σ=0.05

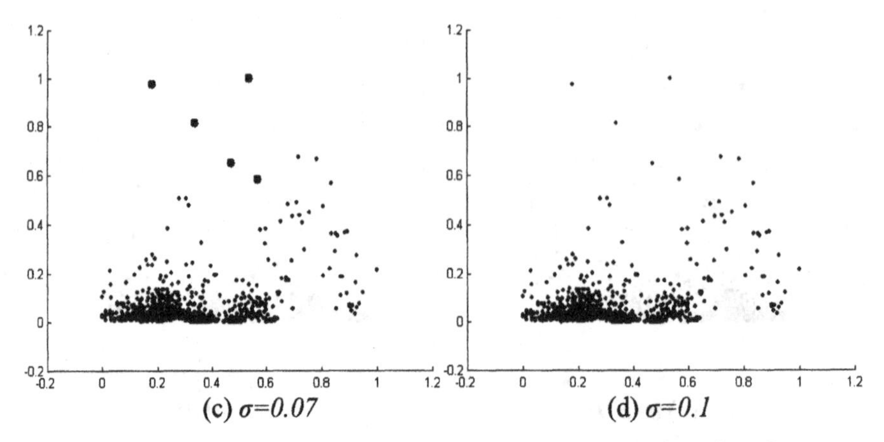

Fig. 9. Outliers (bigger points) detected by ν-SVM are shown in these four figures

5 Discussion

While Section 3 explains why it is reasonable for SVM-OD to select the support vector with the maximal coefficient as an outlier, SVM-OD does not tell us the number of outliers in a given data set and the stopping criteria of the algorithm. In fact, the number of outliers depends on the user's prior knowledge about the fraction of outliers. Actually, the number of outliers is a more intuitive concept than the penalty factor ν in ν-SVM and the user can more easily know the approximate ratio of outliers compared with ν.

ν-SVM does not provide a relationship between the fraction of outliers and ν although ν is proved to be the upper bound of the fraction of outliers [11]. It is still difficult for the user to decide ν even if they know the fraction of outliers in advance. For example, Table 1 in [11] showed that when ν is *4%*, the fraction of outliers is *0.6%* and when ν is *5%*, the fraction of outliers is *1.4%*. However, when the user knows the fraction of outliers (e.g. *1.4%*) before detecting outliers, which value of ν (*4%*, *5%*, or another upper bound of *1.4%*) should be selected to obtain *1.4%* samples as outliers? SVM-OD can avoid this problem when the user knows the approximate fraction of outliers. There is the similar problem in another work about detecting outliers based on support vector clustering [2].

In addition, Table 1 in [11] pointed out that the training time of ν-SVM increases as the fraction of outliers detected becomes more. There is a similar property for the training cost of SVM-OD. This paper does not compare the training costs of these two methods, which is a topic for the future work.

6 Conclusion

This paper has proposed a new method to detect outliers called SVM-OD. Compared to ν-SVM, SVM-OD can be used by data mining users more easily since it avoids the penalty factor ν required in ν-SVM. We have verified the effectiveness of SVM-OD according to the theoretical analysis based on SLT and some experiments on both toy and real-world data. The experiment on a synthetic data set shows that when the kernel parameter is fixed, SVM-OD can detect some local outliers while ν-SVM cannot do that. In the experiment on stock data, it is found that SVM-OD is insensitive for some values of the kernel parameter compared with ν-SVM. In the future work, we will try to give a theoretical explanation to this phenomenon and compare SVM-OD with more methods to detect outliers on more real-world data from the different sides, e.g. the training cost and the effectiveness.

References

1. Barnett V, Lewis T (1984) Outliers in statistical data. John Wiley & Sons, New York
2. Ben-Hur A, Horn D, Siegelmann HT, Vapnik V (2001) Support vector clustering. Journal of Machine Learning Research 2: 125-137
3. Breunig MM, Kriegel H-P, Ng RT, Sander J (2000) LOF: identifying density-based local outliers. Proceedings of ACM SIGMOD Conference
4. Ferguson TS (1961) On the rejection of outliers. Proceedings of the Fourth Berkeley Symposium on Mathematical Statistics and Probability 1: 253-287
5. Gilbert EG (1966) Minimizing the quadratic form on a convex set. SIAM Journal of Control 4: 61-79
6. Han J, Kamber M (2000) Data mining: concepts and techniques. Morgan Kaufmann Publishers, Inc.
7. Hawkins D (1980) Identification of outliers. Chapman and Hall, London
8. Keerthi SS, Shevade SK, Bhattacharyya C, Murthy KRK (2000) A fast iterative nearest point algorithm for support vector machine classifier design. IEEE Transactions on Neural Networks 11(1): 124-136
9. Mitchell BF, Dem'yanov VF, Malizemov VN (1974) Finding the point of a polyhedron closet to the origin. SIAM Journal of Control 12: 19-26
10. Platt JC (1999) Fast training of support vector machines using sequential minimal optimization. In: Advances in kernel methods – support vector learning. MIT press, Cambridge, MA, pp 185-208
11. Schölkopf B, Platt JC, Shawe-Taylor J, Smola AJ, Williamson RC (2001) Estimating the support of a high-dimensional distribution. Neural Computation 13 443-1471
12. Vapnik V (1999) The nature of statistical learning theory (2nd edn). Springer-Verlag, New York

Extracting Rules from Incomplete Decision Systems: System ERID

Agnieszka Dardzińska[1] and Zbigniew W. Raś[2,3]

[1] Bialystok Technical University, Department of Mathematics, ul. Wiejska 45A, 15-351, Bialystok, Poland adardzin@uncc.edu
[2] University of North Carolina, Department of Computer Science, Charlotte, N.C. 28223, USA
[3] Polish Academy of Sciences, Institute of Computer Science, ul. Ordona 21, 01-237 Warsaw, Poland ras@uncc.edu

Summary. We present a new method for extracting rules from incomplete Information Systems (IS) which are generalizations of information systems introduced by Pawlak [7]. Namely, we allow to use a set of weighted attribute values instead of a single value to describe objects in IS. The proposed strategy has some similarities with system LERS [3]. It is a bottom-up strategy, guided by two thresholds values (minimum support and minimum confidence) and generating sets of weighted objects with descriptions of minimal length. The algorithm starts with identifying sets of objects having descriptions of length one (values of attributes). Some of these sets satisfy both thresholds values and they are used for constructing rules. They are marked as successful. All sets having a number of supporting objects below the threshold value are marked as unsuccessful. Pairs of descriptions of all remaining sets (unmarked) are used to construct new sets of weighted objects having descriptions of length 2. This process is continued recursively by moving to sets of weighted objects having k-value properties. In [10], [1], ERID is used as a null value imputation toll for knowledge discovery based chase algorithm.

1 Introduction

There is a number of strategies which can be used to extract rules describing values of one attribute (called decision attribute) in terms of other attributes (called classification attributes) available in the system. For instance, we can mention here such systems like *LERS* [3], *AQ*19 [5], *Rosetta* [6] and, *C*4.5 [8]. In spite of the fact that the number of rule discovery methods is still increasing, most of them are developed under the assumption that the information about objects in information systems is either precisely known or not known at all. This implies that either one value of an attribute is assigned to an object as its property or no value is assigned to it (instead of no value we use the term *null value*). Problem of inducing rules from information systems with attribute values represented as sets of possible values was discussed for instance by Kryszkiewicz and Rybinski [4], Greco, Matarazzo, and Slowinski [2], and by Ras and Joshi [9].

In this paper, we present a new strategy for discovering rules in information systems when data are partially incomplete. Namely, we allow to use a set of weighted attribute values as a value of an attribute. A weight assigned to an attribute value represents user confidence in that value. For instance, by assigning a value $\{(brown, \frac{1}{3}), (black, \frac{2}{3})\}$ of the attribute *Color of Hair* to an object x we say that the confidence in object x having brown hair is $\frac{1}{3}$ whereas the confidence that x has black hair is $\frac{2}{3}$. Similar assumption was used, for instance, in papers by Greco [2] or Slowinski [11] but their approach to rule extraction from incomplete information systems is different than ours.

2 Discovering Rules in Incomplete IS

In this section, we give the definition of an incomplete information system which can be called *probabilistic* because of the requirement placed on values of attributes assigned to its objects. Namely, we assume that value of an attribute for a given object is a set of weighted attribute values and the sum of these weights has to be equal to one. Next, we present an informal strategy for extracting rules from incomplete information systems. Finally, we propose a method of how to compute confidence and support of such rules.

We begin with a definition of an incomplete information system which is a generalization of an information system given by Pawlak [7].

By an incomplete Information System we mean $S = (X, A, V)$, where X is a finite set of objects, A is a finite set of attributes and $V = \bigcup\{V_a : a \in A\}$ is a finite set of values of attributes from A. The set V_a is a domain of attribute a, for any $a \in A$.

We assume that for each attribute $a \in A$ and $x \in X$, $a(x) = \{(a_i, p_i) : i \in J_{a(x)} \wedge (\forall i \in J_{a(x)})[a_i \in V_a] \wedge p_i = 1\}$.

Null value is interpreted as the set all possible values of an attribute with equal confidence assigned to all of them. Table 1 gives an example of an incomplete information system $S = (\{x_1, x_2, x_3, x_4, x_5, x_6, x_7, x_8\}, \{a, b, c, d, e\}, V\}$.

Clearly, $a(x_1) = \{(a_1, \frac{1}{3}), (a_2, \frac{2}{3})\}, a(x_2) = \{(a_2, \frac{1}{4}), (a_3, \frac{3}{4})\}, ..$

Let us begin to extract rules from S describing attribute e in terms of attributes $\{a, b, c, d\}$ following a strategy similar to *LERS* [3]. We start with identifying sets of objects in X having properties $a_1, a_2, a_3, b_1, b_2, c_1, c_2, c_3, d_1, d_2$ and next for each of these sets we check in what relationship it is with a set of objects in X having property e_1, next property e_2, and finally property e_3. Attribute value a_1 is interpreted as a set $a_1^\star$, equal to $\{(x_1, \frac{1}{3}), (x_3, 1), (x_5, \frac{2}{3})\}$. The justification of this interpretation is quite simple. Only 3 objects in S

may have property a_1. Object x_3 has it for sure. The confidence that x_1 has property a_1 is $\frac{1}{3}$ since $(a_1, \frac{1}{3}) \in a(x_1)$. In a similar way we justify the property a_1 for object x_5.

Table 1. Incomplete Information System S

X	a	b	c	d	e
x_1	$\{(a_1,\frac{1}{3}),(a_2,\frac{2}{3})\}$	$\{(b_1,\frac{2}{3}),(b_2,\frac{1}{3})\}$	c_1	d_1	$\{(e_1,\frac{1}{2}),(e_2,\frac{1}{2})\}$
x_2	$\{(a_2,\frac{1}{4}),(a_3,\frac{3}{4})\}$	$\{(b_1,\frac{1}{3}),(b_2,\frac{2}{3})\}$		d_2	e_1
x_3	a_1	b_2	$\{(c_1,\frac{1}{2}),(c_3,\frac{1}{2})\}$	d_2	e_3
x_4	a_3		c_2	d_1	$\{(e_1,\frac{2}{3}),(e_2,\frac{1}{3})\}$
x_5	$\{(a_1,\frac{2}{3}),(a_2,\frac{1}{3})\}$	b_1	c_2		e_1
x_6	a_2	b_2	c_3	d_2	$\{(e_2,\frac{1}{3}),(e_3,\frac{2}{3})\}$
x_7	a_2	$\{(b_1,\frac{1}{4}),(b_2,\frac{3}{4})\}$	$\{(c_1,\frac{1}{3}),(c_2,\frac{2}{3})\}$	d_2	e_2
x_8	a_3	b_2	c_1	d_1	e_3

Values of classification attributes are seen as atomic terms or initial descriptions:

$$a_1^\star = \{(x_1,\tfrac{1}{3}),(x_3,1),(x_5,\tfrac{2}{3})\},$$
$$a_2^\star = \{(x_1,\tfrac{2}{3}),(x_2,\tfrac{1}{4}),(x_5,\tfrac{1}{3}),(x_6,1),(x_7,1)\},$$
$$a_3^\star = \{(x_2,\tfrac{3}{4}),(x_4,1),(x_8,1)\},$$
$$b_1^\star = \{(x_1,\tfrac{2}{3}),(x_2,\tfrac{1}{3}),(x_4,\tfrac{1}{2}),(x_5,1),(x_7,\tfrac{1}{4})\},$$
$$b_2^\star = \{(x_1,\tfrac{1}{3}),(x_2,\tfrac{2}{3}),(x_3,1),(x_4,\tfrac{1}{2}),(x_6,1),(x_7,\tfrac{3}{4}),(x_8,1)\},$$
$$c_1^\star = \{(x_1,1),(x_2,\tfrac{1}{3}),(x_3,\tfrac{1}{2}),(x_7,\tfrac{1}{3}),(x_8,1)\},$$
$$c_2^\star = \{(x_2,\tfrac{1}{3}),(x_4,1),(x_5,1),(x_7,\tfrac{2}{3})\},$$
$$c_3^\star = \{(x_2,\tfrac{1}{3}),(x_3,\tfrac{1}{2}),(x_6,1)\},$$
$$d_1^\star = \{(x_1,1),(x_3,1),(x_5,\tfrac{1}{2}),(x_6,1),(x_7,1)\}.$$

Values of the decision attribute represent the second group of atomic descriptions:

$$e_1^\star = \{(x_1,\tfrac{1}{2}),(x_2,1),(x_4,\tfrac{2}{3}),(x_5,1)\},$$
$$e_2^\star = \{(x_1,\tfrac{1}{2}),(x_4,\tfrac{1}{3}),(x_6,\tfrac{1}{3}),(x_7,1)\},$$
$$e_3^\star = \{(x_3,1),(x_6,\tfrac{2}{3}),(x_8,1)\}.$$

Now, we will propose a new approach for checking the relationship between classification attributes and the decision attribute. Namely, for any two sets $c_i^\star = \{(x_i,p_i)\}_{i\in N}$, $e_j^\star = \{(y_j,q_j)\}_{j\in M}$, where $p_i > 0$ and $q_j > 0$, we say that:

$\{(x_i,p_i)\}_{i\in N} \preceq \{(y_j,q_j)\}_{j\in M}$ iff the support of the rule $[c_i \to e_j]$ is above some threshold value.

So, the relationship between $c_i^\star, e_j^\star$ depends only on the support of the corresponding rule $[c_i \to e_j]$.

Coming back to our example, let us take the rule $a_1 \to e_3$. How can we calculate its confidence and support?

Let us observe that:

- object x_1 supports a_1 with a confidence $\frac{1}{3}$ and it supports e_3 with a confidence 0.
- object x_3 supports a_1 with a confidence 1 and it supports e_3 with a confidence 1.
- object x_5 supports a_1 with a confidence $\frac{2}{3}$ and it supports e_3 with a confidence 0.

We propose to calculate the support of the rule $[a_1 \to e_3]$ as: $\frac{1}{3} \cdot 0 + 1 \cdot 1 + \frac{2}{3} \cdot 0 = 1$.

Similarly, the support of the term a_1 can be calculated as $\frac{1}{3} + 1 + \frac{2}{3} = 2$. Applying the standard strategy for calculating the confidence of association rules, the confidence of $a_1 \to e_3$ will be $\frac{1}{2}$.

Now, let us follow a procedure which has some similarity with *LERS*. Relation $\preceq$ will be automatically marked if the confidence and support of the rule, associated with that relation, are above two corresponding threshold values (given by user). In the our example we take $\frac{1}{2}$ as the threshold for minimum confidence and 1 as the threshold for minimum support.

Assume again that $c_i^\star = \{(x_i, p_i)\}_{i \in N}$, $e_j^\star = \{(y_j, q_j)\}_{j \in M}$. Our algorithm, in its first step, will calculate the support of the rule $c_i \to c_j$. If it is below the threshold value for minimum support, then the corresponding relationship $\{(x_i, p_i)\}_{i \in N} \preceq \{(y_j, q_j)\}_{j \in M}$ does not hold (it is marked as negative) and it is not considered in later steps. Otherwise, the confidence of the rule $c_i \to e_j$ has to be checked. If it is not below the given threshold value, the rule is acceptable and the corresponding relationship $\{(x_i, p_i)\}_{i \in N} \preceq \{(y_j, q_j)\}_{j \in M}$ is marked positive. Otherwise this relationship remains unmarked.

In our example we take $M_{sup} = 1$ as the threshold value for minimum support and $M_{conf} = \frac{1}{2}$ as the threshold for minimum confidence.

$a_1^\star \preceq e_1^\star$ ($sup = \frac{5}{6} < 1$) - marked negative
$a_1^\star \preceq e_2^\star$ ($sup = \frac{1}{6} < 1$) - marked negative
$a_1^\star \preceq e_3^\star$ ($sup = 1 \geq 1$ and $conf = \frac{1}{2}$) - **marked positive**

$a_2^\star \preceq e_1^\star$ ($sup = \frac{11}{12} < 1$) - marked negative
$a_2^\star \preceq e_2^\star$ ($sup = \frac{5}{3} \geq 1$ and $conf = \frac{51}{100}$) - **marked positive**
$a_2^\star \preceq e_3^\star$ ($sup = \frac{2}{3} < 1$) - marked negative

$a_3^\star \preceq e_1^\star$ ($sup = \frac{17}{12} \geq 1$ and $conf = \frac{51}{100}$) - **marked positive**
$a_3^\star \preceq e_2^\star$ ($sup = \frac{1}{3} < 1$) - marked negative
$a_3^\star \preceq e_3^\star$ ($sup = 1 \geq 1$ but $conf = \frac{36}{100}$) - not marked

$b_1^\star \preceq e_1^\star$ ($sup = 2 \geq 1$ and $conf = \frac{72}{100}$) - **marked positive**
$b_1^\star \preceq e_2^\star$ ($sup = \frac{3}{4} < 1$) - marked negative
$b_1^\star \preceq e_3^\star$ ($sup = 0 < 1$) - marked negative

$b_2^\star \preceq e_1^\star$ ($sup = \frac{7}{6} \geq 1$ but $conf = \frac{22}{100}$) - not marked
$b_2^\star \preceq e_2^\star$ ($sup = \frac{17}{12} \geq 1$ but $conf = \frac{27}{100}$) - not marked
$b_2^\star \preceq e_3^\star$ ($sup = \frac{8}{3} \geq 1$ and $conf = \frac{51}{100}$) - **marked positive**

$c_1^\star \preceq e_1^\star$ ($sup = \frac{5}{6} < 1$) - marked negative
$c_1^\star \preceq e_2^\star$ ($sup = \frac{5}{6} < 1$) - marked negative
$c_1^\star \preceq e_3^\star$ ($sup = \frac{3}{2} \geq 1$ but $conf = \frac{47}{100}$) - not marked

$c_2^\star \preceq e_1^\star$ ($sup = 2 \geq 1$ and $conf = \frac{66}{100}$) - **marked positive**
$c_2^\star \preceq e_2^\star$ ($sup = 1 \geq 1$ but $conf = \frac{33}{100}$) - not marked
$c_2^\star \preceq e_3^\star$ ($sup = 0 < 1$) - marked negative

$c_3^\star \preceq e_1^\star$ ($sup = \frac{1}{3} < 1$) - marked negative
$c_3^\star \preceq e_2^\star$ ($sup = \frac{1}{3} < 1$) - marked negative
$c_3^\star \preceq e_3^\star$ ($sup = \frac{7}{6} \geq 1$ and $conf = \frac{64}{100}$) - **marked positive**

$d_1^\star \preceq e_1^\star$ ($sup = \frac{5}{3} \geq 1$ but $conf = \frac{48}{100}$) - not marked
$d_1^\star \preceq e_2^\star$ ($sup = \frac{5}{6} < 1$) - marked negative
$d_1^\star \preceq e_3^\star$ ($sup = 1 \geq 1$ but $conf = \frac{28}{100}$) - not marked

$d_2^\star \preceq e_1^\star$ ($sup = \frac{3}{2} \geq 1$ but $conf = \frac{33}{100}$) - not marked
$d_2^\star \preceq e_2^\star$ ($sup = \frac{1}{3} < 1$) - marked negative
$d_2^\star \preceq e_3^\star$ ($sup = \frac{5}{3} \geq 1$ but $conf = \frac{37}{100}$) - not marked

The next step is to build descriptions of length 2 from descriptions of length 1 interpreted as sets of objects having support ≥ 1, and confidence $< \frac{1}{2}$. We propose the following definition for concatenating such two corresponding sets $c_i^\star, e_j^\star$:

$(c_i \cdot e_j)^\star = \{(x_i, p_i \cdot q_i)\}^{i \in K}$, where $c_i^\star = \{(x_i, p_i)\}_{i \in N}$, $e_j^\star = \{(y_j, q_j)\}_{j \in M}$ and $K = M \cap N$.

Following this definition, we get:

$(a_3 \cdot c_1)^\star \preceq e_3^\star$ ($sup = 1 \geq 1$ and $conf = \frac{8}{10}$) - **marked positive**
$(a_3 \cdot d_1)^\star \preceq e_3^\star$ ($sup = 1 \geq 1$ and $conf = \frac{1}{2}$) - **marked positive**
$(a_3 \cdot d_2)^\star \preceq e_3^\star$ ($sup = 0 < 1$ - marked negative
$(b_2 \cdot d_2)^\star \preceq e_1^\star$ ($sup = \frac{2}{3} < 1$ - marked negative

$(b_2 \cdot c_2)^\star \preceq e_2^\star$ ($sup = \frac{2}{3} < 1$ - marked negative
$(c_1 \cdot d_1)^\star \preceq e_3^\star$ ($sup = 1 \geq 1$ and $conf = \frac{1}{2}$) - **marked positive**
$(c_1 \cdot d_2)^\star \preceq e_3^\star$ ($sup = \frac{1}{2} < 1$) - marked negative

So, from the above example, we can easily see the difference between *LERS* and our proposed strategy for rules induction. *LERS* is considering only one type of marks. This mark in our method corresponds to positive marking. *LERS* is controlled by one threshold value whereas our strategy is controlled by two thresholds (for minimum confidence and minimum support).

3 Algorithm *ERID* (Extracting Rules from Incomplete Decision Systems)

In this section, we present an algorithm for extracting rules when the data in an information system are incomplete. Also, we evaluate its time complexity.

To simplify our presentation, new notations (only for the purpose of this algorithm) will be introduced.

Let us assume that $S = (X, A \cup \{d\}, V)$ is a decision system, where X is a set of objects, $A = \{a[i] : 1 \leq i \leq I\}$ is a set of classification attributes, $V_i = \{a[i,j] : 1 \leq j \leq J(i)\}$ is a set of values of attribute $a[i]$, $i \leq I$. We also assume that d is a decision attribute, where $V_d = \{d[m] : 1 \leq m \leq M\}$.

Finally, we assume that $a(i_1, j_1) \cdot a(i_2, j_2) \cdot ... \cdot a(i_r, j_r)$ is denoted by term $[a(i_k, j_k)]_{k \in \{1,2,...,r\}}$, where all $i_1, i_2, ..., i_r$ are distinct integers and $j_p \leq J(i_p)$, $1 \leq p \leq r$.

Algorithm for Extracting Rules from Incomplete Decision System S

ERID$(S, \lambda_1, \lambda_2, L(D))$;
$S = (X, A, V)$ - incomplete decision system,
λ_1 - threshold for minimum support,
λ_2 - threshold for minimum confidence,
$L(D)$ set of rules discovered from S by *ERID*.
begin
$i := 1$;
while $i \leq I$ **do**
begin
$j := 1$; $m := 1$
while $j \leq J[i]$ **do**
begin
if $sup(a[i,j] \rightarrow d(m)) < \lambda_1$ **then** $mark[a[i,j]^\star \preceq d(m)^\star] = negative$;
if $sup(a[i,j] \rightarrow d(m)) \geq \lambda_1$ and $conf(a[i,j] \rightarrow d(m)) \geq \lambda_2$ **then**
begin

```
      mark[a[i,j]* ⪯ d(m)*] = positive;
      output[a[i,j] → d(m)]
      end
   j := j + 1
   end
 end
 I_k := {i_k}; /i_k - index randomly chosen from {1,2,...,I}/
 for all j_k ≤ J(i_k) do a[(i_k, j_k)]_{i_k∈I_k} := a(i_k, j_k);
 for all i, j such that both rules
    a[(i_k, j_k)]_{i_k∈I_k} → d(m),
    a[i,j] → d(m) are not marked and i ∉ I_k do
    begin
    if sup(a[(i_k, j_k)]_{i_k∈I_k} · a[i,j] → d(m)) < λ1
       then mark[(a[(i_k, j_k)]_{i_k∈I_k} · a[i,j])* ⪯ d(m)*] = negative;
    if sup(a[(i_k, j_k)]_{i_k∈I_k} · a[i,j] → d(m)) ≥ λ1 and
    conf(a[(i_k, j_k)]_{i_k∈I_k} · a[i,j] → d(m)) ≥ λ2 then
       begin
       mark[(a[(i_k, j_k)]_{i_k∈I_k} · a[i,j])* ⪯ d(m)*] = positive;
       output[a[(i_k, j_k)]_{i_k∈I_k} · a[i,j] → d(m)]
       end
    else
       begin
       I_k := I_k ∪ {i};
       a[(i_k, j_k)]_{i_k∈I_k} := a[(i_k, j_k)]_{i_k∈I_k} · a[i,j]
       end
 end
```

Now, we will evaluate the time complexity (*T-Comp*) of the algorithm *ERID*.

Let us assume that $S = (X, A \cup \{d\}, V)$, $N = card(X)$, $I = card(A)$, $V = \bigcup\{V_a : a \in A\}$, $card(V) = k$.

$T\text{-}Comp(ERID) = O[I \cdot N + N + [\sum_{i=1}^{I-1}[card[u(V)] \cdot card[u(V^i)] \cdot N]]] = O[\sum_{i=1}^{I-1}[card[u(V)] \cdot card[u(V^i)] \cdot N]] \leq O[(k + k^2 + ... + k^I) \cdot N] = O[k^{I+1} \cdot N]$.

By $u(V^i)$ we mean non-marked terms built from elements in V^i.

By assigning smaller values to λ_2 we are significantly decreasing the number of elements in $u(V^i)$ when i is getting larger. So, the time complexity of *ERID* is also decreasing when λ_2 is decreasing. It means that instead of k^{I+1} used in $O[k^{I+1} \cdot N]$ much smaller number is usually needed.

4 Testing Results

In this section, we show the results obtained from testing the rule discovery algorithm (given in Section 3) which is implemented as the main module of *ERID*. The testing strategies which include fixed cross-validation, stratified cross-validation, and bootstrap are implemented as another module in *ERID*. The user can chose the number of folds in cross-validation. Also, quantization strategy for numerical attributes (which is a bottom-up strategy based on a nearest distance between clusters represented as cubes) is also implemented as a module in *ERID*.

System *ERID* was tested on the data-set *Adult* which is available at:

http://www.cs.toronto.edu/ delve/data/adult/desc.html.

Since *ERID* accepts data-sets not only incomplete (with null values) but also partially incomplete (as introduced in this paper) as its input, we decided to modify slightly the data-set *Adult* by replacing its missing values by two weighted attribute values following two steps given below:

- Randomly select two attribute values from the corresponding attribute's domain.
- Randomly assign a weight to the first selected attribute value and then *one* minus that weight to the second attribute value.

Table 2. Cross-Validation (Fixed)

a	b	c	d
$folds$	sup	$Conf$	$ErrorRate$
10	[$\geq$ 10]	[$\geq$ 90%]	[11.25%]
10	[$\geq$ 15]	[$\geq$ 90%]	[10.22%]
5	[$\geq$ 10]	[$\geq$ 90%]	[11.18%]
4	[$\geq$ 10]	[$\geq$ 90%]	[11.49%]
3	[$\geq$ 10]	[$\geq$ 90%]	[10.69%]
3	[$\geq$ 10]	[$\geq$ 95%]	[6.59%]
3	[$\geq$ 10]	[= 100%]	[**4.2**%]

Clearly, for each incomplete attribute, we could easily follow some null value imputation strategy to replace its missing values by sets of weighted attribute values. These values can be taken from the domain of that attribute and the corresponding weights can be computed by following either one of the probabilistic methods or one of the rule based methods. When this is accomplished, we can run *ERID* on this new modified data-set. But, for testing *ERID* we do not need to use any sophisticated imputation method for replacement of null values in the testing data-set. Partially incomplete data-set created by using random selection of two weighted attribute values (taken from the same attribute domain) for each missing value and using them as a replacement for each of these missing values will work as well.

Table 3. Cross-Validation (Stratified)

a	b	c	d
folds	*sup*	*Conf*	*ErrorRate*
10	[$\geq$ 10]	[$\geq$ 90%]	[10.5%]
10	[$\geq$ 15]	[$\geq$ 90%]	[9.63%]
3	[$\geq$ 10]	[$\geq$ 90%]	[11.63%]
3	[$\geq$ 10]	[= 100%]	[**3.6**%]
3	[$\geq$ 20]	[$\geq$ 90%]	[9.6%]
3	[$\geq$ 20]	[$\geq$ 95%]	[5.35%]
3	[$\geq$ 20]	[= 100%]	[**1.2**%]

The results of testing *ERID* on the modified data-set *Adult* using Bootstrap and both fixed and stratified 10-fold and 3-fold Cross Validation methods are given in Table 2, Table 3, and Table 4. The last figure shows two snapshots from *ERID*. The first one was taken when the threshold for maximum support was set up as 20 and the threshold for maximum confidence as 90%. In the second one, the minimum support is equal to 10 and the minimum confidence to 100%. In both cases, Bootstrap strategy was used for testing.

Table 4. Bootstrap

a	b	c
sup	*Conf*	*ErrorRate*
[$\geq$ 10]	[$\geq$ 90%]	[9.09%]
[$\geq$ 10]	[$\geq$ 95%]	[5.79%]
[$\geq$ 10]	[= 100%]	[4.54%]
[$\geq$ 20]	[$\geq$ 90%]	[7.14%]
[$\geq$ 20]	[$\geq$ 95%]	[6.91%]
[$\geq$ 20]	[= 100%]	[**0.82**%]
[$\geq$ 10]	[$\geq$ 70%]	[14.0%]

5 Conclusion

Missing data values cause problems both in knowledge extraction from information systems and in query answering systems. Quite often missing data are

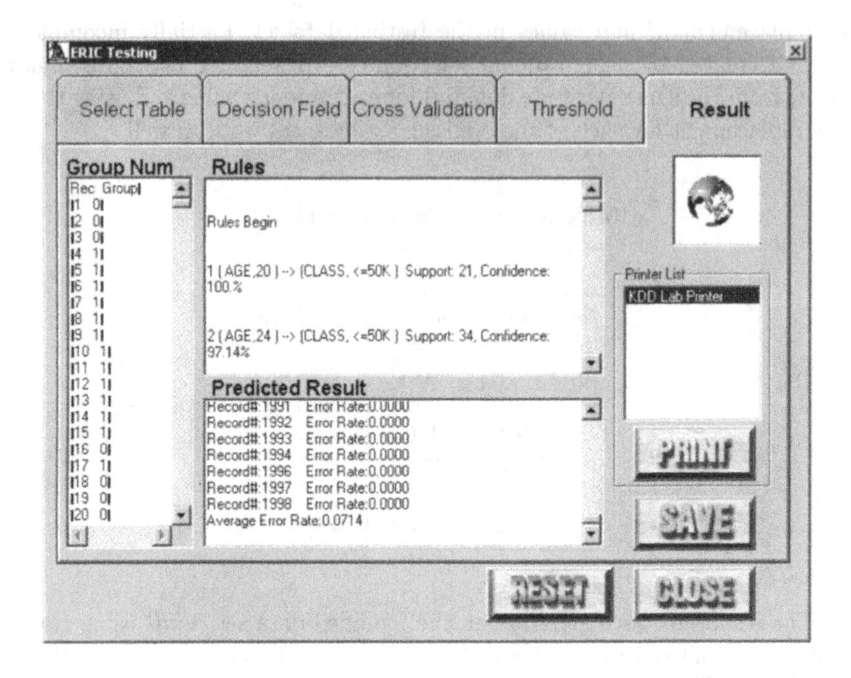

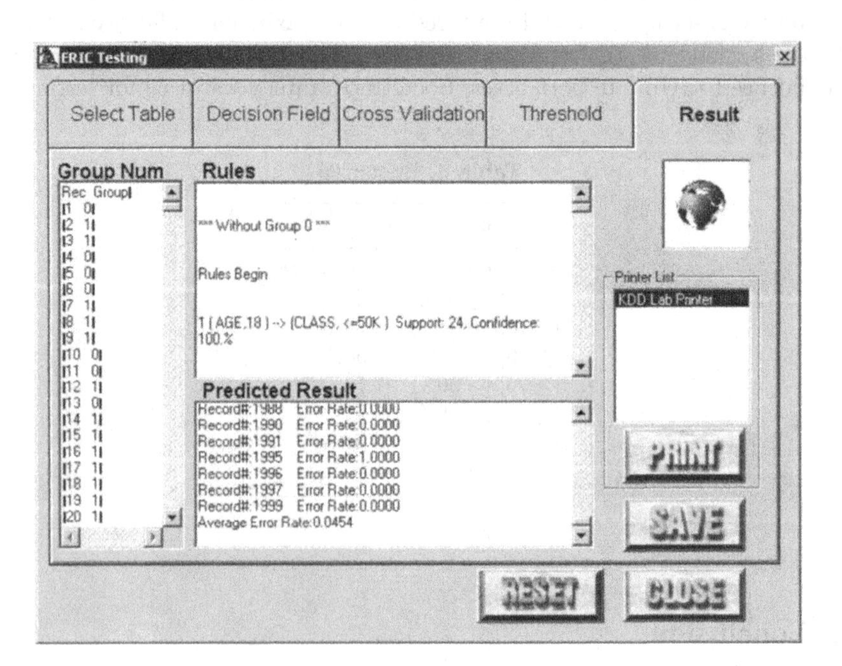

Fig. 1. Snapshots from *ERID* Interface - Bootstrap (Testing)

partially specified. It means that the missing value is an intermediate value between the unknown value and the completely specified value. Clearly there is a number of intermediate values possible for a given attribute. Algorithm *ERID* presented in this paper can be used as a new tool for chasing values in incomplete information systems with rules discovered from them. We can keep repeating this strategy till a fix point is reached. To our knowledge, there are no such tools available for incomplete information systems presented in this paper.

References

1. Dardzinska A, Ras Z W (2003) Rule-based Chase algorithm for partially incomplete information systems, Proceedings of the Second International Workshop on Active Mining (AM'2003), Maebashi City, Japan, October, 42-51
2. Greco S, Matarazzo B, Slowinski R (2001) Rough sets theory for multicriteria decision analysis, European Journal of Operational Research, Vol. 129, 1–47
3. Grzymala-Busse J (1997) A new version of the rule induction system LERS, Fundamenta Informaticae, Vol. 31, No. 1, 27–39
4. Kryszkiewicz M, Rybinski H (1996) Reducing information systems with uncertain attributes, Proceedings of ISMIS'96, LNCS/LNAI, Springer-Verlag, Vol. 1079, 285–294
5. Michalski R S, Kaufman K (2001) The AQ19 System for Machine Learning and Pattern Discovery: A General Description and User's Guide, Reports of the Machine Learning and Inference Laboratory, MLI 01-2, George Mason University, Fairfax, VA
6. Ohrn A, Komorowski J, Skowron A, Synak P (1997) A software systemfor rough data analysis, Bulletin of the International Rough Sets Society, Vol. 1, No. 2, 58–59
7. Pawlak Z (1991) Rough sets-theoretical aspects of reasoning about data, Kluwer, Dordrecht
8. Quinlan J R (1993) C4.5: Programs for Machine Learning, Morgan Kaufmann Publishers, San Mateo, CA
9. Ras Z W, Joshi S (1997) Query approximate answering system for an incomplete DKBS, Fundamenta Informaticae Journal, IOS Press, Vol. 30, No. 3/4, 313–324
10. Ras Z W, Dardzinska A (2004) Query Answering based on Collaboration and Chase, Proceedings of FQAS'04 Conference, Lyon, France, LNCS/LNAI, No. 3055, Springer-Verlag, 125–136
11. Slowinski R, Stefanowski J, Greco S, Matarazzo B (2000) Rough sets processing of inconsistent information, Control and Cybernetics, Vol. 29 No. 1, 379–404

Mining for Patterns Based on Contingency Tables by *KL-Miner* – First Experience

Jan Rauch[1,3], Milan Šimůnek[2,4], and Václav Lín[3]

[1] EuroMISE Centrum – Cardio,
[2] Department of Information Technology,
[3] Department of Information and Knowledge Engineering,
[4] Laboratory for Intelligent Systems,
University of Economics, Prague
nám. W. Churchilla 4, 130 67 Praha 3, Czech Republic
rauch@vse.cz, simunek@vse.cz, xlinv05@vse.cz

Summary. A new datamining procedure called *KL-Miner* is presented. The procedure mines for various patterns based on evaluation of two–dimensional contingency tables, including patterns of statistical or information theoretic nature. The procedure is a result of continued development of the academic system LISp-Miner for KDD.

Key words: Data mining, contingency tables, system LISp–Miner, statistical patterns

1 Introduction

Goal of this paper is to present first experience with data mining procedure *KL-Miner*. The procedure mines for patterns of the form

$$R \sim C/\gamma\,.$$

Here R and C are categorial attributes, the attribute R has *categories* (possible values) $r_1, \ldots, r_K$, the attribute C has categories $c_1, \ldots, c_L$. Further, γ is a Boolean attribute.

The *KL-Miner* works with data matrices. We suppose that R and C correspond to columns of the analyzed data matrix. We further suppose that the Boolean attribute γ is somehow derived from other columns of the analyzed data matrix and thus that it corresponds to a Boolean column of the analyzed data matrix.

The intuitive meaning of the expression $R \sim C/\gamma$ is that the attributes R and C are in relation given by the symbol $\sim$ when the condition given by the derived Boolean attribute γ is satisfied.

The symbol $\sim$ is called *KL-quantifier* . It corresponds to a condition imposed by the user on the contingency table of R and C. There are several restrictions that the user can choose to use (e.g. minimal value, sum over the table, value of the χ^2 statistic, and other).

We call the expression $R \sim C/\gamma$ a *KL-hypothesis* or simply *hypothesis.* The KL-hypothesis $R \sim C/\gamma$ is *true* in the data matrix $\mathcal{M}$ if the condition corresponding to the KL-quantifier $\sim$ is satisfied for the contingency table of R and C on the data matrix $\mathcal{M}/\gamma$. The data matrix $\mathcal{M}/\gamma$ consists of all rows of data matrix $\mathcal{M}$ satisfying the condition γ (i.e. of all rows in which the value of γ is TRUE).

Input of the *KL-Miner* consists of the analyzed data matrix and of several parameters defining a set of potentially interesting hypotheses. Such a set can be very large. The *KL-Miner* automatically generates all potentially interesting hypotheses and verifies them in the analyzed data matrix. The output of the *KL-Miner* consists of all hypotheses that are true in the analyzed data matrix (i.e. supported by the analyzed data). Some details about input of the *KL-Miner* procedure are given in Sect. 2. KL-quantifiers are described in Sect. 3.

The *KL-Miner* procedure is one of data mining procedures of the LISp-Miner system that are implemented using bit string approach [4]. It means that the analysed data is represented using suitable strings of bits. Software modules for dealing with strings of bits developed for the LISp-Miner system are used for *KL-Miner* implementation, see Sect. 4.

Let us remark that the *KL-Miner* is a GUHA procedure in the sense of the book [1]. Therefore, we will use the terminology introduced in [1]. The potentially interesting hypotheses will be called *relevant questions*, and the hypotheses that are true in the analyzed data matrix will be called *relevant truths.* Also, the use of the term *quantifier* for the symbol $\sim$ in the expression $R \sim C/\gamma$ is inspired by [1]. The cited book contains rich enough theoretical framework to build a formal logical theory for the *KL-Miner*–style data mining; however, we will not do it here. Furthermore, *KL-Miner* is related to a GUHA procedure from the 1980's called *CORREL*, see [2]. Let us also remark that the analysis of contingency tables by *KL-Miner* is similar in spirit to that of the procedure *49er* [7].

2 Input of *KL-Miner*

Please recall that the *KL-Miner* mines for hypotheses of the form $R \sim C/\gamma$, where R and C are categorial attributes, γ is a Boolean attribute, and $\sim$ is a KL-quantifier. The attribute R is called the *row attribute*, the attribute C is called the *column attribute.*

Input of the *KL-Miner* procedure consists of

- the analyzed data matrix

- a set $\mathcal{R} = \{R_1, \ldots, R_u\}$ of row attributes
- a set $\mathcal{C} = \{C_1, \ldots, C_v\}$ of column attributes
- specification of the KL-quantifier $\sim$
- several parameters defining set Γ of *relevant conditions* (i.e. derived Boolean attributes) γ, see below.

The *KL-Miner* automatically generates and verifies all relevant questions

$$R \sim C/\gamma,$$

such that $R \in \mathcal{R}$, $C \in \mathcal{C}$ and $\gamma \in \Gamma$. Conventionally, the set of all relevant questions is denoted RQ.

Each relevant condition γ is a conjunction of *partial conditions* $\gamma_1, \ldots, \gamma_t$, and each partial condition γ_i is a conjunction of *literals*. Literal is an expression of the form $B(\omega)$ or $\neg B(\omega)$, where B is an attribute (derived column of the analyzed data matrix) and ω is a proper subset of the set of all categories (possible values) of B. The subset ω is called a *coefficient* of the literal $B(\omega)$ (or $\neg B(\omega)$). The literal $B(\omega)$ is called *positive literal*, $\neg B(\omega)$ is called *negative literal.*

$B(\omega)$ is in fact a Boolean attribute that is true in the row o of analyzed data matrix if, and only if, the value in the column B in the row o belongs to the set ω. $\neg B(\omega)$ is a Boolean attribute that is a negation of $B(\omega)$.

An example of a relevant question is expression

$$R_1 \sim C_1/A_1(1,2) \wedge A_2(3,4) \ .$$

Here R_1 is the row attribute, C_1 is the column attribute and the condition is $A_1(1,2) \wedge A_2(3,4)$. This condition means that value of the attribute A_1 is either 1 or 2, and value of the attribute A_2 is either 3 or 4. An example of *KL-Miner* input is in Sect. 5.

The set Γ of relevant conditions to be automatically generated is given in the same way as the set of relevant antecedents for the procedure *4ft-Miner*, see [4]. The procedure 4ft-Miner mines for association rules of the form

$$\varphi \approx \psi$$

and for conditional association rules of the form

$$\varphi \approx \psi/\chi$$

where φ, ψ and χ are derived Boolean attributes (conjunctions of literals). Intuitive meaning of $\varphi \approx \psi$ is that φ and ψ are in relation given by the symbol $\approx$. Intuitive meaning of $\varphi \approx \psi/\chi$ is that φ and ψ are in relation given by the symbol $\approx$ when the condition χ is satisfied.

Symbol $\approx$ is called *4ft-quantifier.* It corresponds to a condition concerning four fold contingency table of φ and ψ. Various types of dependencies of φ and ψ can be expressed this way. An example is the classical association rule with confidence and support, another example is a relation corresponding to the χ^2-test of independence.

The left part of the association rule, φ, is called *antecedent*, the right part of the association rule, ψ, is called *succedent*, and χ is called *condition*.

The input of the *4ft-Miner* procedure consists of

- the analyzed data matrix
- several parameters defining set of relevant antecedents φ
- several parameters defining set of relevant succedents ψ
- several parameters defining set of relevant conditions χ
- the 4ft-quantifier $\approx$.

Parameters defining set of relevant antecedents, set of relevant succedents and set of relevant conditions have the same structure as the parameters defining the set of relevant conditions in the input of the *KL-Miner* procedure (see above). This fact is very useful from the point of view of implementation of the *KL-Miner*, see Sect. 4.

3 KL-quantifiers

As noted above, the *KL-Miner* mines for hypotheses of the form

$$R \sim C/\gamma$$

where R and C are categorial attributes with admissible values $r_1, \ldots, r_K$ and $c_1, \ldots, c_L$, respectively. γ is a relevant condition.

Hypothesis $R \sim C/\gamma$ is true in the data matrix $\mathcal{M}$ if a condition corresponding to the KL-quantifier $\sim$ is satisfied for the contingency table of R and C on the data matrix $\mathcal{M}/\gamma$. The data matrix $\mathcal{M}/\gamma$ consists of all rows of $\mathcal{M}$, for which the value of γ is TRUE.

We suppose that the contingency table of attributes R and C on the data matrix $\mathcal{M}/\gamma$ has the form of Table 1, where:

- $n_{k,l}$ denotes the number of rows in data matrix $\mathcal{M}/\gamma$ for which $R = r_k$ and $C = c_l$
- $n_{k,*} = \sum_l n_{k,l}$ denotes the number of rows in data matrix $\mathcal{M}/\gamma$ for which $R = r_k$
- $n_{*,l} = \sum_k n_{k,l}$ denotes the number rows in data matrix $\mathcal{M}/\gamma$ for which $C = c_l$
- $n = \sum_k \sum_l n_{k,l}$ denotes the number of all rows in data matrix $\mathcal{M}/\gamma$

We will also use relative frequencies:

- $f_{k,l} = n_{k,l}/n$
- $f_{k,*} = \sum_l f_{k,l} = n_{k,*}/n$
- $f_{*,l} = \sum_k f_{k,l} = n_{*,l}/n$

Table 1. Contingency table of R and C on $\mathcal{M}/\gamma$ - absolute frequencies

$\mathcal{M}/\gamma$	c_1	$\ldots$	c_L	Σ_l
r_1	$n_{1,1}$	$\ldots$	$n_{1,L}$	$n_{1,*}$
$\vdots$	$\vdots$		$\vdots$	$\vdots$
r_K	$n_{K,1}$	$\ldots$	$n_{K,L}$	$n_{K,*}$
Σ_k	$n_{*,1}$	$\ldots$	$n_{*,L}$	n

Semantics of the KL-quantifier $\sim$ is determined by the user, who can choose to set lower / upper threshold values for several functions on the table of absolute or relative frequencies. We list some of these functions below.

Simple Aggregate Functions

$$\min_{k,l}\{n_{k,l}\}\ ,\ \max_{k,l}\{n_{k,l}\}\ ,\ \sum_{k,l} n_{k,l}\ ,\ \frac{1}{KL}\sum_{k,l} n_{k,l}$$

Simple Non-statistical Measures

For example Fnc_S expresses the fact, that C is a function of R:

$$Fnc_S = \frac{1}{n}\sum_k \max_l\{n_{k,l}\}\ .$$

Fnc_S takes values from $\langle L^{-1}, 1\rangle$. It is $Fnc_S = 1$ iff for each category r_k of R, there is exactly one category c_l of C, such that $n_{k,l}$ is nonzero; it is $Fnc_S = L^{-1}$ iff $(\forall k \forall l) n_{k,l} = \max_j\{n_{k,j}\}$ (i.e. each row's distribution of frequencies is uniform).

Statistical and Information Theoretic Functions

Information dependence [5].

$$ID = 1 - \frac{\sum_k f_{k,*}\log_2 f_{k,*} - \sum_{k,l} f_{k,l}\log_2 f_{k,l}}{-\sum_l f_{*,l}\log_2 f_{*,l}}\ .$$

We note that ID corresponds to

$$1 - \frac{-H(R) + H(C,R)}{H(C)} = 1 - \frac{H(C|R)}{H(C)}\ ,$$

where $H(.)$, $H(.,.)$ and $H(.|.)$ denote entropy, joint entropy and conditional entropy, respectively. ID takes values from $\langle 0, 1\rangle$; $ID = 0$ iff C is independent of R (i.e. $H(C|R) = H(C)$), and $ID = 1$ iff C is a function of R (i.e. $H(C|R) = 0$).

The Pearson χ^2 statistic [5].

$$\chi^2 = \sum_{k,l} \frac{\left(n_{k,l} - n_{k,*}n_{*,l}/n\right)^2}{n_{k,*}n_{*,l}/n} \,.$$

Kendall's coefficient [5].

$$\tau_b = \frac{2(P-Q)}{\sqrt{\left(n^2 - \sum_k n_{k,*}^2\right)\left(n^2 - \sum_l n_{*,l}^2\right)}} \,,$$

where

$$P = \sum_{k,l} n_{k,l} \sum_{i>k} \sum_{j>l} n_{i,j} \,, \; Q = \sum_{k,l} n_{k,l} \sum_{i>k} \sum_{j<l} n_{i,j} \,.$$

τ_b takes values from $\langle -1, 1\rangle$ with the following interpretation: $\tau_b > 0$ indicates positive ordinal dependence[5], $\tau_b < 0$ indicates negative ordinal dependence, $\tau_b = 0$ indicates ordinal independence, $|\tau_b = 1|$ indicates that C is a function of R. We note that the possibility to express ordinal dependence is rather useful in practice.

Mutual information.

$$I_m = \frac{\sum_{k,l} f_{k,l}\left(\log_2 f_{k,l} - \log_2 f_{k,*}f_{*,l}\right)}{\min\left[-\sum_l f_{*,l}\log_2 f_{*,l}, -\sum_k f_{k,*}\log_2 f_{k,*}\right]}$$

corresponds to $\left(H(C) - H(C|R)\right)/\min\left[H(C), H(R)\right]$ and has the following properties: $0 \le I_m \le 1$, $I_m = 0$ indicates independence, $I_m = 1$ indicates that there is a functional dependence of C on R or of R on C.

Values of χ^2, ID, τ_b and I_m have reasonable interpretation only when working with data of statistical nature (i.e. resulting from simple random sampling). For an example of KL-quantifier, see Sect. 5.

4 *KL-Miner* Implementation

The KL-Miner procedure is one of data mining procedures of the LISp-Miner system that are implemented using bit string approach [3, 4]. Software tools developed earlier for the LISp-Miner are utilized. We will use the data matrix shown in Fig. 1 to explain the implementation principles.

We suppose to have only one row attribute R with categories $r_1, \ldots, r_K$, and one column attribute C with categories $c_1, \ldots, c_L$. We also suppose that the relevant conditions will be automatically generated from attributes X, Y and Z with categories $x_1, \ldots, x_p$, $y_1, \ldots, y_q$ and $z_1, \ldots, z_r$, respectively.

[5] i.e. high values of C often coincide with high values of R and low values of C often coincide with low values of R

row	R	C	X	Y	Z
o_1	r_2	c_L	x_2	y_3	z_4
o_2	r_K	c_4	x_7	y_2	z_6
$\vdots$	$\vdots$	$\vdots$	$\vdots$	$\vdots$	$\vdots$
o_m	r_1	c_1	x_p	y_9	z_3

Fig. 1. Data matrix $\mathcal{M}$

Data matrix $\mathcal{M}$ (see Fig. 1) has m rows $o_1, \ldots, o_m$; value of the attribute R in the row o_1 is r_2, value of the attribute C in the row o_1 is c_L, etc.

The *KL-Miner* has to automatically generate and verify relevant questions of the form

$$R \sim C/\gamma$$

where the relevant condition γ is automatically generated from attributes X, Y and Z. The key problem of the *KL-Miner* implementation is fast computation of contingency tables of attributes R and C on the data matrix $\mathcal{M}/\gamma$ for particular relevant conditions γ, see also Table 1.

It means that we have to compute the frequencies $n_{k,l}$ for $k = 1, \ldots, K$ and $l = 1, \ldots, L$. Let us remember that $n_{k,l}$ denotes the number of rows in data matrix $\mathcal{M}/\gamma$ for which $R = r_k$ and $C = c_l$, see Sect. 3. In other words the frequency $n_{k,l}$ is the number of rows in data matrix $\mathcal{M}$ for which the Boolean attribute

$$R(r_k) \wedge C(c_l) \wedge \gamma$$

is true. Recall that $R(r_k)$ and $C(c_l)$ are Boolean attributes – literals, see Sect. 2.

We use a bit-string representation of the analyzed data matrix $\mathcal{M}$. Each attribute is represented by *cards* of its categories, i.e. the attribute R is represented by cards of categories $r_1, \ldots, r_K$.

For $k = 1, \ldots, K$, the card of the category r_k of the attribute R is denoted $R[r_k]$. $R[r_k]$ is a string of bits. Each row of $\mathcal{M}$ corresponds to one bit in $R[r_k]$. There is "1" in such a bit if, and only if, there is the value r_k in the corresponding row of the column R. Cards of the categories of the attribute R are shown in Fig. 2. Cards of categories of attributes C, X, Y and Z are

	attribute	cards of categories of R			
row	R	$R[r_1]$	$R[r_2]$	$\ldots$	$R[r_K]$
o_1	r_2	0	1	$\ldots$	0
o_2	r_K	0	0	$\ldots$	1
$\vdots$	$\vdots$	$\vdots$	$\vdots$	$\vdots$	$\vdots$
o_m	r_1	1	0	$\ldots$	0

Fig. 2. Cards of categories of the attribute R

denoted analogously.

Card $\mathcal{C}(\gamma)$ of the Boolean attribute γ is a string of bits that is analogous to a card of category. Each row of data matrix corresponds to one bit of $\mathcal{C}(\gamma)$ and there is "1" in this bit if and only if the Boolean attribute γ is true in the corresponding row.

Clearly, for arbitrary Boolean attributes γ_1 and γ_2,

$$\mathcal{C}(\gamma_1 \wedge \gamma_2) = \mathcal{C}(\gamma_1) \dot{\wedge} \mathcal{C}(\gamma_2) .$$

Here $\mathcal{C}(\gamma_1)\dot{\wedge}\mathcal{C}(\gamma_2)$ is a bit-wise conjunction of bit-strings $\mathcal{C}(\gamma_1)$ and $\mathcal{C}(\gamma_2)$. Similarly it is

$$\mathcal{C}(\neg \gamma_1) = \dot{\neg} \mathcal{C}(\gamma_1) .$$

It is important that the bit-wise Boolean operations $\dot{\wedge}$ and $\dot{\vee}$ are realised by very fast computer instructions. Very fast computer instructions are also used to realise a bit-string function $Count(\xi)$ returning number of values "1" in the bit-string ξ. These bit-string operations and function are used to compute the frequency $n_{k,l}$:

$$n_{k,l} = Count(R[r_k] \dot{\wedge} C[c_l] \dot{\wedge} \mathcal{C}(\gamma)) .$$

The only task we have to solve is the task of fast computation of $\mathcal{C}(\gamma)$. It is done in the same way as fast computation of antecedents for the 4ft-Miner procedure, see [4].

Apart of efficient implementation, the generating algorithm employs an optimization similar in spirit to support-based pruning known from association rule discovery. When the quantifier $\sim$ contains a condition $\sum n_{k,l} \geq S$ (S a constant), the set *RQ* of relevant questions can be substantially pruned. This is due to the obvious fact that with R and C fixed and γ varying, $\sum n_{k,l}$ is non-increasing in the length of γ. Thus we may avoid generating conditions γ that are sure to result in tables with $\sum n_{k,l} < S$. The algorithm is quite fast, results of some experiments are in Sect. 6.

5 *KL-Miner* Application Example

We will present an application example concerning the STULONG data set (see [6] for details). The data set consists of several data matrices, comprising data from cardiology research. We will work with data matrix called ENTRY. The matrix results from observing 219 attributes on 1 419 middle-aged men upon their entry examination.

Our example task concerns relation between patients' body mass index (BMI) and patients' level of cholesterol, conditioned by patients' vices (smoking, alcohol consumption, etc.). Definition of the task in *KL-Miner*'s GUI is shown in Fig. 3.

The set *RQ* of relevant questions is given by:

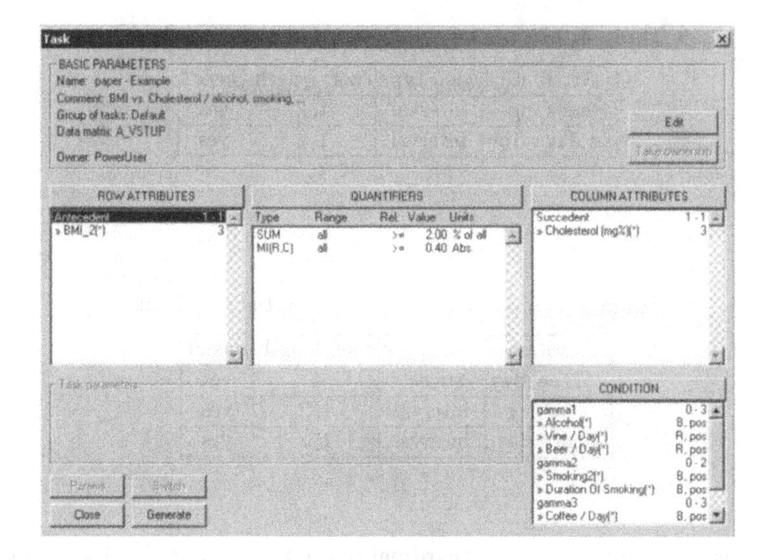

Fig. 3. Task definition in the *KL-Miner*

- a set of row attributes: $\mathcal{R} = \{\texttt{BMI}\}$,
- a set of column attributes: $\mathcal{C} = \{\texttt{cholesterol}\}$,
- a specification of KL-quantifier (see Sect. 3): we set conditions $I_m > 0.4$ & $\sum_{k,l} n_{k,l} \geq 0.02 * m$ (m is number of rows of the ENTRY data matrix),
- definition of the set Γ of relevant conditions, see below.

Recall (see Sect. 2) that each relevant condition is a conjunction of partial conditions, each partial condition being a conjunctions of literals. In our task, we have defined the following three sets of partial conditions:

- set Γ_1 of conjunctions of length $0, \ldots, 3$ of literals defined in Table 2,
- set Γ_2 of conjunctions of length $0, \ldots, 2$ of literals defined in Table 3,
- set Γ_3 of conjunctions of length $0, \ldots, 3$ of literals defined in Table 4.

Only positive literals were allowed. The set Γ of all relevant conditions is defined as

$$\Gamma = \{\gamma_1 \wedge \gamma_2 \wedge \gamma_3 | \gamma_1 \in \Gamma_1, \gamma_2 \in \Gamma_2, \gamma_3 \in \Gamma_3\} .$$

Table 2. Literals for the Γ_1 set of partial conditions

Attribute	coef. type	coef. length	Basic?
`alcohol`	subset	1	yes
`vine/day`	subset	1	no
`beer/day`	subset	1	no

Table 3. Literals for the Γ_2 set of partial conditions

Attribute	coef. type	coef. length	Basic?
`smoking`	interval	$1,2$	yes
`sm_duration`	interval	$1,2$	yes

Table 4. Literals for the Γ_3 set of partial conditions

Attribute	coef. type	coef. length	Basic?
`coffee/day`	subset	1	yes
`tea/day`	interval	$1,2$	yes
`sugar/day`	interval	$1,2$	yes

Let us give a brief account of cardinalities of the above defined sets: $|\Gamma_1| = 49$, $|\Gamma_2| = 64$, $|\Gamma_3| = 288$, $|RQ| = |\Gamma| = |\Gamma_1| \times |\Gamma_2| \times |\Gamma_3| = 903\,168$. (See Table 5 to confirm these numbers). To solve the task, *KL-Miner* searched the set

Table 5. Numbers of admissible values of attributes

Attribute	No. of admissible values
`alcohol`	3
`vine/day`	3
`beer/day`	3
`smoking`	4
`sm_duration`	4
`coffee/day`	3
`tea/day`	3
`sugar/day`	6

RQ and found one relevant truth. Due to pruning, only 26 239 contingency tables were actually evaluated. Execution of the task took 3 seconds, see Table 6. The true hypothesis is

$$\texttt{BMI} \sim \texttt{cholesterol}/\gamma^* \ ,$$

here γ^* is conjunction of the following literals:

- `tea/day(0 cups)`
- `smoking(> 14 cigarettes/day)`.
- `sugar/day(1 - 4 lumps)`
- `coffee/day(1 - 2 cups)`
- `alcohol(occasionally)`

This means that strength of the correlation between attributes BMI and cholesterol measured by I_m exceeds 0.4 among those observed patients, who satisfy the condition γ^* (i.e. they do not drink tea, smoke more than 14 cigarettes a day, etc.). Furthermore, the number of patients who satisfy γ^* exceeds $0.02 * m$. The output of this hypothesis in the procedure *KL-Miner* is shown in Fig. 4. To further examine the dependence, we used the *4ft-Miner*

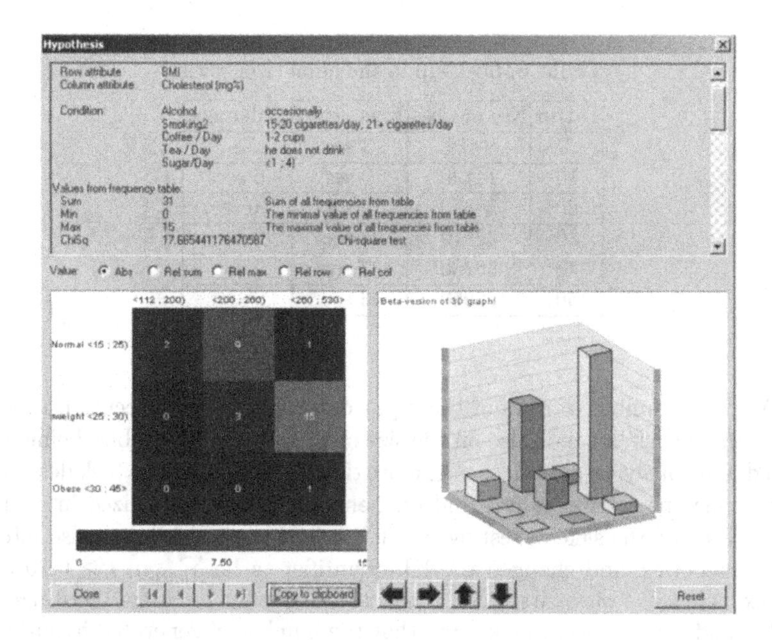

Fig. 4. Hypothesis output in the *KL-Miner*

procedure (see Sect. 2) and found association rules

$$\text{BMI}(<25, 30)) \rightarrow \text{cholesterol}(<260, 530>)/\gamma^* ,$$

with $confidence = 0.83$ and $support = 0.48$, and

$$\text{BMI}(<15, 25)) \rightarrow \text{cholesterol}(<200, 260))/\gamma^* ,$$

with $confidence = 0.75$ and $support = 0.29$. Here $confidence$ and $support$ are computed with respect to the data matrix ENTRY$/\gamma^*$.

6 An Example on Scalability

We have conducted some preliminary experiments concerning scalability. All the experiments were conducted on PC with Pentium III on 1133 MHz, with

256 MB of operational memory. We have run the task from the preceding section on the ENTRY data matrix magnified by the factor of 10, 20, etc.. Please recall that execution of this particular task consists of generation and evaluation of more than 26 thousands of contingency tables. See Table 6 for results. Apparently, the execution time is approximately linear in the number of rows (see the "Difference" column). However, a more detailed examination of *KL-Miner*'s performance on large data sets is yet to be done.

Table 6. Scale-up in the number of rows

Factor	No. of rows	Exec. time	Difference
1	1419	3 sec.	–
10	14190	9 sec.	6 sec.
20	28380	19 sec.	10 sec.
30	42570	28 sec.	9 sec.
40	56760	37 sec.	9 sec.
50	70950	43 sec.	6 sec.

Apart of number of rows of the input data matrix, the execution time of a *KL-Miner* task depends also on the size of the contingency tables being generated. Obviously, larger tables take more time to compute. The task described above concerns attributes `BMI` and `cholesterol`, both discretized into three categories. For the sake of testing, we discretized these attributes also into 4, 5, ... categories, and changed the KL-quantifier to be $\sum_{k,l} n_{k,l} \geq 0.10 * m$. We excluded I_m, since its value tends to be greater on larger contingency tables, and we wanted to make sure that the number of generated hypotheses remains the same across all tasks. We run the modified tasks on the ENTRY data matrix. Each time, 1 092 hypotheses were generated and tested (note the drastic pruning!), and 332 hypotheses were returned as true. Results show a good scale-up, see Table 7.

Table 7. Scale-up in the size of contingency table

Size of table	Exec. time	Difference
3×3	12 sec.	–
4×4	17 sec.	5 sec.
5×5	21 sec.	4 sec.
6×6	26 sec.	5 sec.
7×7	34 sec.	8 sec.
8×8	40 sec.	6 sec.
9×9	48 sec.	8 sec.
10×10	67 sec.	9 sec.

7 Conclusions and Further Work

We have presented a new data mining procedure called *KL-Miner*. Purpose of the procedure is to mine for patterns based on evaluation of two-dimensional contingency tables. To evaluate a contingency table, combinations of several functions can be used – these functions range from simple conditions on frequencies to information theoretic measures.

We have also outlined principles of the bit string technique used to efficiently implement the mining algorithm. Application of the procedure was shown on a simple example.

As for the further work, we plan to apply the procedure *KL-Miner* to further data sets, and to explore the possibilities of combining *KL-Miner* with other data mining procedures. Also, the study of statistical properties of KL-quantifiers is of importance.

Acknowledgement: The work described here has been supported by project LN00B107 of the Ministry of Education of the Czech Republic and by project IGA 17/04 of University of Economics, Prague.

References

1. Hájek P, Havránek T (1978) Mechanising Hypothesis Formation – Mathematical Foundations for a General Theory. Springer, Berlin Heidelberg New York
2. Hájek P, Havránek T, Chytil M (1983) The GUHA Method. Academia, Prague (in Czech)
3. Rauch, J (1978) Some Remarks on Computer Realisations of GUHA Procedures. International Journal of Man-Machine Studies 10, 23–28.
4. Rauch J, Šimůnek M (2005) An Alternative Approach to Mining Association Rules. In: Lin T Y, Ohsuga S, Liau C J, Tsumoto S, and Hu X (eds) Foundations of Data Mining and Knowledge Discovery, Springer-Verlag, 2005, pp. 219 - 238.
5. Řehák J, Řeháková B (1986) Analysis of Categorized Data in Sociology. Academia, Prague (in Czech)
6. Tomečková M, Rauch J, Berka P (2002) STULONG – Data from Longitudinal Study of Atherosclerosis Risk Factors. In: Berka P (ed) ECML/PKDD-2002 Discovery Challenge Workshop Notes. Universitas Helsingiensis, Helsinki
7. Zembowicz R, Zytkow J (1996) From Contingency Tables to Various Forms of Knowledge in Databases. In: Fayyad U M and all (eds.) Advances in Knowledge Discovery and Data Mining. AAAI Press, Menlo Park (CA)

Knowledge Discovery in Fuzzy Databases Using Attribute-Oriented Induction

Rafal A. Angryk, Frederick E. Petry
Electrical Engineering and Computer Science Department
Tulane University
New Orleans, LA 70118
USA
{angryk, fep}@eecs.tulane.edu

Abstract

In this paper we analyze an attribute-oriented data induction technique for discovery of generalized knowledge from large data repositories. We employ a fuzzy relational database as the medium carrying the original information, where the lack of precise information about an entity can be reflected via multiple attribute values, and the classical equivalence relation is replaced with relation of the fuzzy proximity. Following a well-known approach for exact data generalization in the ordinary databases [1], we propose three ways in which the original methodology can be successfully implemented in the environment of fuzzy databases. During our investigation we point out both the advantages and the disadvantages of the developed tactics when applied to mine knowledge from fuzzy tuples.

1. Introduction

The majority of current works on data mining describes the construction or application of algorithms performing complex analyses of stored data. Despite the predominant attention on this phase of analysis, because of the extensive volume of data in databases, techniques allowing conversion of raw data into condensed representations has become a practical necessity in many data-mining projects [2-3].

Attribute-Oriented Induction (AOI) [1, 4-12] is a descriptive database mining technique allowing such a transformation. It is an iterative process of grouping of data, enabling hierarchical transformation of similar itemsets stored originally in a database at the low (primitive) level, into more abstract conceptual representations. It allows compression of the original data set (i.e. initial relation) into a generalized relation, which provides concise and summarative information about the massive set of task-relevant data.

To take advantage of computationally expensive analyses in practice it is often indispensable to start by pruning and compressing the voluminous sets of the original data. Continuous processing of the original data is excessively time consuming and might be expendable, if we are actually interested only in information on abstraction levels much higher than directly reflected by the technical details stored usually in large databases (e.g. serial numbers, time of transactions with precision in seconds, detailed GPS locations, etc.). Simultaneously, the data itself represents information at multiple levels (e.g. Tulane University represents an academic institutions of Louisiana, the university is located in the West South, which is a part of the North American educational system, etc.), and is naturally suitable for generalization (i.e. transformation to a preferred level of abstraction).

Despite the attractive myth of fully automatic data-mining applications, detailed knowledge about the analyzed areas remains indispensable in avoiding many fundamental pitfalls of data mining. As appropriately pointed out in [6], there are actually three foundations of effective data mining projects: (1) the set of data relevant to a given data mining task, (2) the expected form of knowledge to be discovered and (3) the background knowledge, which usually supports the whole process of knowledge acquisition.

Generalization of database records in the AOI approach is performed on an attribute-by-attribute basis, applying a separate concept hierarchy for each of the generalized attributes included in the relation of task-relevant data. The concept hierarchy, which in the original AOI approach is considered to be a part of background knowledge (which makes it the third element of the above mentioned primitives), is treated as an indispensable and crucial element of this data mining technique. Here we investigate the character of knowledge available in the similarity and proximity relations of fuzzy databases and analyze possible ways of its application to the generalization of the information originally stored in fuzzy tuples.

In the next sections we introduce attribute-oriented induction, and briefly characterize crisp and fuzzy approaches to the data generalization; we will also discuss the unique features of fuzzy database schemas that were utilized in our approach to attribute-oriented generalization. In the third part we will present three techniques allowing convenient generalization of records stored in fuzzy databases. The increase in efficiency of these methods over the originally proposed solutions is achieved by taking full advantage of the knowledge about generalized domains stored implicitly in fuzzy database models. Then we will propose a method that allows multi-contextual generalization of tuples in the analyzed database environments. Finally we will also introduce a simple algorithm allowing generalization of imprecise information stored in fuzzy tuples.

2. Background

2.1. Attribute-Oriented Induction

AOI is a process transforming similar itemsets, stored originally in a database, into more abstract conceptual representations. The transformation has an iterative character, based on the concept hierarchies provided to data analysts. Each concept hierarchy reflects background knowledge about the domain which is going to be generalized. The hierarchy permits gradual, similarity-based, aggregation of attribute values stored in the original tuples. Typically, the hierarchy is built in a bottom-up manner progressively increasing the abstraction of the generalization concepts introduced at each new level (ideally a *0-abstraction-level*, located at the bottom of the hierarchy, includes all attribute values, which occurred in the mined dataset for the particular attribute). To guarantee the reduction of the size of original dataset, each new level of hierarchy has a more coarse-grained structure. In other words, at each new level, there is a smaller number of descriptors, but they have a broaden (i.e. general) character. The new concepts, though reduced in number, because of their more general meaning are still able to represent all domain values from the lower abstraction level.

Typically in the AOI technique we initially retrieve a set of task-relevant data (e.g. using a relational database query). Then, we gradually perform actual generalization on the dataset by replacing stored attribute values with more general descriptors. This replacement has an iterative character, where at each stage we are replacing low-level values only with their direct abstracts (i.e. the concepts which are placed at the next abstraction level in the concept hierarchy). After each phase of replacements we merge the tuples that appear identical at the particular abstraction level. To preserve original dependencies among the data at each stage of AOI, we have to accumulate a count of the merged tuples. To be able to keep track of the number of original records that are represented by identical abstract concept, we extend the generalized relation with a new attribute COUNT. At each stage of induction the number stored in the field COUNT informs us about the number of original tuples (i.e. *votes*), which are characterized by the particular abstract entity.

The AOI approach seems to be much more appropriate for detailed analysis of datasets than commonly used simplified, non-hierarchical summarization. The hierarchical character of induction provides analysts with the opportunity to view the original information at multiple levels of abstraction, allowing them to progressively discover interesting data aggregations. In contrast to the flat summarization a gradual process of AOI through concept hierarchies allows detailed tracking of all records and guarantees discovering significant clusters of data at the lowest abstraction level of their occurrence. For that reason the analysts, who use the AOI technique, are able to avoid unnecessary loss of information due to over-generalization. Moreover, in this approach the tuples, which did not achieve significant aggregation at a low abstraction level, rather than being removed from analysis, are gradually further aggregated and so given a chance to reach a meaningful count at one of higher abstraction levels.

Originally the hierarchical grouping proposed by Han and his co-workers [4] was based on tree-like generalization hierarchies, where each of the concepts at the lower level of the generalization hierarchy was allowed to have just one abstract concept at the level directly above it. An example of such a concept hierarchy characterizing a generalization of students' status (*{master of art, master of science, doctorate}* $\subset$ *graduate*, *{freshman, sophomore, junior, senior}* $\subset$ *undergraduate*, *{undergraduate, graduate}* $\subset$ *ANY*) is presented in Figure 1, which appeared in [6].

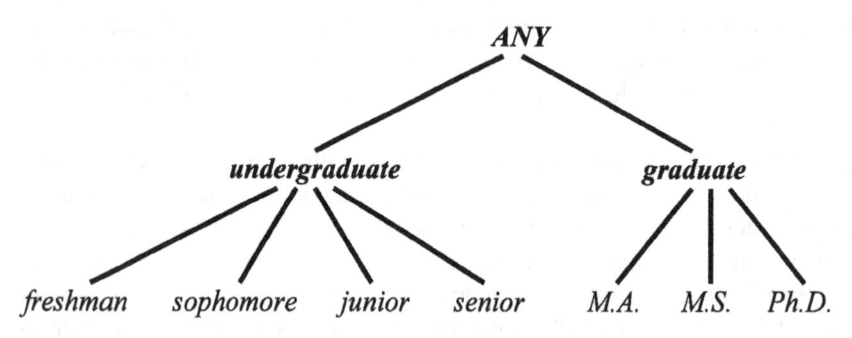

Fig. 1. A concept tree for attribute STUDENT_STATUS [6].

Han [5-6] developed a few elementary guidelines (strategies), which are the basis of effective attribute-oriented induction for knowledge discovery purposes:

1. ***Generalization on the smallest decomposable components.*** Generalization should be performed on the smallest decomposable components (or attributes) of a data relation.

2. ***Attribute removal.*** If there is a large set of distinct values for an attribute but there is no higher level concept provided for the attribute, the attribute should be removed in the generalization process.
3. ***Concept tree ascension.*** If there exists a higher-level concept in the concept tree for an attribute value of a tuple, the substitution of the value by its higher-level concept generalizes the tuples. Minimal generalization should be enforced by ascending the tree one level at a time.
4. ***Vote propagation.*** The value of the vote of a tuple should be carried to its generalized tuple and the votes should be accumulated when merging identical tuples in generalization.

The work initiated by Han and his co-researchers [4-8] was extended further by Hamilton, Hilderman with their co-workers [9-11], and also by Hwang and Fu in [12]. The attribute-oriented induction methods have been implemented in commercially used systems (DBLearn/DB-Miner [10] and DB-Discover [11]), and applied to a number of research and commercial databases to produce interesting and useful results [5].

Fuzzy hierarchies of concepts were applied to tuples summarization in late nineties by four groups of independent researchers. Lee and Kim [13] used ISA hierarchies, from area of data modeling, to generalize database records to more abstract concepts. Lee [14] applied fuzzy generalization hierarchies to mine generalized fuzzy quantitative association rules. Cubero, Medina, Pons and Vila [15] presented fuzzy gradual rules for data summarization and Raschia, Ughetto and Mouaddib [16-17] implemented SaintEtiq system for data summarization through extended concept hierarchies.

A fuzzy hierarchy of concepts reflects the degree with which a concept belongs to its direct abstract. In addition, more than one direct abstract of a single concept is allowed during fuzzy induction. Each link ℓ in the fuzzy concept hierarchy is a bottom-up directed arc (edge) between two nodes with a certain weight assigned to it. Such a structure reflects a fuzzy induction triple (c^k, c^{k+1}, $\mu_{c^k c^{k+1}}$), where c^k (i.e. an attribute value generalized to the kth-abstraction level), and c^{k+1} (one of its direct generalizers) are endpoints of ℓ, and $\mu_{c^k c^{k+1}}$, being the weight of the link ℓ, represents the strength of conviction that the concept c^k (a source of ℓ) should be qualified as concept c^{k+1}(a target of ℓ) when the data generalization process moves to the next abstraction level. During attribute-oriented fuzzy induction (AOFI) the value of $\mu_{c^k c^{k+1}}$ dictates actually what fraction of a vote, representing original attribute values generalized already to the concept c^k, is to be propagated to its higher-level representation, c^{k+1}.

An example of a fuzzy concept hierarchy (FCH) presented in the Fig. 2. comes from [13]. According to the authors, utilization of the four popular text editors could be generalized as follows (we denote fuzzy generalization of concept a to its direct abstract b with membership degree c as $a \subset b \mid c$):

1st level of abstraction: *{emacs* $\subset$ *editor| 1.0; emacs* $\subset$ *documentation| 0.1; vi* $\subset$ *editor| 1.0; vi* $\subset$ *documentation| 0.3; word* $\subset$ *documentation | 1.0; word* $\subset$ *spreadsheet| 0.1; wright* $\subset$ *spreadsheet| 1.0}*

2nd level of hierarchy: *{editor* $\subset$ *engineering| 1.0; documentation* $\subset$ *engineering | 1.0; documentation* $\subset$ *business | 1.0; spreadsheet* $\subset$ *engineering | 0.8; spreadsheet* $\subset$ *business| 1.0}*

3rd level of hierarchy: *{engineering* $\subset \omega$ *| 1.0; business* $\subset \omega$ *| 1.0}*

Fuzzy hierarchies of concepts allow a more flexible representation of real life dependencies. A significant weakness of the above-mentioned approaches is a lack of automatic preservation of *exact vote propagation* at each stage of attribute-oriented fuzzy induction (AOFI). To guarantee exactness of the data summarization via AOFI we need to assure that each record from the original database relation will be counted exactly once at each of the levels of the fuzzy hierarchy.

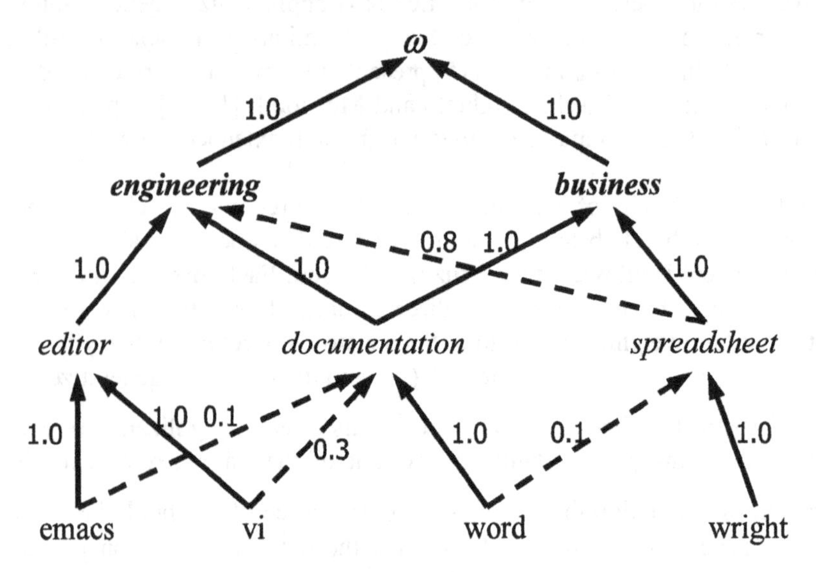

Fig. 2. A fuzzy ISA hierarchy on computer programs [13].

This issue was successfully resolved by utilization of consistent fuzzy concept hierarchies [18-19], which preserve *exact vote propagation* due to their completeness and consistency. Speaking formally, to preserve completeness and consistency of the AOFI process for all adjacent levels, C^k and C^{k+1}, in fuzzy concept hierarchy the following relationship must be satisfied:

$$\sum_{j=1}^{\|C^{k+1}\|} \mu_{c^k c_j^{k+1}} = 1.0,\ c^k \in C^k, c_j^{k+1} \in C^{k+1}$$

In other words, the sum of weights assigned to the links leaving any single node in a fuzzy concept hierarchy needs to be 1.0. Preservation of the above property prevents any attribute value (or its abstract) from being counted more or less during the process of AOFI (i.e. consistency of the fuzzy induction model). It also guarantees that a set of abstracts concepts at each level of hierarchy will cover all of the attribute values that occurred in the original dataset (i.e. completeness of the model). In effect, we are guaranteed to not lose count of the original tuples when performing fuzzy induction on their attribute values.

Formally, each non-complete fuzzy concept hierarchy can be transformed to the complete generalization model via simple normalization of membership values in all outgoing links of the hierarchical induction model.

Summarizing, for the purpose of formal classification, we can distinguish three basic types of generalization hierarchies that have been used to date:

1. *Crisp Concept Hierarchy* [1, 4-12], where each attribute variable (concept) at each level of hierarchy can have only one direct abstract (its direct generalization) to which it fully belongs (there is no consideration of the degree of relationship).
2. *Fuzzy Concept Hierarchy* [13-17], which allows the reflection of degree with which one concept belongs to its direct abstract and more than one direct abstract of a single concept is allowed. Because of the lack of guarantee of exact vote propagation, such a hierarchy seems to be more appropriate for approximate summarizations of data, or to the cases where subjective results are to be emphasized (we purposely want to modify the role or influence of certain records).
3. *Consistent Fuzzy Concept Hierarchy* [18-19], where each degree of membership is normalized to preserve an exact vote propagation of each tuple when being generalized.

2.2. Similarity- and Proximity-based Fuzzy Databases

The similarity-based fuzzy model of a relational database [20-23] is actually a formal generalization of the ordinary relational database [24]. The fuzzy model, based on the max-min composition of a fuzzy similarity relation, which replaces the classical equivalence relation coming from the theory of crisp sets, was further extended by Shenoi and Melton [25-28] with the concept of the proximity relation.

Important aspects of fuzzy relational databases are: (1) allowing non-atomic domain values, when characterizing attributes of a single entity and (2) generation of equivalence classes based on the specific fuzzy relations applied in the place of traditional identity relation.

As mentioned above, each attribute value of the fuzzy database record is allowed to be a subset of the whole base set of attribute values describing a particular domain. Formally, if we denote a set of acceptable attribute values as D_j, and we let d_{ij} to symbolize a particular (j^{th}) attribute value, characterizing the ith entity. Instead of $d_{ij} \in D_j$ the more general case $d_{ij} \subseteq D_j$ is allowed, i.e. any member of the power set of accepted domain values can be used as an attribute value except the null set. A fuzzy database relation is a subset of the cross product of all power sets of its constituent attributes $2^{D_1} \times 2^{D_2} \times \ldots \times 2^{D_m}$. This allows representation of inexactness arising from the original source of information. When a particular entity's attribute cannot be clearly characterized by a single descriptor, this aspect of uncertainty can be reflected by multiple attribute values.

Another feature characterizing proximity fuzzy databases is substitution of the ordinary equivalence relation, defining the notion of redundancy in the ordinary database, with an explicitly declared proximity relation of which both the identity and similarity relations are actually special cases. Since the original definition of fuzzy proximity relations (also called tolerance relations) was only reflexive and symmetric, which is not sufficient to effectively replace the classical equivalence relation, the transitivity of proximity relation was added [27]. This was achieved by extending the original definition of a fuzzy proximity relation to allow transitivity via similarity paths (sequences of similarities), using Tamura chains [29]. So the α-proximity relation used in proximity databases has the following structure:

If P is a proximity relation on D_j, then given an $\alpha \in [0, 1]$, two elements $x, z \in D_j$ are *α-similar* (denoted by $xP_\alpha z$) if and only if $P(x,z) \geq \alpha$, and are said to be *α-proximate* (denoted by $x P_\alpha^+ z$) if and only if they are (1) either α-similar or (2) there exists a sequence $y_1, y_2, \ldots, y_m \in D_j$, such that $xP_\alpha y_1 P_\alpha y_2 P_\alpha \ldots P_\alpha y_m P_\alpha z$.

Table 1. Proximity table for a domain COUNTRY.

	Canada	USA	Mexico	Colombia	Venezuela	Australia	N. Zealand
Canada	1.0	0.8	0.5	0.1	0.1	0.0	0.0
USA	0.8	1.0	0.8	0.3	0.2	0.0	0.0
Mexico	0.5	0.8	1.0	0.4	0.2	0.0	0.0
Colombia	0.1	0.3	0.4	1.0	0.8	0.0	0.0
Venezuela	0.1	0.2	0.2	0.8	1.0	0.0	0.0
Australia	0.0	0.0	0.0	0.0	0.0	1.0	0.8
N. Zealand	0.0	0.0	0.0	0.0	0.0	0.8	1.0

Each of the attributes in the fuzzy database has its own *proximity table*, which includes the *degrees of proximity* (*α-similarity*) between all values occurring for the particular attribute. A proximity table for the domain of COUNTRIES, which we will use as an example for our further analysis, is presented in the Table 1.

The proximity table can be transformed by Tamura chains to represent such an *α-proximity relation*. Results of this transformation are seen in Table 2.

Table 2. α-proximity table for a domain COUNTRY.

	Canada	USA	Mexico	Colombia	Venezuela	Australia	N. Zealand
Canada	1.0	0.8	0.8	0.4	0.4	0.0	0.0
USA	0.8	1.0	0.8	0.4	0.4	0.0	0.0
Mexico	0.8	0.8	1.0	0.4	0.4	0.0	0.0
Colombia	0.4	0.4	0.4	1.0	0.8	0.0	0.0
Venezuela	0.4	0.4	0.4	0.8	1.0	0.0	0.0
Australia	0.0	0.0	0.0	0.0	0.0	1.0	0.8
N. Zealand	0.0	0.0	0.0	0.0	0.0	0.8	1.0

Now the disjoint classes of attribute values, considered to be equivalent at a specific *α-level*, can be extracted from the table. They are marked by shadings in Table 2.

Such separation of the equivalence classes arises mainly due to the sequential similarity proposed by Tamura. For instance, despite the fact that the proximity degree, presented in Table 1, between the concepts *Canada* and *Venezuela* is *0.1*, the α-proximity is *0.4*. Using the sequence of the original proximity degrees: $CanadaP_{\alpha}Mexico = 0.7 \wedge MexicoP_{\alpha}Colombia = 0.4 \wedge ColombiaP_{\alpha}Venezuela = 0.8$, we obtain $Canada\, P_{\alpha}^{+}\, Venezuela = 0.4$, as presented in Table 2.

The transformation based on the sequences of proximities converts the original proximity table back to a similarity relation [30] as in a similarity database model. The practical advantage of the proximity approach comes from the lack of necessity to preserve the *max-min transitivity* when defining the proximity degrees. This makes a proximity table much easier for a user to define. The α-proximity table, although based on the proximity table, is generated dynamically only for the attribute values which were actually used in the fuzzy database.

3. Attribute-Oriented Induction in Fuzzy Databases

In view of the fact that a fuzzy database model is an extension of the ordinary relational database model, the generalization of fuzzy tuples through the concept hierarchies can always be performed by the same procedure as presented by Han [6, 8] for ordinary relational databases. In this work however we focus on the utilization of the unique features of fuzzy databases in the induction process. First of all, due to the nature of fuzzy databases we have an ability to re-organize the original data to the required level of detail (by the merge of records, considered to be identical at a certain *α-cut level*, according to the given similarity relation), so then we can start attribute-oriented generalization from the desired level of detail. This approach, while valuable in removing unnecessary detail, must be applied with caution. When merging the tuples in fuzzy databases according to the equivalence at the given similarity level (e.g. by using SELECT queries with a high threshold level), we are not able to keep the track of the number of original data records to be merged to a single tuple. Lack of such information may result in significant change of balance among the tuples in the database and lead to the erroneous (not reflecting reality) information presented later in the form of support and confidence of the extracted knowledge. This problem, which we call a *count dilemma* is derived from the principle of *vote propagation* and can be easily avoided by performing extraction of initial data table at the very detailed level (i.e. α=*1.0*), where only identical values are merged (e.g. values *England* and *Great Britain* will be unified). So then no considerable number of records would be lost as the result of such redundancy removal, which one way or another is a self-contained part of the data pre-processing phase in many of data mining activities.

3.1. Extraction of multi-level concept hierarchies

The generation of an *α-proximity relation* for a particular domain D_j, in addition to providing a fuzzy alternative to an equivalence relation, also allows extraction of a partition tree, which can then be successfully employed by non-experts to perform attribute-oriented induction.

From the placement of shadings in the Table 2, we can easily observe that the equivalence classes marked in the table have a nested character. As in [30], each *α-cut* (where $\alpha \in (0, 1]$) of the fuzzy binary relation (Table 2) creates disjoint equivalence classes in the domain D_j. If we let Π_α denote a single equivalence class partition induced on the domain D_j by a single *α-level-set*, through the increase of the value of α to α' we are able to extract the subclass of Π_α, denoted $\Pi_{\alpha'}$ (a refinement of the previous equivalence class partition). A nested sequence of partitions Π_{α^1}, $\Pi_{\alpha^2}, \ldots, \Pi_{\alpha^k}$, where $\alpha^1 < \alpha^2 < \ldots < \alpha^k$, may be represented in the form of a partition tree, as in Figure 3.

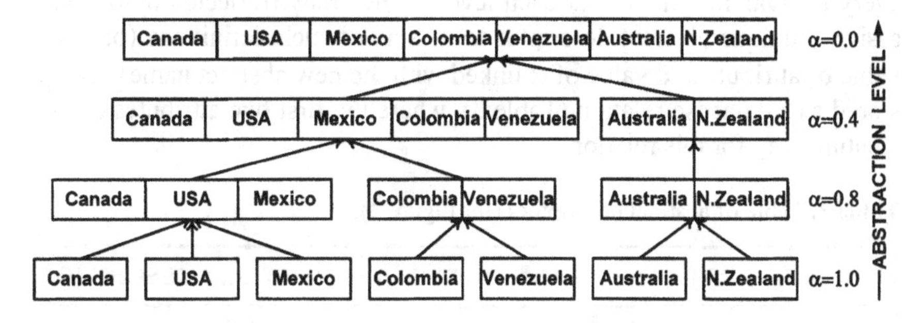

Fig. 3. Partition tree of domain COUNTRY, built on the basis of Table 2.

This nested sequence of partitions in the form of a tree has a structure identical with that of a crisp concept hierarchy applicable for AOI. Since the AOI approach is based on the reasonable assumption that an increase of degree of abstraction makes particular attribute values appear identical, it seems to be appropriate to allow the increase of conceptual abstraction in the concept hierarchy to be reflected by the decrease of α values (representing degrees of similarity between original attribute values), as long as the context of similarity (i.e. dimensions it is measured in) is in full agreement with the context (directions) of performed generalization (e.g. Figure 4). The lack of abstraction (0-abstraction level) at the bottom of generalization hierarchy complies with the 1-cut of the α-proximity relation ($\alpha=1.0$) from the fuzzy model and can be denoted as $P^{+}_{1.0}$. In other words, it is the level where only those attribute values that have identical meaning (i.e. synonyms) are aggregated.

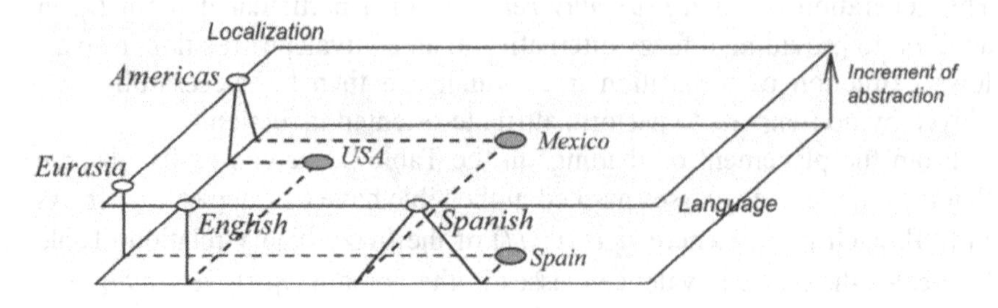

Fig. 4. Example of two different contexts of generalization.

The only thing differentiating the hierarchy in the Figure 3 from the crisp concept hierarchies applicable for AOI is the lack of abstract concepts, which are used as the labels characterizing the sets of generalized (grouped) concepts. To create a complete set of the abstract labels it is sufficient to choose only one member (the original value of the attribute) of every equivalence class Π at each level of hierarchy (reflected by α), and assign a unique abstract descriptor to it. Sets of such definitions (original value of attribute and value of α linked with the new abstract name) can be stored as a database relation (Table 3), where the first two attributes create a natural key for this relation.

Table 3. Table of abstract descriptors (for Figure 3).

ATTRIBUTE VALUE	ABSTRACTION LEVEL (α)	ABSTRACT DESCRIPTOR
Canada	0.8	N. America
Colombia	0.8	S. America
Australia	0.8	Oceania
Canada	0.4	Americas
Australia	0.4	Oceania
Canada	0.0	Any

The combination of partition tree in Figure 3 and the relation of abstract descriptors (Table 3) allows us to create the classical generalization hierarchy in the form of Figure 5.

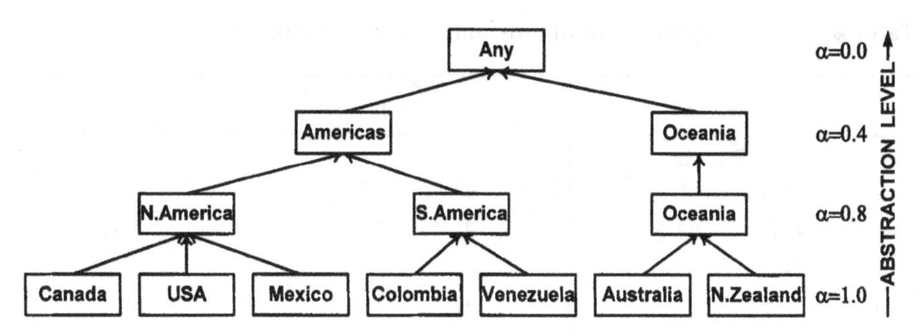

Fig. 5. Crisp generalization hierarchy formed using Tables 4 and 7.

The disjoint character of equivalence classes generated from the α-proximity (i.e. similarity) table does not allow any concept in the hierarchy to have more than one direct abstract at every level of generalization hierarchy. Therefore this approach can be utilized only to form a crisp generalization hierarchy. Such a hierarchy, however, can be then successfully applied as a foundation to the development of a fuzzy concept hierarchy – by extending it with additional edges to represent partial membership of the lower level concepts in their direct abstract descriptors. Depending on the values of assigned memberships, data-analysts can generate consistent or inconsistent fuzzy concept hierarchies.

A partition tree generated directly from the originally defined proximity table (without transforming it first to the α-proximity table as presented in section 2.2) has a structure much less useful for AOI than the solution discussed above. Such a graph (since it does not preserve a tree-like structure), although extractable, usually has a rather complicated structure. Due to the lack of the max-min transitivity requirement in fuzzy proximity relation, partitions of proximity generated by α-cuts (when we increment α) do not always have a nested character. Generated equivalence classes may overlap at the same α-level (e.g. Table 4). In consequence, the extracted graph of partitions often contains the same attribute value assigned multiple times to different equivalence classes at the same abstraction level (α). Furthermore, links between nodes (i.e. proximity partitions, i.e. equivalence classes) may cross the neighboring levels of the hierarchy, which is in contradiction with requirement of minimal tree ascension introduced by Han. When utilizing a partition hierarchy extracted directly from proximity table, we are taking a risk that some tuples would not be counted at the particular stage of induction, which in result may cancel correctness of the obtained results.

Table 4. Overlapping partitions of proximity generated by 0.4-cut.

	Canada	USA	Mexico	Colombia	Venezuela	Australia	N. Zealand
Canada	1.0	0.8	0.5	0.1	0.1	0.0	0.0
USA	0.8	1.0	0.8	0.3	0.2	0.0	0.0
Mexico	0.5	0.8	1.0	0.4	0.2	0.0	0.0
Colombia	0.1	0.3	0.4	1.0	0.8	0.0	0.0
Venezuela	0.1	0.2	0.2	0.8	1.0	0.0	0.0
Australia	0.0	0.0	0.0	0.0	0.0	1.0	0.8
N. Zealand	0.0	0.0	0.0	0.0	0.0	0.8	1.0

An original proximity table can be however successfully utilized when applying other approaches to AOI, as presented in the remaining part of section 3.

3.2. Generation of single-level concept hierarchies

Both similarity and proximity relations can be interpreted in terms of fuzzy classes P(x) [31], where memberships of other elements (in our case other attribute values) in each of the fuzzy classes P(x) are derived from the rows in the proximity (or similarity) relations. In other words the grade of membership of attribute value y in the fuzzy class P(x) (e.g. a fuzzy representative of attribute value x, or of the set of such values), denoted by $\mu_{P(x)}(y)$, is $x\,P_\alpha\,y$ (or $x\,P_\alpha^+\,y$ for the similarity relation). This alternative view on fuzzy binary relations provides motivation for development of two other approaches. These may be more convenient for a user, who does not have the expertise to initiate the whole induction process from the basics and would like to use some support derived from existing proximity tables. Both of the techniques to be described require some knowledge about generalized domains, but not at an expert's level.

3.2.1. Induction based on typical elements

In practice, when generalizing attribute values, there are usually distinguished subsets of lower-level concepts that can be easily assigned to the particular abstract concepts even by users who have only ordinary knowledge about the generalized domain. At the same time the assignment of the other lower-level values is problematic even among experts. Using the terminology presented in [32] we will say that each attribute has a domain (allowed values), a range (actually occurring values) and a typical range (most common values). In this section we apply this approach to the generalization process. When we have an abstract concept we can often identify its *typical direct specializers*, which are the elements clearly belonging to it (e.g. we all would probably agree that the country *Holland* can be generalized to the concept *Europe* with 100% surety). This can be represented as a core of the fuzzy set (abstract concept) characterized by a membership of *1.0*. However, there are usually also lower-level concepts, which cannot be definitely assigned to only one of their direct abstracts (e.g. assigning the *Russian Federation* fully to the abstract concept *Asia* would probably raise some doubts, as almost one fifth of this country lies west of the Urals). We call such cases *possible direct specializers* and define them as the concepts occurring in the group of lower level concepts characterized by the given abstract descriptor (fuzzy set) with the membership $0<\mu\leq 1$. Such elements create the support of a fuzzy set and are to be interpreted as the range of the abstract concept.

In short, the idea behind this approach is to define initially the abstract concepts via choosing their basic representative attribute values (i.e. typical representative specializers) and then to use the proximity table to extract a more precise definition of the abstract class. For such extraction we assume a certain level of similarity (α), which should be interpreted as a level of precision reflected in our abstract concept definition.

Using this approach we initially characterize each new abstract concept as a set of its typical original attribute values with the level of doubt about its other possible specializers reflected by the value of α. Then we select the fuzzy similarity class created from the α-cut of similarity relation for these pre-defined typical specializers and assess if this class fits well our expectations. Obviously some background knowledge about a generalized domain is now necessary so that the data analyst could point out typical direct specializers. However, there is still a significant support provided by the proximity table allowing smooth generalization of the most problematical cases. When we characterize an abstract concept by more than one typical element, we must also choose an intersection operator which best fits our preferences. This is necessary so that we would be able to extract a unified definition in the case when the typical elements (i.e. fuzzy classes) are overlapping.

For instance, if we predefine the abstract concept *North America* by its two typical countries *Canada* and *USA* with the level of similarity $\alpha=0.6$, and with the operator *MAX* to reflect our optimistic approach, we can derive from Table 1 that:

North America = *MAX* ($Canada_{0.6}$, $USA_{0.6}$)= {*MAX* (*Canada*|*1.0*, *USA*|*0.8*, *Mexico*|*0.5*; *Canada*|*0.8*; *USA*|*1.0*; *Mexico*|*0.8*} = {*Canada*|*1.0*; *USA*|*1.0*; *Mexico*|*0.8*}.

Defining now *Middle America* as the cut of fuzzy class *Colombia* on the *0.4*-level we can extract from the proximity table:

Middle America = $Colombia_{0.4}$ =

= {*Mexico*|*0.4*; *Colombia*|*1.0*; *Venezuela*|*0.8*}.

As a result we can derive the fuzzy concept hierarchy and even modify the generalization model to become consistent through the normalization of derived memberships:

$$North\ America = \{Canada|1.0;\ USA|1.0;\ Mexico|\frac{0.8}{0.8+0.4}\} =$$

$$= \{Canada|1.0;\ USA|1.0;\ Mexico|0.67\}$$

$$Middle\ America = \{Mexico|\frac{0.4}{0.8+0.4};\ Colombia|1.0;\ Venezuela|\frac{0.8}{0.8}\} =$$

$$= \{Mexico|0.33;\ Colombia|1.0;\ Venezuela|1.0\}.$$

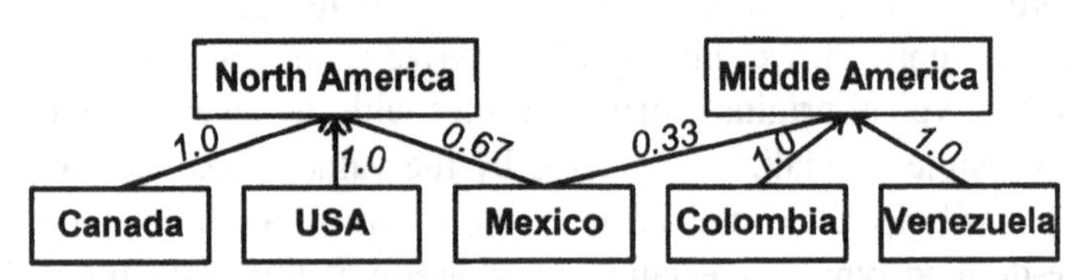

Fig. 6. Consistent fuzzy generalization hierarchy built by extraction of partial knowledge stored in the proximity relation for the attribute COUNTRY.

As distinct from the previous approach, here users are given much more freedom in defining abstracts. It is totally their decision as to which lower-level concepts will be aggregated and to what abstraction level. They can modify two elements to control results of the generalization process: (1) sets of typical elements, and (2) level of acceptable proximity between these values (α). Depending on the choice the extracted definitions of abstract concepts may have different characteristics (Table 5).

Table 5. Characteristics of abstract definitions extracted from the proximity table.

Quantity of typical elements \ *Similarity level* (α)	*LOW (close to 0)*	*HIGH (close to 1)*
SMALL (close to 1)	**Spread widely**	**Precise** *(Short)*
LARGE (close to the total number of attribute values)	**Confusing** *(Error suspected)*	***Precise*** *(Confirmed)*

When extracting overlapping definitions of abstract concepts from a proximity table, a fuzzy concept hierarchy results and one must be extremely careful to keep the definitions semantically meaningful. For instance it makes no sense to pre-define two or more general concepts at a level of abstraction so high, that they are interpreted almost as identical. Some basic guidelines are necessary when utilizing this approach:

1. We need to assure that the intuitively assumed value of α extracts the cut (subset) of attribute values that corresponds closely to the definition of the abstract descriptor we desired. The strategy for choosing the most appropriate level of *α-cut* when extracting the abstract concept definitions comes from the principle of minimal generalization (minimal concept tree ascension strategy in [6]), which translates to minimal proximity level decrease in our approach. Accepting this strategy we would recommend always choosing the definition extracted at the highest possible level of proximity (biggest α), where all pre-defined typical components of desired abstract descriptor are already considered (i.e. where they occur first time).
2. The problem of selecting appropriate representative elements without external knowledge about a particular attribute still remains, however it can now be supported by the analysis of the values stored in the proximity table. Choosing typical values and then extracting detailed definition from the similarity (or proximity) table will make AOFI more accessible for non-experts. By using common knowledge they may be able to point out a few typical elements of a generalized concept even while lacking expert knowledge necessary to characterize a particular abstract in detail.
3. Moreover, we should be aware that if the low-level concepts, predefined as typical components of the particular abstract descriptor, do not occur in the common proximity class, then the contexts of the generalized descriptor and the proximity relation may be not in agreement

and revision of the proximity table (or the abstract concepts) is necessary.

This approach allows us to place at the same level of the concept hierarchy, abstract concepts that were extracted with different levels of proximity (different α-values). As a result this achieves more effective (i.e. compressed) induction. However, allowing such a situation especially when using a similarity (i.e. α-proximity) table, we must always remember that the abstract concepts derived from the similarity relation have a nested character. Placement of one abstract concept at the same level of abstraction with another, which is its actual refinement, may lead to the partial overlapping of partitions. This is in contradiction with the character of a similarity relation. This restriction may not always appear in the case of a proximity relation, as it has an intransitive character.

3.2.2. Induction with context oriented on the single attribute values

As in the previous approach, this technique also leads to the generation of one-level hierarchies. It is derived from a slightly different interpretation of the extracted fuzzy similarity (or proximity) classes. The biggest advantage of this method is its ability to perform generalization of all original attribute values in the context reflected in the proximity table, but from the point of view of the relevance of all of these attribute values with respect to the distinguished one (i.e. the one being used as a basis of the extracted fuzzy class).

Let us use a simple example to illustrate this. From the proximity relation (Table 1) we can extract a single row, representing a fuzzy proximity class for the attribute value Canada (as presented in Table 6).

Table 6. Fuzzy class extracted from Table1; it represents proximity of different countries to Canada.

	Canada	USA	Mexico	Colombia	Venezuela	Australia	N. Zealand
Canada	1.0	0.8	0.5	0.1	0.1	0.0	0.0

Now we can generate subsets of this fuzzy class's domain (which is actually the whole set of acceptable attribute values), defining disjoint ranges of acceptable membership values:

Canada-similar = $\{Canada_{[0.6,\ 1.0]}\}$ = *{Canada, USA}*
Canada -semi-similar = $\{Canada_{[0.1,\ 0.6)}\}$ = *{Mexico, Colombia, Venezuela}*
Canada -un-similar = $\{Canada_{[0.0,\ 0.1)}\}$ = *{Australia, N. Zealand}*.

The size of the assumed ranges is dependent on the preferences of data analyst. Smaller ranges generate larger number of abstract classes, but may reduce the number of lower-level concepts in each of the extracted abstracts. The classes presented above provide us sufficient information to perform generalization of all values occurring in the attribute COUNTRY from the point of view of the similarity of these values to the country *Canada* (Figure 7). Obviously values *Canada* and *USA* will be generalized as *Canada-similar* concepts, countries more distant as *semi-similar* and *unsimilar*. Since degrees of proximity were already utilized to extract these three subsets, we have not inserted them in the concept hierarchy.

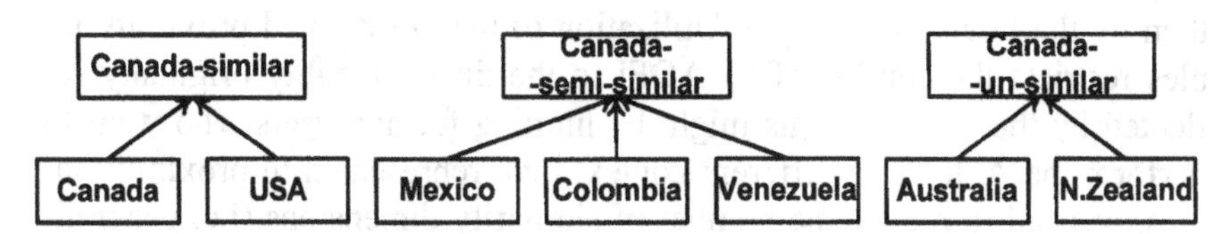

Fig. 7. Value-similarity based generalization.

Technically speaking, we simply sliced the single fuzzy class extracted from the proximity relation into layers reflecting levels of the relevance of the attribute values to the original attribute value, which then became a center of our attribute-oriented induction process.

These two approaches we have discussed above allow us to form only one-level generalization hierarchies or to derive the generalized concepts at the first level of abstraction in the concept hierarchy. Each of abstract concepts defined with this method is a generalization of original attribute values, and therefore cannot be placed at the higher level of the concept hierarchy.

The inability to derive multi-level hierarchical structures does not stop this approach from being appropriate and actually very convenient for rapid data summarization or something we call – selective attribute-oriented generalization. To quickly summarize the given data set we may actually prefer to not perform gradual (hierarchical) generalization, but to replace it with a one-level hierarchy covering whole domain of attribute values. Such "flat hierarchies" still represent dependencies between the attribute values which were originally characterized with the proximity relation; although these dependencies are now restricted to a certain extent, as reflected by the user's preferences. These flat hierarchies can also be successfully utilized as a foundation for further hierarchical induction, as discussed in the next section.

3.3. Multi-context AOI

The previously discussed approaches, despite being very convenient (and almost automatic) to use, have one common limitation. As described both similarity and proximity relations can be successfully utilized when building concept hierarchies. However the nature of the generalization based on these particular hierarchies can only reflect the context represented by the original similarity or proximity tables. Each of the attribute values can be considered basically as a point in a multi-dimensional space, reflecting multiple contexts of possible inductions (e.g. countries can be aggregated by province, and then by continent, or by the official native language, and then by the linguistic groups). Utilization of the pre-defined proximity tables restricts the context of an AOFI to the dimensionality originally reflected by these tables. This might be limiting for any users who want to perform the AOFI in a different context than represented in proximity tables, or need to reduce the number of similarity dimensions (i.e. generalization contexts) at a certain stage of hierarchical induction. The techniques presented above do not provide flexibility in allowing such modifications. Obviously, the data-mining analysts can be allowed to modify the values in proximity table in their own user-views of the fuzzy database to represent the similarity between the concepts (attribute values) in the context of their interests. The advantage of such modifications is that now these new proximity tables can be used to merge the records that are similar in the context of their interests.

The last two methods presented in the previous section allowed solely one-level generalization. To extend AOI to a higher abstraction level we need to simply define new similarity or proximity tables reflecting similarity between the generalized concepts. There is nothing preventing users from defining such tables in the new context, as long as they have sufficient knowledge about generalized domain to be able to correctly describe such dependencies. We can even halt the induction process performed with the multilevel hierarchy, presented in the Figure 5, at any level and build new similarity (or proximity) tables reflecting the distance between abstract concepts in a totally different context than those primarily represented by the original hierarchy. In Table 7 we present such a similarity relation, reflecting similarities at the higher level of abstraction and in a context different than the one originally represented by Table 1.

Table 7. New similarity table reflecting relationship between concepts from *1st*-abstraction level, according to the placement on Northern and Southern hemisphere of the Earth.

	N. America	S. America	Oceania
N. America	1.0	0.5	0.5
S. America	0.5	1.0	0.7
Oceania	0.5	0.7	1.0

By cutting the hierarchy from Figure 5 after the first level of generalization and introducing new abstract names, which better fit the relation presented in Table 7, we can generate a new concept hierarchy allowing gradual induction from the *1st*-abstraction level to the *3rd*-level in the new context. Then it will be merged with layers cut from Figure 5 to perform AOI based on the modification of generalization context at the *1st* level of the abstraction. The hierarchy constructed in this manner is seen in Figure 8.

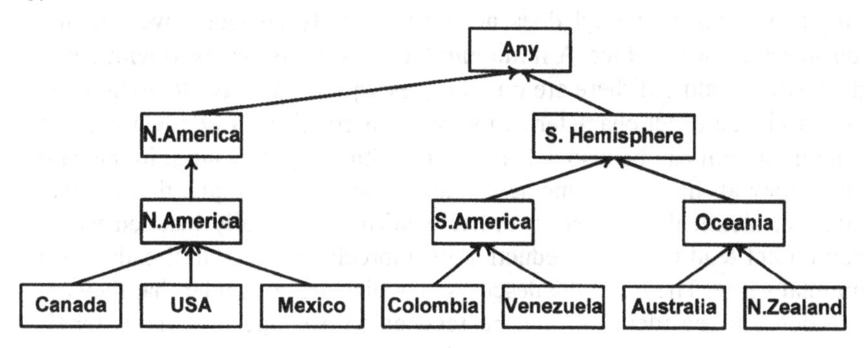

Fig. 8. Multi-contextual generalization hierarchy built from merge of two similarity tables starting at different abstraction levels.

4. AOI over imprecise data

There are two actual carriers of imprecision's representation in the fuzzy database schema. First, as already mentioned in background section, is the occurrence of multiple attribute values. Obviously, the more descriptors we insert to characterize a particular record in the database, the more imprecise is its depiction. The uncertainty about the description is also implicitly reflected in the similarity of values characterizing any particular entity. When the reports received by Department of Homeland Security state that a particular individual was seen recently in the following countries {*Iran, Germany, Australia*}, they cause more doubt about the person's current location than in the case where he/she was seen in {*Iran, Iraq, Afghanistan*}, since this information would be rather immediately interpreted as "*Middle East*". There are exactly the same number of attribute values in each case and they all are at the same level of abstraction, however the higher similarity of reports provided in the second set results in the higher informativeness carried by the second example. So the imprecision of the original information is actually reflected both in the number of inserted descriptors for particular attributes and in the similarity of these values.

A simplified characterization of data imprecision can be provided by an analysis of the boundary values. The range of imprecision degree ranges between 0 (i.e. the lack of uncertainty about stored information) and infinity (maximum imprecision). Following the common opinion that even flawed information is better than the lack of information, we then say that imprecision is at a maximum when there are no data values at all. Since our fuzzy database model does not permit empty attributes we will not consider this case further. A minimum (zero value) is achieved with a single attribute value. If there are no other descriptors, we have to assume the value to be an exact characterization of the particular entity's feature. The same minimum can be also accomplished with multiple values, in the case where they all have the same semantics (synonyms). Despite the fact that multiple, identical descriptors further confirm the initially inserted value, they do not lead to further reduction of imprecision, since it already has a minimal value. Therefore the descriptors, which are so similar that they are considered to be identical, can be reduced to a single descriptor. Obviously, some attribute values, initially considered as different, may be treated as identical at the higher abstraction level. Therefore we can conclude that the practically achievable minimum of imprecision depends on the abstraction level of employed descriptors, and can reach its original (absolute) 0 value only at the lowest level of abstraction (for $\alpha = 1.0$ in our fuzzy database model).

During attribute-oriented induction we usually employ concept hierarchies that have single attribute values in their leaves (at the 0-abstraction level). Therefore the generalization of tuples with single descriptors is straightforward. The problem arises however when there are multiple attribute values describing a single entity. Where should we expect to find a drug dealer who, as not-confirmed reports say, was seen recently in {*Canada, Colombia, Venezuela*}? Our solution is based on partial vote propagation, where a single vote, corresponding to one database tuple, is partitioned before being assigned to the concepts placed in the leaves of the concept hierarchy. Now the fractions of vote are assigned to separate 0-level concepts to represent each of the originally inserted attribute values. During AOI all fractions of this vote propagate gradually through multiple levels of generalization hierarchy, the same way as the regular (precise) records do. The only difference is that the tuple with uncertainty has multiple entries to the generalization paths (different leaves of concept hierarchy for different vote's fractions), whereas each of the precise tuples has only one beginning of its generalization path.

The most trivial solution would be to split the vote equally among all inserted descriptors: {*Canada*|*0.(3), Colombia*|*0.(3), Venezuela*|*0.(3)*}. This approach however does not take into consideration real life dependencies, which are reflected not only in the number of inserted descriptors, but also in their similarity. We propose replacement of the even distribution of vote with a nonlinear spread, dependent on the similarity and the number of inserted values. Using the partition tree built from the similarity table, we can extract from the set of the originally inserted values those concepts which are more similar to each other than to the remaining descriptors. We call these subsets of resemblances (e.g. {*Colombia, Venezuela*} from the above-mentioned example). Then we use these as a basis for calculating a distribution of vote's fractions. An important aspect of this approach is extraction of these subsets of similarities at the lowest possible level of their occurrence, since the nested character of α-proximity relation guarantees that above this α-level they are going to co-occur every time. Repetitive extraction of such subsets could unbalance the original dependencies among inserted values.

The algorithm to achieve it is straightforward. Given (1) a set of attribute values inserted as a description of particular entity, and (2) a hierarchical structure reflecting Zadeh's partition tree for the particular attribute; we want to extract a table, which includes (a) the list of all subsets of resemblances from the given set of descriptors, with (b) the highest level of α-proximity of their common occurrence. The algorithm uses preorder recursive traversal for searching the partition tree. The partition tree is searched starting from its root and if any subset of the given set of descriptors occurs at the particular node of the concept hierarchy we store the values that were recognized as similar, and the adequate value of α. An example of such a search for subsets of resemblances in a tuple with values {*Canada, Colombia, Venezuela*} is depicted in Figure 9. Numbers on the links in the tree represent the order in which the particular subsets of similarities were extracted.

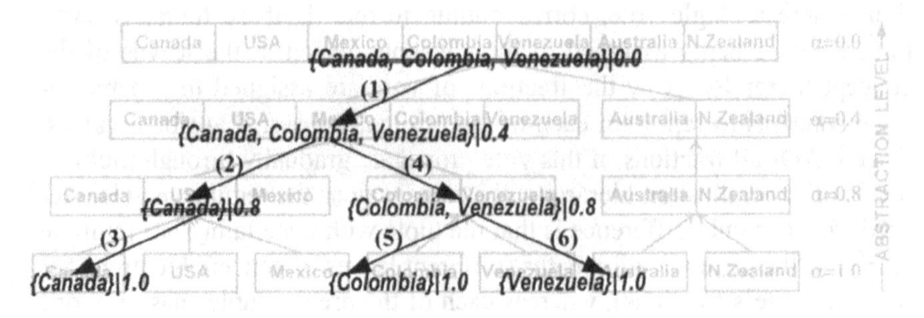

Fig. 9. Subsets of similar values extracted from the original set of descriptors.

An output with subsets of resemblance generated for this example is presented in Table 8.

Table 8. Subsets of resemblances and their similarity levels for the analyzed example.

OUTPUT	Comments
{Canada, Colombia, Venezuela} | 0.0	STORED
{Canada, Colombia, Venezuela} | ~~0.0~~ 0.4	UPDATED
{Canada} | 0.8	STORED
{Canada} | ~~0.8~~ 1.0	UPDATED
{Colombia, Venezuela} | 0.8	STORED
{Colombia} | 1.0	STORED
{Venezuela} | 1.0	STORED

After extracting the subsets of similarities (i.e. *subsets of resemblances*), we apply a summarization of α values as a measure reflecting both the frequency of occurrence of the particular attribute values in the subsets of similarities, as well as the abstraction level of these occurrences. Since the country *Canada* was reported only twice, we assigned it a grade *1.4 (1.0+0.4)*. The remaining attribute values were graded as follows:
Colombia|(1.0 + 0.8 + 0.4) = Colombia|2.2
Venezuela|(1.0 + 0.8 + 0.4) = Venezuela|2.2

In the next step, we added all generated grades (*1.4+2.2 + 2.2 = 5.8*) to normalize grades finally assigned to each of the participating attribute values:
Canada |(1.4/5.8) = Canada |0.24
Colombia |(2.2/5.8) = Colombia |0.38
Venezuela |(2.2/5.8) = Venezuela |0.38

This leads to the new distribution of the vote's fractions, which more accurately reflects real life dependencies than a linear weighting approach. The results obtained are presented in the Figure 10.

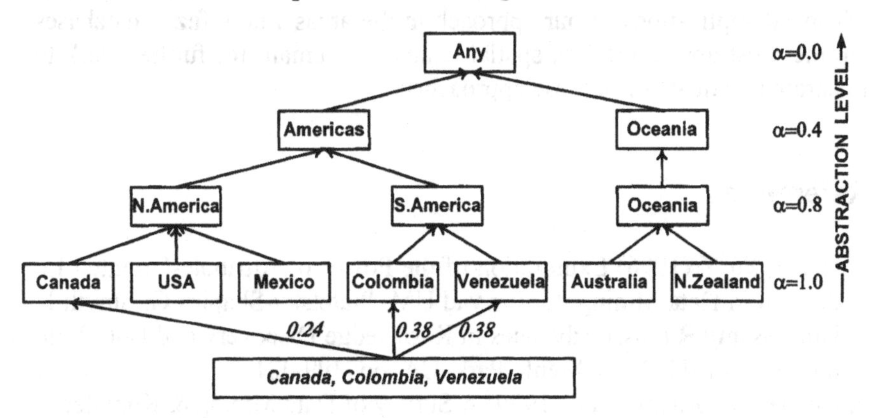

Fig. 10. Partial Vote Propagation for records with uncertainty.

Normalization of the initial grades has a crucial meaning for preservation of the generalization model's completeness. It guarantees that each of the records is represented as a unity, despite being variously distributed at each of the generalization levels.

During an AOI process all fractions of the vote may gradually merge to finally become unity at a level of abstraction high enough to overcome the originally occurring imprecision. In such a case, we observe that there is a removal of imprecision from the data due to its generalization. Such connection between the precision and certainty seems to be natural and has already been noted by researchers [33-34]. In general, very abstract statements have a greater likelihood of being valid than more detailed ones.

4. Conclusions

In this paper we discussed three possible ways that similarity and proximity relations, implemented as the essential parts of fuzzy databases, can be successfully applied to knowledge discovery via attribute-oriented generalization. We proved that both of these relations, due to their basic properties, could be successfully utilized when building bottom-up oriented concept hierarchies, assuming that both the context of an intended generalization and the perspective represented in the similarity or proximity table are in the agreement. Moreover, we also considered how both of these fuzzy relations could be employed to building one-level concept hierarchies, involving the generalization of original attribute values into a representative group of concepts via single layer fuzzy hierarchies. Finally, we presented an algorithm allowing generalization of imprecise data. More advanced applications of our approach in the areas where fuzzy databases are the most applicable (i.e. spatial databases) remain for further work to illustrate practical use of these approaches.

References

1. Han J, Fu Y (1996) Exploration of the Power of Attribute-Oriented Induction in Data Mining. In: Fayyad UM, Piatetsky-Shapiro G, Smyth P, Uthurusamy R (eds.) Advances in Knowledge Discovery and Data Mining. AAAI/MIT Press, Menlo Park, CA, pp. 399-421.
2. Goebel M, Gruenwald L (1999) A Survey of Data Mining & Knowledge Discovery Software Tools. ACM SIGKDD Explorations Newsletter 1(1), pp. 20-33.
3. Feelders A, Daniels H, Holsheimer M (2000) Methodological and practical aspects of data mining, Information & Management 37, pp. 271-281.
4. Cai Y, Cercone N, Han J (1989) Attribute-Oriented Induction in Relational Databases, In: Proc. IJCAI-89 Workshop on Knowledge Discovery in Databases, Detroit, MI, pp. 26-36.
5. Cai Y, Cercone N, Han J (1991) Attribute-Oriented Induction in Relational Databases. In: Piatetsky-Shapiro G, Frawley WJ (eds.) Knowledge Discovery in Databases, AAAI/MIT Press, Menlo Park, CA, pp. 213-228.
6. Han J, Cai Y, Cercone N (1992) Knowledge discovery in databases: An attribute-oriented approach. In: Proc. 18th Int. Conf. Very Large Data Bases, Vancouver, Canada, 1992, pp. 547-559.
7. Chen MS, Han J, and Yu PS (1996) Data Mining: An Overview from a Database Perspective. IEEE Transactions on Knowledge and Data Engineering 8(6), pp. 866-883.
8. Han J, Kamber M (2000) Data Mining: Concepts and Techniques. Morgan Kaufmann, New York, NY.
9. Hamilton HJ, Hilderman RJ, Cercone N (1996) Attribute-oriented induction using domain generalization graphs. In: Proc. 8th IEEE Int'l Conf. on Tools with Artificial Intelligence, Toulouse, France, pp. 246-253.

10. Carter CL, Hamilton HJ (1998) Efficient Attribute-Oriented Generalization for Knowledge Discovery from Large Databases. IEEE Transactions on Knowledge and Data Engineering 10(2), pp. 193-208.
11. Hilderman RJ, Hamilton HJ, Cercone N (1999) Data mining in large databases using domain generalization graphs. Journal of Intelligent Information Systems 13(3), pp. 195-234.
12. Hwang HY, Fu WC(1995) Efficient Algorithms for Attribute-Oriented Induction. In: Proc. 1st International Conference on Knowledge Discovery and Data Mining, Montreal, Canada, pp. 168-173.
13. Lee DH & Kim MH (1997) Database summarization using fuzzy ISA hierarchies. IEEE Transactions on Systems, Man, and Cybernetics - part B 27(1), pp. 68-78.
14. Lee KM (2001) Mining generalized fuzzy quantitative association rules with fuzzy generalization hierarchies. In: Proc. Joint 9th IFSA World Congress and 20th NAFIPS International Conference, Vancouver, Canada, pp. 2977-2982.
15. Cubero JC, Medina JM, Pons O, Vila MA (1999) Data Summarization in Relational Databases Through Fuzzy Dependencies. Information Sciences 121(3-4), pp. 233-270.
16. Raschia G, Ughetto L, Mouaddib N (2001) Data summarization using extended concept hierarchies. In: Proc. Joint 9th IFSA World Congress and 20th NAFIPS International Conference, Vancouver, Canada, pp. 2289 -2294.
17. Raschia G, Mouaddib N (2002) SAINTETIQ:a fuzzy set-based approach to database summarization. Fuzzy Sets and Systems 129(2), pp. 137-162.
18. Angryk RA, Petry FE (2003) Consistent fuzzy concept hierarchies for attribute generalization. In: Proc. IASTED International Conference on Information and Knowledge Sharing (IKS '03), Scottsdale, AZ, pp. 158-163.
19. Angryk RA, Petry FE (2003) Data Mining Fuzzy Databases Using Attribute-Oriented Generalization. In: Proc. 3rd IEEE International Conference on Data Mining (ICDM '03), Workshop on Foundations and New Direction in Data Mining, Melbourne, FL, pp. 8-15.
20. Buckles BP, Petry FE (1982) A fuzzy representation of data for relational databases. Fuzzy Sets and Systems 7(3), pp. 213-226.
21. Buckles BP, Petry FE (1983) Information-theoretic characterization of fuzzy relational databases. IEEE Transactions on Systems, Man, and Cybernetics 13(1), pp. 74-77.
22. Buckles BP, Petry FE (1984) Extending the fuzzy database with fuzzy numbers. Information Sciences 34(2), pp. 145-155.

23.Petry FE (1996) Fuzzy Databases: Principles and Applications. Kluwer Academic Publishers, Boston, MA.
24.Codd FE (1970) A relational model of data for large share data banks. Communications of the ACM, 13(6), pp. 377-387.
25.Shenoi S, Melton A (1989) Proximity Relations in the Fuzzy Relational Database Model. International Journal of Fuzzy Sets and Systems, 31(3), pp. 285-296.
26.Shenoi S, Melton A (1990) An Extended Version of the Fuzzy Relational Database Model. Information Sciences 52 (1), pp. 35-52.
27.Shenoi S, Melton A (1991) Fuzzy Relations and Fuzzy Relational Databases. International Journal of Computers and Mathematics with Applications 21 (11/12), pp. 129-138.
28.Shenoi S, Melton A, Fan LT (1992) Functional Dependencies and Normal Forms in the Fuzzy Relational Database Model. Information Sciences 60, pp. 1-28.
29.Tamura S, Higuchi S, Tanaka K (1971) Pattern Classification Based on Fuzzy Relations. IEEE Transactions on Systems, Man, and Cybernetics 1(1), pp. 61-66.
30.Zadeh LA (1970) Similarity relations and fuzzy orderings. Information Sciences, 3(2), pp. 177-200.
31.Dubois D, Prade H (1980) Fuzzy Sets and Systems: Theory and Applications. Academic Press, New York, NY.
32.Dubois D, Prade H, Rossazza JP (1991) Vagueness, typicality and uncertainty in class hierarchies. International Journal of Intelligent Systems 6, pp. 167-183.
33.Bosc P, Prade H (1993) An introduction to fuzzy set and possibility theory based approaches to the treatment of uncertainty and imprecision in database management systems. In: Proc. 2nd Workshop on Uncertainty Management in Information Systems (UMIS '94): From Needs to Solutions, Catalina, CA.
34.Parsons S (1996) Current approaches to handling imperfect information in data and knowledge bases. IEEE Transactions on Knowledge and Data Engineering 8(3), pp. 353-372.

Rough Set Strategies to Data with Missing Attribute Values

Jerzy W. Grzymala-Busse

Department of Electrical Engineering and Computer Science, University of Kansas, Lawrence, KS 66045, USA
and
Institute of Computer Science, Polish Academy of Sciences, 01-237 Warsaw, Poland
`Jerzy@ku.edu`
`http://lightning.eecs.ku.edu/index.html`

Summary. In this paper we assume that a data set is presented in the form of the incompletely specified decision table, i.e., some attribute values are missing. Our next basic assumption is that some of the missing attribute values are lost (e.g., erased) and some are "do not care" conditions (i.e., they were redundant or not necessary to make a decision or to classify a case). Incompletely specified decision tables are described by characteristic relations, which for completely specified decision tables are reduced to the indiscernibility relation. It is shown how to compute characteristic relations using an idea of block of attribute-value pairs, used in some rule induction algorithms, such as LEM2. Moreover, the set of all characteristic relations for a class of congruent incompletely specified decision tables, defined in the paper, is a lattice. Three definitions of lower and upper approximations are introduced. Finally, it is shown that the presented approach to missing attribute values may be used for other kind of missing attribute values than lost values and "do not care" conditions.

Key words: Data mining, rough set theory, incomplete data, missing attribute values

1 Introduction

Usually all ideas of rough set theory are explored using decision tables as a starting point [10], [11]. The decision table describes cases (also called examples or objects) using attribute values and a decision. Attributes are independent variables while the decision is a dependent variable. In the majority of papers on rough set theory it is assumed that the information is complete, i.e., that for all cases all attribute values and decision values are specified. Such a decision table is said to be completely specified.

In practice, however, input data, presented as decision tables, may have missing attribute and decision values, i.e., decision tables are incompletely specified. Since our main concern is learning from examples, and an example with a missing decision value, (i.e., not classified) is useless, we will assume that only attribute values may be missing.

There are two main reasons why an attribute value is missing: either the value was lost (e.g., was erased) or the value was not important. In the former case attribute value was useful but currently we have no access to it. In the latter case the value does not matter, so such values are also called "do not care" conditions. In practice it means that originally the case was classified (the decision value was assigned) in spite of the fact that the attribute value was not given, since the remaining attribute values were sufficient for such a classification or to make a decision. For example, a test, represented by that attribute, was redundant.

The first rough set approach to missing attribute values, when all missing values were lost, was described in 1997 in [7], where two algorithms for rule induction, LEM1 and LEM2, modified to deal with such missing attribute values, were presented. In 1999 this approach was extensively described in [13], together with a modification of the original idea in the form of a valued tolerance based on a fuzzy set approach.

The second rough set approach to missing attribute values, in which the missing attribute value is interpreted as a "do not care" condition, was used for the first time in 1991 [4]. A method for rule induction was introduced in which each missing attribute value was replaced by all possible values. This idea was further developed and furnished with theoretical properties in 1995 [8].

In this paper a more general rough set approach to missing attribute values is presented. In this approach, in the same decision table, some missing attribute values are assumed to be lost and some are "do not care" conditions. A simple method for computing a characteristic relation describing the decision table with missing attribute values of either of these two types is presented. The characteristic relation for a completely specified decision table is reduced to the ordinary indiscernibility relation. It is shown that the set of all characteristic relations, defined by all possible decision tables with missing attribute values being one of the two types, together with two defined operations on relations, forms a lattice.

Furthermore, three different definitions of lower and upper approximations are introduced. Similar three definitions of approximations were studied in [15], [16], [17]. Some of these definitions are better suited for rule induction. Examples of rules induced from incompletely specified decision tables are provided. The paper ends up with a discussion of other approaches to missing attribute values.

A preliminary version of this paper was presented at the Workshop on Foundations and New Directions in Data Mining, associated with the third IEEE International Conference on Data Mining [6].

2 Missing attribute values and characteristic relations

In the sequel we will assume that information about any phenomena are presented by *decision tables.* Rows of the table are labeled by *cases* and columns by *variables.* The set of all cases will be denoted by U. Independent variables are called *attributes* and a dependent variable is called a *decision* and is denoted by d. The set of all attributes will be denoted by A. An example of such a decision table is presented in Table 1.

Table 1. An example of a completely specified decision table

	Attributes			Decision
Case	Location	Basement	Fireplace	Value
1	good	yes	yes	high
2	bad	no	no	small
3	good	no	no	medium
4	bad	yes	no	medium
5	good	no	yes	medium

Obviously, any decision table defines a function ρ that maps the set of ordered pairs (case, attribute) into the set of all values. For example, $\rho(1, Location) = good.$

Rough set theory is based on the idea of an *indiscernibility relation.* Let B be a nonempty subset of the set A of all attributes. The indiscernibility relation $IND(B)$ is a relation on U defined for $x, y \in U$ as follows

$$(x, y) \in IND(B) \; if \; and \; only \; if \; \rho(x, a) = \rho(y, a) \; for \; all \; a \in B.$$

For completely specified decision tables the indiscernibility relation $IND(B)$ is an equivalence relation. Equivalence classes of $IND(B)$ are called *elementary sets* of B. For example, for Table 1, elementary sets of *IND({Location, Basement})* are {1}, {2}, {3, 5} and {4}.

Function ρ describing Table 1 is completely specified (total). In practice, input data for data mining are frequently affected by missing attribute values. In other words, the corresponding function ρ is incompletely specified (partial).

In the sequel we will assume that all decision values are specified, i.e., are not missing. Also, we will assume that all missing attribute values are denoted either by "?" or by "*", lost values will be denoted by "?", "do not care" conditions will be denoted by "*". Additionally, we will assume that for each case at least one attribute value is specified.

Incompletely specified tables are described by characteristic relations instead of indiscernibility relations. An example of an incompletely specified table is presented in Table 2, where all missing attribute values are lost.

Table 2. An example of an incompletely specified decision table, in which all missing attribute values are lost

	Attributes			Decision
Case	Location	Basement	Fireplace	Value
1	good	yes	yes	high
2	bad	?	no	small
3	good	no	?	medium
4	bad	yes	no	medium
5	?	?	yes	medium

For decision tables, in which all missing attribute values are lost, a special characteristic relation was defined by J. Stefanowski and A. Tsoukias in [13], see also [12], [14]. In this paper that characteristic relation will be denoted by $LV(B)$, where B is a nonempty subset of the set A of all attributes. For $x, y \in U$ characteristic relation $LV(B)$ is defined as follows:

$$(x, y) \in LV(B) \; if \; and \; only \; if \; \rho(x, a) = \rho(y, a)$$
$$for \; all \; a \in B \; such \; that \; \rho(x, a) \neq ?.$$

For any case x, the characteristic relation $LV(B)$ may be presented by the characteristic set $I_B(x)$, where

$$I_B(x) = \{y | (x, y) \in LV(B)\}.$$

For Table 2, characteristic sets $I_A(x)$, where $x \in U$, are the following sets:

$I_A(1) = \{1\}$,
$I_A(2) = \{2, 4\}$,
$I_A(3) = \{3\}$,
$I_A(4) = \{4\}$, and
$I_A(5) = \{1, 5\}$.

For any decision table in which all missing attribute values are lost, characteristic relation $LV(B)$ is reflexive, but—in general—does not need to be symmetric or transitive.

Another example of a decision table with all missing attribute values, this time with only "do not care" conditions, is presented in Table 3.

For decision tables where all missing attribute values are "do not care" conditions a special characteristic relation, in this paper denoted by $DCC(B)$, was defined by M. Kryszkiewicz in [8], see also, e.g., [9]. For $x, y \in U$ characteristic relation $DCC(B)$ is defined as follows:

$$(x, y) \in DCC(B) \; if \; and \; only \; if \; \rho(x, a) = \rho(y, a)$$
$$or \; \rho(x, a) = * \; or \; \rho(y, a) = * \; for \; all \; a \in B.$$

Table 3. An example of an incompletely specified decision table, in which all missing attribute values are "do not care" conditions

	Attributes			Decision
Case	Location	Basement	Fireplace	Value
1	good	yes	yes	high
2	bad	*	no	small
3	good	no	*	medium
4	bad	yes	no	medium
5	*	*	yes	medium

Similarly, for a case $x \in U$, the characteristic relation $DCC(B)$ may be presented by the characteristic set $J_B(x)$, where

$$J_B(x) = \{y|(x,y) \in DCC(B)\}.$$

For Table 3, characteristic sets $J_A(x)$, where $x \in U$, are the following sets:

$J_A(1) = \{1,5\}$,
$J_A(2) = \{2,4\}$,
$J_A(3) = \{3,5\}$,
$J_A(4) = \{2,4\}$, and
$J_A(5) = \{1,3,5\}$.

Relation $DCC(B)$ is reflexive and symmetric but—in general—not transitive.

Table 4 presents a more general case, a decision table with missing attribute values of both types: lost values and "do not care" conditions.

Table 4. An example of an incompletely specified decision table, in which some missing attribute values are lost and some are "do not care" conditions

	Attributes			Decision
Case	Location	Basement	Fireplace	Value
1	good	yes	yes	high
2	bad	?	no	small
3	good	no	?	medium
4	bad	yes	no	medium
5	*	*	yes	medium

In a similar way we may define a characteristic relation $R(B)$ on U for an incompletely specified decision table with both types of missing attribute values: lost values and "do not care" conditions:

$$(x,y) \in R(B) \; if \; and \; only \; if \; \rho(x,a) = \rho(y,a) \; or \; \rho(x,a) = * \; or \rho(y,a) = * \\ for \; all \; a \in B \; such \; that \; \rho(x,a) \neq ?.$$

where $x, y \in U$ and B is a nonempty subset of the set A of all attributes. For a case x, the characteristic relation $R(B)$ may be also presented by its characteristic set $K_B(x)$, where

$$K_B(x) = \{y | (x, y) \in R(B)\}.$$

For Table 4, characteristic sets $K_A(x)$, where $x \in U$, are the following sets:

$K_A(1) = \{1, 5\}$,
$K_A(2) = \{2, 4\}$,
$K_A(3) = \{3, 5\}$,
$K_A(4) = \{4\}$, and
$K_A(5) = \{1, 5\}$.

Obviously, characteristic relations $LV(B)$ and $DCC(B)$ are special cases of the characteristic relation $R(B)$. For a completely specified decision table, the characteristic relation $R(B)$ is reduced to $IND(B)$. The characteristic relation $R(B)$ is reflexive but—in general—does not need to be symmetric or transitive.

3 Computing characteristic relations

The characteristic relation $R(B)$ is known if we know characteristic sets $K(x)$ for all $x \in U$. Thus we may concentrate on computing characteristic sets $K(x)$. We need a few definitions. For completely specified decision tables if $t = (a, v)$ is an attribute-value pair then a block of t, denoted $[t]$, is a set of all cases from U that for attribute a have value v [1], [5]. For incompletely specified decision tables the definition of a block of an attribute-value pair must be modified. If for an attribute a there exists a case x such that $\rho(x, a) = ?$, i.e., the corresponding value is lost, then the case x is not included in the block $[(a, v)]$ for any value v of attribute a. If for an attribute a there exists a case x such that the corresponding value is a "do not care" condition, i.e., $\rho(x, a) = *$, then the corresponding case x should be included in blocks $[(a, v)]$ for all values v of attribute a. The characteristic set $KB(x)$ is the intersection of blocks of attribute-value pairs (a, v) for all attributes a from B for which $\rho(x, a)$ is specified and $\rho(x, a) = v$.

For decision table from Table 4,

$[(Location, good)] = \{1, 3, 5\}$,
$[(Location, bad)] = \{2, 4, 5\}$,
$[(Basement, yes)] = \{1, 4, 5\}$,
$[(Basement, no)] = \{3, 5\}$,
$[(Fireplace, yes)] = \{1, 5\}$, and
$[(Fireplace, no)] = \{2, 4\}$.

Thus

$K_A(1) = \{1,3,5\} \cap \{1,4,5\} \cap \{1,5\} = \{1,5\}$,
$K_A(2) = \{2,4,5\} \cap \{2,4\} = \{2,4\}$,
$K_A(3) = \{1,3,5\} \cap \{3,5\} = \{3,5\}$,
$K_A(4) = \{2,4,5\} \cap \{1,4,5\} \cap \{2,4\} = \{4\}$, and
$K_A(5) = \{1,5\}$.

4 Lattice of characteristic relations

For the sake of simplicity, in this section all characteristic relations will be defined for the entire set A of attributes instead of its subset B and we will write R instead of $R(A)$. By the same token, in characteristic sets $K_A(x)$, the subscript A will be omitted.

Two decision tables with the same set U of all cases, the same attribute set A, the same decision d, and the same specified attribute values will be called *congruent*. Thus, two congruent decision tables may differ only by missing attribute values * and ?. Decision tables from Tables 2, 3, and 4 are all pairwise congruent.

Table 5. Decision table indistinguishable from decision table presented in Table 6

	Attributes			Decision
Case	Location	Basement	Fireplace	Value
1	good	yes	yes	high
2	bad	*	no	small
3	good	no	*	medium
4	bad	yes	no	medium
5	?	*	yes	medium

Table 6. Decision table indistinguishable from decision table presented in Table 5

	Attributes			Decision
Case	Location	Basement	Fireplace	Value
1	good	yes	yes	high
2	bad	*	no	small
3	good	no	*	medium
4	bad	yes	no	medium
5	*	?	yes	medium

Two congruent decision tables that have the same characteristic relations will be called *indistinguishable*. For example, decision tables, presented in

Tables 5 and 6 are indistinguishable, both have the same characteristic relation with the following characteristic sets:

$K(1) = \{1\}$,
$K(2) = \{2, 4\}$,
$K(3) = \{3\}$,
$K(4) = \{2, 4\}$, and
$K(5) = \{1, 3, 5\}$.

On the other hand, if the characteristic relations for two congruent decision tables are different, the decision tables will be called *distinguishable*. Obviously, there is 2^n congruent decision tables, where n is the total number of all missing attribute values in a decision table.

Let D_1 and D_2 be two congruent decision tables, let R_1 and R_2 be their characteristic relations, and let $K_1(x)$ and $K_2(x)$ be their characteristic sets for some $x \in U$, respectively. We say that $R_1 \leq R_2$ if and only if $K_1(x) \subseteq K_2(x)$ for all $x \in U$. We will use also notation that $D_1 \leq D_2$.

For two congruent decision tables D_1 and D_2, $D_1 \leq D_2$ if for every missing attribute value "?" in D_2, say $\rho_2(x, a)$, the missing attribute value for D_1 is also "?", i.e., $\rho_1(x, a)$, where ρ_1 and ρ_2 are functions defined by D_1 and D_2, respectively.

Two subsets of the set of all congruent decision tables are special: set E of n decision tables such that every decision table from E has exactly one missing attribute value "?" and all remaining missing attribute values equal to "*" and the set F of n decision tables such that every decision table from E has exactly one missing attribute value "*" and all remaining missing attribute values equal to "?". In our example, decision tables presented in Tables 5 and 6 belong to the set E.

Let G be the set of all characteristic relations associated with the set E and let H be the set of all characteristic relations associated with the set F. In our example, the set G has three elements, say R_1, R_2, and R_3, defined by the following family of characteristic sets K_1, K_2, and K_3, respectively:

$$\begin{array}{lll}
K_1(1) = \{1\}, & K_2(2) = \{2,4\}, & K_3(1) = \{1,5\}, \\
K_1(2) = \{2,4\}, & K_2(1) = \{1,5\}, & K_3(2) = \{2,4\}, \\
K_1(3) = \{3\}, & K_2(3) = \{3,5\}, & K_3(3) = \{3,5\}, \\
K_1(4) = \{2,4\}, & K_2(4) = \{4\}, & K_3(4) = \{2,4\}, \\
K_1(5) = \{1,3,5\}, & K_2(5) = \{1,3,5\}, & K_3(5) = \{1,5\},
\end{array}$$

where R_1 is the characteristic relation of the decision table D_1 from Table 5, R_2 is the characteristic relation of the decision table D_2 congruent with D_1 and with $\rho_2(2, Basement) =?$ and all remaining missing attribute values equal to "*", and R_3 is the characteristic relation of the decision table D_3

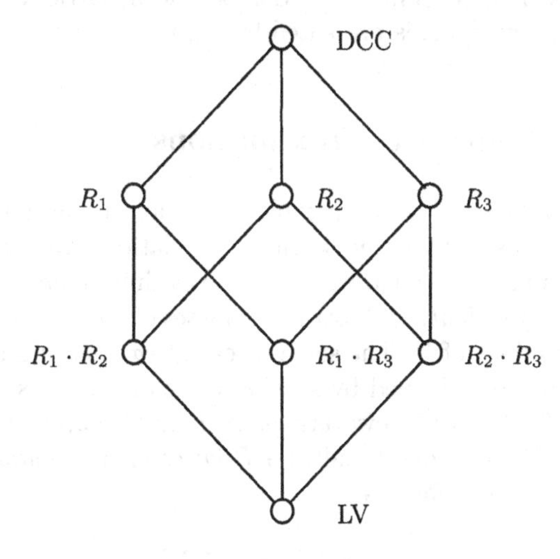

Fig. 1. Diagram of the lattice of all characteristic relations

congruent with D_1 and with $\rho_3(3, Fireplace) =?$ and all remaining missing attribute values equal to "*".

Let D and D' be two congruent decision tables with characteristic relations R and R', and with characteristic sets $K(x)$ and $K'(x)$, respectively, where $x \in U$. We define a characteristic relation $R + R'$ as defined by characteristic sets $K(x) \cup K'(x)$, for $x \in U$, and a characteristic relation $R \cdot R'$ as defined by characteristic sets $K(x) \cap K'(x)$. The set of all characteristic relations for the set of all congruent tables, together with operations $+$ and $\cdot$, is a lattice L (i.e., operations $+$ and $\cdot$ satisfy the four postulates of idempotent, commutativity, associativity, and absorption laws [2]).

Each characteristic relation from L can be represented (using the lattice operations $+$ and $\cdot$) in terms of characteristic relations from G (and, similarly for H). Thus G and H are sets of generators of L. In our example, set G, together with the operation $\cdot$, generates all characteristic relations from L, except for DCC, which may be computed as $R_1 + R_2$, for any two distinct characteristic relations R_1 and R_2 from G. Similarly, set $H = \{R_1 \cdot R_2, R_1 \cdot R_3, R_2 \cdot R_3\}$, together with the operation $+$, generates all characteristic relations from L, except for LV, which may be computed as $R_1 \cdot R_2$, for any two distinct characteristic relations R_1 and R_2 from H.

A characteristic relation R_1 covers another characteristic relation R_2 if and only if $R_1 \neq R_2$, $R_1 \geq R_2$, and there is no R with $R_1 \neq R \neq R_2$ and $R_1 \geq R \geq R_2$. A *diagram* of the lattice L represents elements of L by circles; the characteristic relation R_1 will be placed higher than R_2 if and only if $R_1 \geq R_2$, the circles represented by R_1 and R_2 are connected by a straight line if and only if R_1 covers R_2. The diagram of the lattice of all characteristic relations for our example is presented by Figure 1.

5 Lower and upper approximations

For completely specified decision tables lower and upper approximations are defined on the basis of the indiscernibility relation. An equivalence class of $IND(B)$ containing x is denoted by $[x]_B$. Any finite union of elementary sets of B is called a *B-definable set.* Let U be the set of all cases, called an universe. Let X be any subset of U. The set X is called *concept* and is usually defined as the set of all cases defined by specific value of the decision. In general, X is not a B-definable set. However, set X may be approximated by two B-definable sets, the first one is called a *B-lower approximation* of X, denoted by $\underline{B}X$ and defined as follows

$$\{x \in U | [x]_B \subseteq X\}.$$

The second set is called a *B-upper approximation* of X, denoted by $\overline{B}X$ and defined as follows

$$\{x \in U | [x]_B \cap X \neq \emptyset\}.$$

The B-lower approximation of X is the greatest B-definable set, contained in X. The B-upper approximation of X is the least B-definable set containing X.

For incompletely specified decision tables lower and upper approximations may be defined in a few different ways. In this paper we suggest three different definitions. Again, let X be a concept, let B be a subset of the set A of all attributes, and let $R(B)$ be the characteristic relation of the incompletely specified decision table with characteristic sets $K(x)$, where $x \in U$. Our first definition uses a similar idea as in the previous articles on incompletely specified decision tables [8], [9], [12], [13], [14], i.e., lower and upper approximations are sets of singletons from the universe U satisfying some properties. We will call these definitions *singleton.* A singleton B-lower approximation of X is defined as follows:

$$\underline{B}X = \{x \in U | K_B(x) \subseteq X\}.$$

A singleton B-upper approximation of X is

$$\overline{B}X = \{x \in U | K_B(x) \cap X \neq \emptyset\}.$$

In our example of the decision presented in Table 2 let us say that $B = A$, hence $R(A) = LV(A)$. Then the singleton A-lower and A-upper approximations are:

$$\underline{A}\{1\} = \{1\},$$

$$\underline{A}\{2\} = \emptyset,$$

$$\underline{A}\{3,4,5\} = \{3,4\},$$

$$\overline{A}\{1\} = \{1,5\},$$

$$\overline{A}\{2\} = \{2\},$$

$$\overline{A}\{3,4,5\} = \{2,3,4,5\}.$$

In our example of the decision presented in Table 3 let us say that $B = A$, hence $R(A) = DCC(A)$. Then the singleton A-lower and A-upper approximations are:

$$\underline{A}\{1\} = \emptyset,$$

$$\underline{A}\{2\} = \emptyset,$$

$$\underline{A}\{3,4,5\} = \{3\}$$

$$\overline{A}\{1\} = \{1,5\},$$

$$\overline{A}\{2\} = \{2,4\},$$

$$\overline{A}\{3,4,5\} = \{1,2,3,4,5\} = U.$$

The second definition uses another idea: lower and upper approximations are unions of characteristic sets, subsets of U. We will call these definitions *subset*. A subset B-lower approximation of X is defined as follows:

$$\underline{B}X = \cup\{K_B(x) | x \in U, K_B(x) \subseteq X\}.$$

A *subset* B-upper approximation of X is

$$\overline{B}X = \cup\{K_B(x) | x \in U, K_B(x) \cap X \neq \emptyset\}.$$

In our example of the decision table presented in Table 2 and $R(A) = LV(A)$, the subset A-lower and A-upper approximations are

$$\underline{A}\{1\} = \{1\},$$

$$\underline{A}\{2\} = \emptyset,$$

$$\underline{A}\{3,4,5\} = \{3,4\},$$

$$\overline{A}\{1\} = \{1,5\},$$

$$\overline{A}\{2\} = \{2,4\},$$

$$\overline{A}\{3,4,5\} = \{1,2,3,4,5\} = U.$$

In our example of the decision table presented in Table 3 and $R(A) = DCC(A)$, the subset A-lower and A-upper approximations are

$$\underline{A}\{1\} = \emptyset,$$

$$\underline{A}\{2\} = \emptyset,$$

$$\underline{A}\{3,4,5\} = \{3,5\}$$

$$\overline{A}\{1\} = \{1,3,5\},$$

$$\overline{A}\{2\} = \{2,4\},$$

$$\overline{A}\{3,4,5\} = \{1,2,3,4,5\} = U.$$

The next possibility is to modify the subset definition of upper approximation by replacing the universe U from the previous definition by a concept X. A *concept* B-lower approximation of the concept X is defined as follows:

$$\underline{B}X = \cup\{K_B(x)|x \in X, K_B(x) \subseteq X\}.$$

Obviously, the subset B-lower approximation of X is the same set as the concept B-lower approximation of X. A concept B-upper approximation of the concept X is defined as follows:

$$\overline{B}X = \cup\{K_B(x)|x \in X, K_B(x) \cap X \neq \emptyset\} = \cup\{K_B(x)|x \in X\}.$$

In our example of the decision presented in Table 2 and $R(A) = LV(A)$, the concept A-upper approximations are

$$\overline{A}\{1\} = \{1\},$$

$$\overline{A}\{2\} = \{2,4\},$$

$$\overline{A}\{3,4,5\} = \{1,3,4,5\}.$$

In our example of the decision presented in Table 3 and $R(A) = DCC(A)$, the concept A-upper approximations are

$$\overline{A}\{1\} = \{1,5\},$$

$$\overline{A}\{2\} = \{2,4\},$$

$$\overline{A}\{3,4,5\} = \{1,2,3,4,5\} = U.$$

Note that for completely specified decision tables, all three definitions of lower approximations coalesce to the same definition. Also, for completely specified decision tables, all three definitions of upper approximations coalesce to the same definition. This is not true for incompletely specified decision tables, as the example shows. Since any characteristic relation $R(B)$

is reflexive, singleton lower and upper approximations are subsets of subset lower and upper approximations, respectively.

Also, note that using characteristic relation $LV(A)$, even if we are going to use all three attributes to describe case 2, we cannot describe this case not describing, at the same time, case 4. Thus, the set of rules describing only $\{2\}$ is the empty set. In some rule induction systems the expectation is that the set of all possible rules, induced from an upper approximation cannot be the empty set, so such a system may encounter an infinite loop. This situation cannot happen to the subset or concept definitions of upper approximation. Besides, the concept definition of upper approximation is a subset of the subset definition of upper approximation, so the concept definition of upper approximation is better suited for rule induction. Moreover, it better fits into the idea that the upper approximation should be the *smallest* set containing the concept.

Furthermore, some properties that hold for singleton lower and upper approximations do not hold—in general—for subset lower and upper approximations and for concept lower and upper approximations. For example, as noted in [13], for singleton lower and upper approximations

$$\{x \in U | I_B(x) \subseteq X\} \supseteq \{x \in U | J_B(x) \subseteq X\}$$

and

$$\{x \in U | I_B(x) \cap X \neq \emptyset\} \subseteq \{x \in U | J_B(x) \cap X \neq \emptyset\},$$

where $I_B(x)$ is a characteristic set of $LV(B)$ and $J_B(X)$ is a characteristic set of $DCC(B)$.

In our example, for the subset definition of A-lower approximation, $X = \{3,4,5\}$, and the characteristic relation $LV(A)$ (see Table 2)

$$\cup\{I_B(x) | I_B(x) \subseteq X\} = \{3,4\}$$

while for the subset definition of A-lower approximation, $X = \{3,4,5\}$, and the characteristic relation $DCC(A)$ (see Table 3)

$$\cup\{J_B(x) | J_B(x) \subseteq X\} = \{3,5\},$$

so neither the former set is a subset of the latter nor vice versa.

6 Rule induction

Since all characteristic sets $K(x)$, where $x \in U$, are intersections of blocks of attribute-value pairs, for attributes from B, and for subset and concept definitions of lower and upper approximations, lower and upper approximations are unions of sets of the type $K(x)$, it is natural for rule induction to use an algorithm based on blocks of attribute-value pairs, such as LEM2 [1], [5].

For example, for Table 2, i.e., for the characteristic relation $LV(A)$, the certain rules [3], induced from the concept lower A-approximations are

(Location, good) & (Basement, yes) -> (Value, high),
(Basement, no) -> (Value, medium),
(Location, bad) & (Basement, yes) -> (value, medium).

The possible rules [3], induced from the concept upper A-approximations, for the same characteristic relation $LV(A)$ are

(Location, good) & (Basement, yes) -> (Value, high),
(Location, bad) -> (Value, small),
(Location, good) -> (Value, medium),
(Basement, yes) -> (Value, medium),
(Fireplace, yes) -> (Value, medium).

7 Other approaches to missing attribute values

In this paper two basic approaches to missing attribute values, based on interpretation of a missing attribute value as lost or a "do not care" condition were discussed. Even though the suggested definitions cover the situation in which in the same decision table some missing attribute values are considered to be lost and other are "do not care" conditions, there exist many other possibilities to interpret missing attribute values.

For example, for the attribute *Basement* from our example, we may introduce a special, new value, say *maybe*, for case 2 and we may consider that the missing attribute value for case 5 should be *no*. Neither of these two cases falls into the category of lost values or "do not care" conditions. Nevertheless, such approaches may be studied using the same idea of blocks of attribute-value pairs. More specifically, for attribute *Basement*, the blocks will be

[(Basement, maybe)] = {2},
[(Basement, yes)] = {1, 3}, and
[(Basement, no)} = {3, 5}.

Then we may compute a new characteristic relation, using the technique from Section 3 and define lower and upper approximations using one of the possibilities of Section 5, preferable concept lower and upper approximations, and, eventually, induce certain and possible rules.

8 Conclusions

This paper discusses data with missing attribute values using rough set theory as the main research tool. The existing two approaches to missing attribute values, interpreted as a lost value or as a "do not care" condition are generalized by interpreting every missing attribute value separately as a lost value or as a "do not care" condition. Characteristic relations are introduced to describe incompletely specified decision tables. For completely specified decision tables any characteristic relation is reduced to an indiscernibility relation. It is shown that the basic rough set idea of lower and upper approximations for incompletely specified decision tables may be defined in a variety of different ways. Some of these definitions should have preference of use for rule induction because there is a guarantee that all necessary cases will be described by rules. Again, for completely specified decision tables, all of these definitions of lower and upper approximations are reduced to the standard definition of lower and upper approximations.

References

1. Chan, C.C. and Grzymala-Busse, J.W.: On the attribute redundancy and the learning programs ID3, PRISM, and LEM2. Department of Computer Science, University of Kansas, TR-91-14, December 1991, 20 pp.
2. Birkhoff, G.: Lattice Theory. American Mathematical Society, Providence, RI (1940).
3. Grzymala-Busse, J.W.: Knowledge acquisition under uncertainty—A rough set approach. *Journal of Intelligent & Robotic Systems* **1** (1988), 3–16.
4. Grzymala-Busse, J.W.: On the unknown attribute values in learning from examples. Proc. of the ISMIS-91, 6th International Symposium on Methodologies for Intelligent Systems, Charlotte, North Carolina, October 16–19, 1991. Lecture Notes in Artificial Intelligence, vol. 542, Springer-Verlag, Berlin, Heidelberg, New York (1991) 368–377.
5. Grzymala-Busse, J.W.: LERS—A system for learning from examples based on rough sets. In Intelligent Decision Support. Handbook of Applications and Advances of the Rough Sets Theory, ed. by R. Slowinski, Kluwer Academic Publishers, Dordrecht, Boston, London (1992) 3–18.
6. Grzymala-Busse, J.W.: Rough set strategies to data with missing attribute values. Workshop Notes, Foundations and New Directions of Data Mining, the 3-rd International Conference on Data Mining, Melbourne, FL, USA, November 19–22, 2003, 56–63.
7. Grzymala-Busse, J.W. and A. Y. Wang A.Y.: Modified algorithms LEM1 and LEM2 for rule induction from data with missing attribute values. Proc. of the Fifth International Workshop on Rough Sets and Soft Computing (RSSC'97) at the Third Joint Conference on Information Sciences (JCIS'97), Research Triangle Park, NC, March 2–5, 1997, 69–72.
8. Kryszkiewicz, M.: Rough set approach to incomplete information systems. Proceedings of the Second Annual Joint Conference on Information Sciences, Wrightsville Beach, NC, September 28–October 1, 1995, 194–197.
9. Kryszkiewicz, M.: Rules in incomplete information systems. *Information Sciences* **113** (1999) 271–292.
10. Pawlak, Z.: Rough Sets. *International Journal of Computer and Information Sciences* **11** (1982) 341–356.

11. Pawlak, Z.: Rough Sets. Theoretical Aspects of Reasoning about Data. Kluwer Academic Publishers, Dordrecht, Boston, London (1991).
12. Stefanowski, J.: Algorithms of Decision Rule Induction in Data Mining. Poznan University of Technology Press, Poznan, Poland (2001).
13. Stefanowski, J. and Tsoukias, A.: On the extension of rough sets under incomplete information. Proceedings of the 7th International Workshop on New Directions in Rough Sets, Data Mining, and Granular-Soft Computing, RSFD-GrC'1999, Ube, Yamaguchi, Japan, November 8–10, 1999, 73–81.
14. Stefanowski, J. and Tsoukias, A.: Incomplete information tables and rough classification. *Computational Intelligence* **17** (2001) 545–566.
15. Yao, Y.Y.: Two views of the theory of rough sets in finite universes. *International J. of Approximate Reasoning* **15** (1996) 291–317.
16. Yao, Y.Y.: Relational interpretations of neighborhood operators and rough set approximation operators. *Information Sciences* **111** (1998) 239–259.
17. Yao, Y.Y.: On the generalizing rough set theory. Proc. of the 9th Int. Conference on Rough Sets, Fuzzy Sets, Data Mining and Granular Computing (RSFDGrC'2003), Chongqing, China, October 19–22, 2003, 44–51.

Privacy-Preserving Collaborative Data Mining

Justin Zhan[1], LiWu Chang[2], and Stan Matwin[3]

[1] School of Information Technology & Engineering, University of Ottawa, Canada
zhizhan@site.uottawa.ca
[2] Center for High Assurance Computer Systems, Naval Research Laboratory, USA
lchang@itd.nrl.navy.mil
[3] School of Information Technology & Engineering, University of Ottawa, Canada
stan@site.uottawa.ca

Summary. Privacy-preserving data mining is an important issue in the areas of data mining and security. In this paper, we study how to conduct association rule mining, one of the core data mining techniques, on private data in the following scenario: Multiple parties, each having a private data set, want to jointly conduct association rule mining without disclosing their private data to other parties. Because of the interactive nature among parties, developing a secure framework to achieve such a computation is both challenging and desirable. In this paper, we present a secure framework for multiple parties to conduct privacy-preserving association rule mining.

Key Words:

privacy, security, association rule mining, secure multi-party computation.

1 INTRODUCTION

Business successes are no longer the result of an individual toiling in isolation; rather successes are dependent upon collaboration, team efforts, and partnership. In the modern business world, collaboration becomes especially important because of the mutual benefit it brings. Sometimes, such a collaboration even occurs among competitors, or among companies that have conflict of interests, but the collaborators are aware that the benefit brought by such a collaboration will give them an advantage over other competitors. For this kind of collaboration, data's privacy becomes extremely important: all the parties of the collaboration promise to provide their private data to the collaboration, but neither of them wants each other or any third party to learn much about their private data.

This paper studies a very specific collaboration that becomes more and more prevalent in the business world. The problem is the collaborative data mining. Data mining is a technology that emerges as a means for identifying patterns and trends from a large quantity of data. The goal of our studies is to develop technologies to enable multiple parties to conduct data mining collaboratively without disclosing their private data.

In recent times, the explosion in the availability of various kinds of data has triggered tremendous opportunities for collaboration, in particular collaboration in data mining. The following is some realistic scenarios:

1. Multiple competing supermarkets, each having an extra large set of data records of its customers' buying behaviors, want to conduct data mining on their joint data set for mutual benefit. Since these companies are competitors in the market, they do not want to disclose too much about their customers' information to each other, but they know the results obtained from this collaboration could bring them an advantage over other competitors.
2. Several pharmaceutical companies, each have invested a significant amount of money conducting experiments related to human genes with the goal of discovering meaningful patterns among the genes. To reduce the cost, the companies decide to join force, but neither wants to disclose too much information about their raw data because they are only interested in this collaboration; by disclosing the raw data, a company essentially enables other parties to make discoveries that the company does not want to share with others.

To use the existing data mining algorithms, all parties need to send their data to a trusted central place (such as a super-computing center) to conduct the mining. However, in situations with privacy concerns, the parties may not trust anyone. We call this type of problem the *Privacy-preserving Collaborative Data Mining* (PCDM) problem. For each data mining problem, there is a corresponding PCDM problem. Fig.1 shows how a traditional data mining problem could be transformed to a PCDM problem (this paper only focuses on the heterogeneous collaboration (Fig.1.c))(heterogeneous collaboration means that each party has different sets of attributes. Homogeneous collaboration means that each party has the same sets of attributes.)

Generic solutions for any kind of secure collaborative computing exist in the literature [5]. These solutions are the results of the studies of the Secure Multi-party Computation problem [10, 5], which is a more general form of secure collaborative computing. However, none of the proposed generic solutions is practical; they are not scalable and cannot handle large-scale data sets because of the prohibitive extra cost in protecting data's privacy. Therefore, practical solutions need to be developed. This need underlies the rationale for our research.

Data mining includes a number of different tasks, such as association rule mining, classification, and clustering. This paper studies the association rule mining problem. The goal of association rule mining is to discover meaningful association rules among the attributes of a large quantity of data. For example,

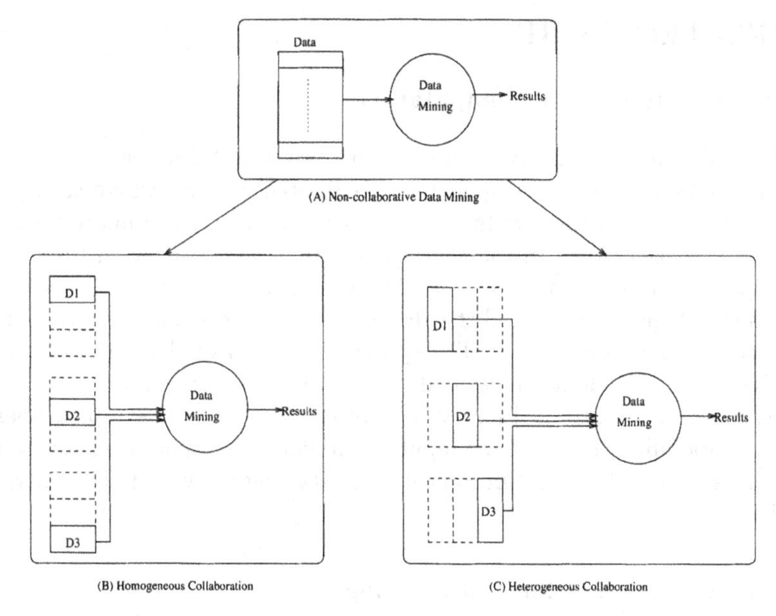

Fig. 1. Privacy Preserving Non-collaborative and Collaborative Data Mining

let us consider the database of a medical study, with each attribute representing a symptom found in a patient. A discovered association rule pattern could be "70% of patients who are drug injection takers also have hepatitis". This information can be useful for the disease-control, medical research, etc. Based on the existing association rule mining technologies, we study the *Mining Association Rules On Private Data* (MAP) problem defined as follows: multiple parties want to conduct association rule mining on a data set that consist all the parties' private data, and neither party is willing to disclose its raw data to other parties.

The existing research on association rule mining [1] provides the basis for the collaborative association rule mining. However, none of those methods satisfy the security requirements of MAP or can be trivially modified to satisfy them. With the increasing needs of privacy-preserving data mining, more and more people are interested in finding solutions to the MAP problem. Vaidya and Clifton proposed a solution [7] for two parties to conduct privacy-preserving association rule mining. However, for the general case where more than two parties are involved, the MAP problem presents a much greater challenge.

The paper is organized as follows: The related work is discussed in Section 2. We describe the association rule mining procedure in Section 3. We then formally define our proposed secure protocol in Section 4. In Section 5, we conduct security and communication analysis. We give our conclusion in Section 6.

2 RELATED WORK

2.1 Secure Multi-party Computation

Briefly, a Secure Multi-party Computation (SMC) problem deals with computing any function on any input, in a distributed network where each participant holds one of the inputs, while ensuring that no more information is revealed to a participant in the computation than can be inferred from that participant's input and output. The SMC problem literature was introduced by Yao [10]. It has been proved that for any function, there is a secure multi-party computation solution [5]. The approach used is as follows: the function F to be computed is first represented as a combinatorial circuit, and then the parties run a short protocol for every gate in the circuit. Every participant gets corresponding shares of the input wires and the output wires for every gate. This approach, although appealing in its generality and simplicity, is highly impractical.

2.2 Privacy-preservation Data Mining

In the early work on such a privacy-preserving data mining problem, Lindell and Pinkas [8] propose a solution to the privacy-preserving classification problem using the oblivious transfer protocol, a powerful tool developed through the secure multi-party computation studies. Another approach for solving the privacy-preserving classification problem was proposed by Agrawal and Srikant [9]. In their approach, each individual data item is perturbed and the distributions of the all data is reconstructed at an aggregate level. The technique works for those data mining algorithms that use the probability distributions rather than individual records. In [7], a solution to the association mining problem for the case of two parties was proposed. In [3], a procedure is provided to build a classifier on private data, where a semi-trusted party was employed to improve the performance of communication and computation. In this paper, we also adopt the model of the semi-trusted party because of the effectiveness and usefulness it brings and present a secure protocol allowing computation to be carried out by the parties.

3 MINING ASSOCIATION RULES ON PRIVATE DATA

Since its introduction in 1993 [1], the association rule mining has received a great deal of attention. It is still one of most popular pattern-discovery methods in the field of knowledge discovery. Briefly, an association rule is an expression $X \Rightarrow Y$, where X and Y are sets of items. The meaning of such rules is as follows: Given a database D of records, $X \Rightarrow Y$ means that whenever a record R contains X then R also contains Y with certain confidence. The rule confidence is defined as the percentage of records containing both X and Y with regard to the overall number of records containing X. The fraction of records R supporting an item X with respect to database D is called the support of X.

3.1 Problem Definition

We consider the scenario where multiple parties, each having a private data set (denoted by D_1, D_2, $\cdots$, and D_n respectively), want to collaboratively conduct association rule mining on the union of their data sets. Because they are concerned about their data's privacy, neither party is willing to disclose its raw data set to others. Without loss of generality, we make the following assumptions on the data sets. The assumptions can be achieved by pre-processing the data sets D_1, D_2, $\cdots$, and D_n, and such a pre-processing does not require one party to send its data set to other parties. (In this paper, we consider applications where the identifier of each data record is recorded. In contrast, for transactions such as the supermarket-buying, customers' IDs may not be needed. The IDs and the names of attributes are known to all parties during the joint computation. A data record used in the joint association rule mining has the same ID in different databases.)

1. D_1, D_2, $\cdots$ and D_n are binary data sets, namely they only contain 0's and 1's, where n is the total number of parties. (Our method is applicable to attributes that are of non-binary value. An attribute of non-binary value will be converted to a binary representation. Detailed implementation includes discretizing and categorizing attributes that are of continuous or ordinal values.)
2. D_1, D_2, $\cdots$ and D_n contain the same number of records. Let N denote the total number of records for each data set.
3. The identities of the ith (for $i \in [1, N]$) record in D_1, D_2, $\cdots$ and D_n are the same.

Mining Association Rules On Private Data problem:

Party 1 has a private data set D_1, party 2 has a private data set D_2, $\cdots$ and party n has a private data set D_n. The data set $[D_1 \cup D_2 \cup \cdots \cup D_n]$ is the union of D_1, D_2, $\cdots$ and D_n (by vertically putting D_1, D_2, $\cdots$ and D_n together so that the concatenation of the ith row in D_1, D_2, $\cdots$ and D_n becomes the ith row in $[D_1 \cup D_2 \cup \cdots \cup D_n]$). The n parties want to conduct association rule mining on $[D_1 \cup D_2 \cup \cdots \cup D_n]$ and to find the association rules with support and confidence being greater than the given thresholds. We say an association rule (e.g., $x_i \Rightarrow y_j$) has confidence $c\%$ in the data set $[D_1 \cup D_2 \cup \cdots \cup D_n]$ if in $[D_1 \cup D_2 \cup \cdots \cup D_n]$ $c\%$ of the records which contain x_i also contain y_j (namely, $c\% = P(y_j \mid x_i)$). We say that the association rule has support $s\%$ in $[D_1 \cup D_2 \cup \cdots \cup D_n]$ if $s\%$ of the records in $[D_1 \cup D_2 \cdots \cup D_n]$ contain both x_i and y_j (namely, $s\% = P(x_i \cap y_j)$).

3.2 Association Rule Mining Procedure

The following is the procedure for mining association rules on $[D_1 \cup D_2 \cdots \cup D_n]$.

1. L_1 = large 1-itemsets
2. **for** (k = 2; $L_{k-1} \neq \phi$; k++) **do begin**
3. C_k = **apriori-gen**(L_{k-1})
4. **for** all candidates $c \in C_k$ **do begin**
5. **Compute** ***c.count*** \\ We will show how to compute it in Section 3.3
6. **end**
7. $L_k = \{c \in C_k | c.count \geq min\text{-}sup\}$
8. **end**
9. Return L = $\cup_k L_k$

The procedure **apriori-gen** is described in the following (please also see [6] for details).

apriori-gen(L_{k-1}: large (k-1)-itemsets)

1. **for** each itemset $l_1 \in L_{k-1}$ **do begin**
2. **for** each itemset $l_2 \in L_{k-1}$ **do begin**
3. **if** $((l_1[1] = l_2[1]) \wedge (l_1[2] = l_2[2]) \wedge \cdots \wedge (l_1[k-1] = l_2[k-1]) \wedge (l_1[k-1] < l_2[k-1]))\{$
4. **then** c = l_1 *join* l_2
5. **for each** (k-1)-subset s of c **do begin**
6. **if** s $\notin L_{k-1}$
7. **then** delete c
8. **else** add c to C_k
9. **end**
10. }
11. **end**
12. **end**
13. return C_k

3.3 How to compute *c.count*

If all the candidates belong to the same party, then *c.count*, which refers to the frequency counts for candidates, can be computed by this party. If the candidates belong to different parties, they then construct vectors for their own attributes and apply our number product protocol, which will be discussed in Section 4, to obtain the *c.count*. We use an example to illustrate how to compute *c.count* among three parties. Party 1, party 2 and party 3 construct vectors X, Y and Z for their own attributes respectively. To obtain *c.count*, they need to compute $\sum_{i=1}^{N} X[i] \cdot Y[i] \cdot Z[i]$ where N is the total number of values in each vector. For instance, if the vectors are as depicted in

Fig.2, then $\sum_{i=1}^{N} X[i] \cdot Y[i] \cdot Z[i] = \sum_{i=1}^{5} X[i] \cdot Y[i] \cdot Z[i] = 3$. We provide an efficient protocol in Section 4 for the parties to compute this value without revealing their private data to each other.

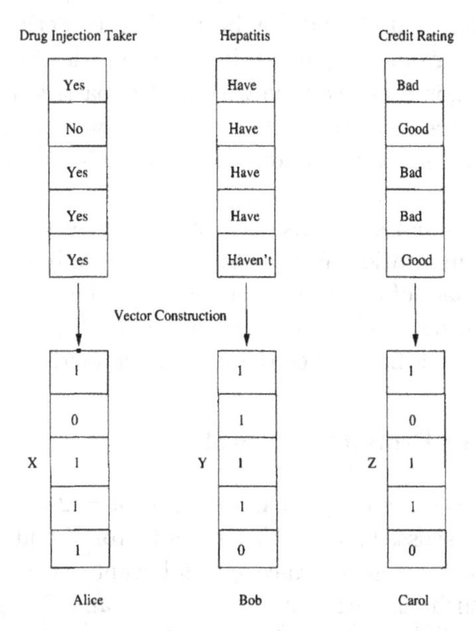

Fig. 2. Raw Data For Alice, Bob and Carol

4 BUILDING BLOCK

How two or multiple parties jointly compute *c.count* without revealing their raw data to each other is the challenge that we want to address. The number product protocol described in this section is the main technical tool used to compute it. We will describe an efficient solution of the number product protocol based on a commodity server, a semi-trusted party.

Our building blocks are two protocols: The first protocol is for two parties to conduct the multiplication operation. This protocol differs from [3] in that we consider the product of numbers instead of vectors. Since vector product can only applied for two vectors, it cannot deal with the computation involved in multiple parties where more than two vectors may participate in the computation. The second protocol, with the first protocol as the basis, is designed for the secure multi-party product operation.

4.1 Introducing The Commodity Server

For performance reasons, we use an extra server, the commodity server [2] in our protocol. The parties could send requests to the commodity server and receive data (called *commodities*) from the server, but the commodities must be independent of the parties' private data. The purpose of the commodities is to help the parties conduct the desired computations.

The commodity server is semi-trusted in the following senses: (1) It should not be trusted; therefore it should not be possible to derive the private information of the data from the parties; it should not learn the computation result either. (2) It should not collude with all the parties. (3) It follows the protocol correctly. Because of these characteristics, we say that it is a semi-trusted party. In the real world, finding such a semi-trusted party is much easier than finding a trusted party.

As we will see from our solutions, the commodity server does not participate in the actual computation among the parties; it only supplies commodities that are independent of the parties' private data. Therefore, the server can generate independent data off-line beforehand, and sell them as commodities to the prover and the verifier (hence the name "commodity server").

4.2 Secure Number Product Protocol

Let's first consider the case of two parties where $n = 2$ (more general cases where $n \geq 3$ will be discussed later). Alice has a vector X and Bob has a vector Y. Both vectors have N elements. Alice and Bob want to compute the product between X and Y such that Alice gets $\sum_{i=1}^{N} U_x[i]$ and Bob gets $\sum_{i=1}^{N} U_y[i]$, where $\sum_{i=1}^{N} U_x[i] + \sum_{i=1}^{N} U_y[i] = \sum_{i=1}^{N} X[i] \cdot Y[i] = X \cdot Y$. $U_y[i]$ and $U_x[i]$ are random numbers. Namely, the scalar product of X and Y is divided into two secret pieces, with one piece going to Alice and the other going to Bob. We assume that random numbers are generated from the integer domain.

Secure Two-party Product Protocol

Protocol 1 *(Secure Two-party Product Protocol)*

1. The Commodity Server generates two random numbers $R_x[1]$ and $R_y[1]$, and lets $r_x[1] + r_y[1] = R_x[1] \cdot R_y[1]$, where $r_x[1]$ (or $r_y[1]$) is a randomly generated number. Then the server sends $(R_x[1], r_x[1])$ to Alice, and $(R_y[1], r_y[1])$ to Bob.
2. Alice sends $\hat{X}[1] = X[1] + R_x[1]$ to Bob.
3. Bob sends $\hat{Y}[1] = Y[1] + R_y[1]$ to Alice.
4. Bob generates a random number $U_y[1]$, and computes $\hat{X}[1] \cdot Y[1] + (r_y[1] - U_y[1])$, then sends the result to Alice.
5. Alice computes $(\hat{X}[1] \cdot Y[1] + (r_y[1] - U_y[1])) - (R_x[1] \cdot \hat{Y}[1]) + r_x[1] = X[1] \cdot Y[1] - U_y[1] + (r_y[1] - R_x[1] \cdot R_y[1] + r_x[1]) = X[1] \cdot Y[1] - U_y[1] = U_x[1]$.
6. Repeat step 1-5 to compute $X[i] \cdot Y[i]$ for $i \in [2, N]$. Alice then gets $\sum_{i=1}^{N} U_x[i]$ and Bob gets $\sum_{i=1}^{N} U_y[i]$.

The bit-wise communication cost of this protocol is $7 * M * N$, where M is the maximum bits for the values involved in our protocol. The cost is approximately 7 times of the *optimal* cost of a two-party scalar product (the optimal cost of a scalar product is defined as the cost of conducting the product of X and Y without the privacy constraints, namely one party simply sends its data in plain to the other party). The cost can be decreased to $3 * M * N$ if the commodity server just sends seeds to Alice and Bob since the seeds can be used to generate a set of random numbers.

Secure Multi-party Product Protocol

We have discussed our protocol of secure number product for two parties. Next, we will consider the protocol for securely computing the number product for multiple parties. For simplicity, we only describe the protocol when $n = 3$. The protocols for the cases when $n > 3$ can be similarly derived. Our concern is that similarly derived solution may not be efficient when the number of parties is large, and more efficient solution is still under research. Without loss of generality, let Alice has a private vector X and a randomly generated vector R_x, Bob has a private vector Y and a randomly generated vector R_y and let $R_y[i] = R'_y[i] + R''_y[i]$ for $i \in [1, N]$ and Carol has a private vector Z and a randomly generated vector R_z. First, we let the parties hide these private numbers by using their respective random numbers, then conduct the product for the multiple numbers.

$$
\begin{aligned}
T[1] &= (X[1] + R_x[1]) * (Y[1] + R_y[1]) * (Z[1] + R_z[1]) \\
&= X[1]Y[1]Z[1] + X[1]R_y[1]Z[1] + R_x[1]Y[1]Z[1] \\
&\quad + R_x[1]R_y[1]Z[1] + X[1]Y[1]R_z[1] + X[1]R_y[1]R_z[1] \\
&\quad + R_x[1]Y[1]R_z[1] + R_x[1]R_y[1]R_z[1] \\
&= X[1]Y[1]Z[1] + X[1]R_y[1]Z[1] + R_x[1]Y[1]Z[1] \\
&\quad + R_x[1]R_y[1]Z[1] + X[1]Y[1]R_z[1] + X[1]R_y[1]R_z[1] \\
&\quad + R_x[1]Y[1]R_z[1] + R_x[1](R'_y[1] + R''_y[1])R_z[1] \\
&= X[1]Y[1]Z[1] + X[1]R_y[1]Z[1] + R_x[1]Y[1]Z[1] \\
&\quad + R_x[1]R_y[1]Z[1] + X[1]Y[1]R_z[1] + X[1]R_y[1]R_z[1] \\
&\quad + R_x[1]Y[1]R_z[1] + R_x[1]R'_y[1]R_z[1] + R_x[1]R''_y[1]R_z[1] \\
&= X[1]Y[1]Z[1] + (X[1]R_y[1] + R_x[1]Y[1] + R_x[1]R_y[1])Z[1] \\
&\quad + (X[1]Y[1] + X[1]R_y[1] + R_x[1]Y[1] + R_x[1]R'_y[1])R_z[1] \\
&\quad + R_x[1]R''_y[1]R_z[1] = T_0[1] + T_1[1] + T_2[1] + T_3[1],
\end{aligned}
$$

where

$T_0[1] = X[1]Y[1]Z[1],$
$T_1[1] = (X[1]R_y[1] + R_x[1]Y[1] + R_x[1]R_y[1])Z[1],$
$T_2[1] = (X[1]Y[1] + X[1]R_y[1] + R_x[1]Y[1] + R_x[1]R'_y[1])R_z[1],$
$T_3[1] = R_x[1]R''_y[1]R_z[1].$

$T_0[1]$ is what we want to obtain. To compute $T_0[1]$, we need to know $T[1]$, $T_1[1]$, $T_2[1]$ and $T_3[1]$. In this protocol, we let Alice get $T[1]$, Bob get $T_3[1]$ and Carol get $T_1[1]$ and $T_2[1]$. Bob separates $R_y[1]$ into $R'_y[1]$ and $R''_y[1]$. If he fails to do so, then his data might be disclosed during the computation of these terms.

To compute $T_1[1]$ and $T_2[1]$, Alice and Bob can use Protocol 1 to compute $X[1]R_y[1]$, $R_x[1]Y[1]$, $R_x[1]R_y[1]$, $X[1]Y[1]$ and $R_x[1]R'_y[1]$. Thus, according to Protocol 1, Alice gets $U_x[1]$, $U_x[2]$, $U_x[3]$, $U_x[4]$ and $U_x[5]$ and Bob gets $U_y[1]$, $U_y[2]$, $U_y[3]$, $U_y[4]$ and $U_y[5]$. Then they compute $(X[1]R_y[1] + R_x[1]Y[1] + R_x[1]R_y[1])$ and $(X[1]Y[1]+X[1]R_y[1]+R_x[1]Y[1]+R_x[1]R'_y[1])$ and send the results to Carol who can then compute $T_1[1]$ and $T_2[1]$. Note that $X[1]R_y[1] = U_x[1]+U_y[1]$, $R_x[1]Y[1] = U_x[2]+U_y[2]$, $R_x[1]R_y[1] = U_x[3]+U_y[3]$, $X[1]Y[1] = U_x[4] + U_y[4]$ and $R_x[1]R'_y[1] = U_x[5] + U_y[5]$.

To compute $T_3[1]$, Alice and Carol use Protocol 1 to compute $R_x[1]R_z[1]$, then send the results to Bob who can then compute $T_3[1]$.

To compute T[1], Bob sends $Y[1] + R_y[1]$ to Alice and Carol sends $Z[1] + R_z[1]$ to Alice.

Repeat the above process to compute $T[i]$, $T_1[i]$, $T_2[i]$, $T_3[i]$ and $T_0[i]$ for $i \in [2, N]$. Then, Alice has $\sum_{i=1}^{N} T[i]$, Bob has $\sum_{i=1}^{N} T_3[i]$ and Carol has $\sum_{i=1}^{N} T_1[i]$ and $\sum_{i=1}^{N} T_2[i]$. Finally, we achieve the goal and obtain $\sum_{i=1}^{N} X[i]Y[i]Z[i] = \sum_{i=1}^{N} T_0[i] = \sum_{i=1}^{N} T[i] - \sum_{i=1}^{N} T_1[i] - \sum_{i=1}^{N} T_2[i] - \sum_{i=1}^{N} T_3[i]$.

Protocol 2 *(Secure Multi-party Product Protocol)*

Step I: Random Number Generation

1. Alice generates a random number $R_x[1]$.
2. Bob generates two random numbers $R'_y[1]$ and $R''_y[1]$.
3. Carol generates a random number $R_z[1]$.

Step II: Collaborative Computing

Sub-step 1: To Compute $T[1]$

1. Carol computes $Z[1] + R_z[1]$ and sends it to Bob.
2. Bob computes $Y[1] + R_y[1]$ and sends it to Alice.
3. Alice computes $T[1] = (X[1] + R_x[1]) * (Y[1] + R_y[1]) * (Z[1] + R_z[1])$.

Sub-step 2: To Compute $T_1[1]$ *and* $T_2[1]$

1. Alice and Bob use Protocol 1 to compute $X[1] \cdot R_y[1]$, $R_x[1] \cdot Y[1]$, $R_x[1] \cdot R_y[1]$, $X[1] \cdot Y[1]$ and $R_x[1] \cdot R'_y[1]$. Then Alice obtains $U_x[1]$, $U_x[2]$, $U_x[3]$, $U_x[4]$ and $U_x[5]$ and Bob obtains $U_y[1]$, $U_y[2]$, $U_y[3]$, $U_y[4]$ and $U_y[5]$.

2. Alice sends $(U_x[1] + U_x[2] + U_x[3])$ and $(U_x[4] + U_x[1] + U_x[2] + U_x[5])$ to Carol.
3. Bob sends $(U_y[1] + U_y[2] + U_y[3])$ and $(U_y[4] + U_y[1] + U_y[2] + U_y[5])$ to Carol.
4. Carol computes
   $T_1[1] = (X[1] \cdot R_y[1] + R_x[1] \cdot Y[1] + R_x[1] \cdot R_y[1]) \cdot Z[1]$, and
   $T_2[1] = (X[1] \cdot Y[1] + X[1] \cdot R_y[1] + R_x[1] \cdot Y[1] + R_x[1] \cdot R'_y[1]) \cdot R_z[1]$.

Sub-step 3: To Compute $T_3[1]$

1. Alice and Carol use Protocol 1 to compute $R_x[1] \cdot R_z[1]$ and send the values they obtained from the protocol to Bob.
2. Bob computes $T_3[1] = R''_y[1] \cdot R_x[1] \cdot R_z[1]$.

Step III: Repeating

1. Repeat the Step I and Step II to compute $T[i]$, $T_1[i]$, $T_2[i]$ and $T_3[i]$ for $i \in [2, N]$.
2. Alice then gets
   $[a] = \sum_{i=1}^{N} T[i] = \sum_{i=1}^{N} (X[i] + R_x[i]) * (Y[i] + R_y[i]) * (Z[i] + R_z[i])$.
3. Bob gets
   $[b] = \sum_{i=1}^{N} T_3[i] = \sum_{i=1}^{N} (Y[i] + R_y[i]) * (Z[i] + R_z[i])$.
4. Carol gets
   $[c] = \sum_{i=1}^{N} T_1[i] = \sum_{i=1}^{N} (X[i] \cdot R_y[i] + R_x[i] \cdot Y[i] + R_x[i] \cdot R_y[i]) \cdot Z[i]$, and
   $[d] = \sum_{i=1}^{N} T_2[i] = \sum_{i=1}^{N} (X[i] \cdot Y[i] + X[i] \cdot R_y[i] + R_x[i] \cdot Y[i] + R_x[i] \cdot R'_y[i]) \cdot R_z[i]$.
   Note that $\sum_{i=1}^{N} X[i] \cdot Y[i] \cdot Z[i] = \sum_{i=1}^{N} T_0[i] = [a] - [b] - [c] - [d]$.

Theorem 1. *Protocol 1 is secure such that Alice cannotlearn Y and Bob cannot learn X either.*

Proof. The number $\hat{X}[i] = X[i] + R_x[i]$ is all what Bob gets. Because of the randomness and the secrecy of $R_x[i]$, Bob cannot find out $X[i]$. According to the protocol, Alice gets (1) $\hat{Y}[i] = Y[i] + R_y[i]$, (2) $Z[i] = \hat{X}[i] \cdot Y[i] + (r_y[i] - U_y[i])$, and (3) $r_x[i]$, $R_x[i]$, where $r_x[i] + r_y[i] = R_x[i] \cdot R_y[i]$. We will show that for any arbitrary $Y'[i]$, there exists $r'_y[i]$, $R'_y[i]$ and $U'_y[i]$ that satisfies the above equations. Assume $Y'[i]$ is an arbitrary number. Let $R'_y[i] = \hat{Y}[i] - Y'[i]$, $r'_y[i] = R_x[i] \cdot R_y[i] - r_x[i]$, and $U'_y[i] = \hat{X}[i] \cdot Y'[i] + r'_y[i]$. Therefore, Alice has (1) $\hat{Y}[i] = Y'[i] + R'_y[i]$, (2) $Z[i] = \hat{X}[i] \cdot Y'[i] + (r'_y[i] - U'_y[i])$ and (3) $r_x[i]$, $R_x[i]$, where $r_x[i] + r'_y[i] = R_x[i] \cdot R'_y[i]$. Thus, from what Alice learns, there exists infinite possible values for $Y[i]$. Therefore, Alice cannot know Y and neither can Bob know X.

Theorem 2. *Protocol 2 is secure such that Alice cannot learn Y and Z, Bob cannot learn X and Z, and Carol cannot learn X and Y.*

Proof. According to the protocol, Alice obtains
(1) $(Y[i] + R_y[i])$, and (2)$(Z[i] + R_z[i])$.
Bob gets $R_x[i] \cdot R_z[i]$.
Carol gets
(1) $(X[i] \cdot R_y[i] + R_x[i] \cdot Y[i] + R_x[i] \cdot R_y[i])$ and
(2) $(X[i] \cdot Y[i] + X[i] \cdot R_y[i] + R_x[i] \cdot Y[i] + R_x[i] \cdot R'_y[i])$.

Since $R_x[i]$, $R_y[i](= (R'_y[i] + R''_y[i]))$ and $R_z[i]$ are arbitrary random numbers. From what Alice learns, there exists infinite possible values for $Y[i]$ and $Z[i]$. From what Bob learns, there also exists infinite possible values for $Z[i]$. From what Carol learns, there still exists infinite possible values for $X[i]$ and $Y[i]$.

Therefore, Alice cannot learn Y and Z, Bob cannot learn X and Z, and Carol cannot learn X and Y either.

5 ANALYSIS

5.1 Security analysis

Malicious Model Analysis

In this paper, our algorithm is based on the semi-honest model, where all the parties behave honestly and cooperatively during the protocol execution. However, in practice, one of the parties (e.g., Bob) may be malicious in that it wants to gain true values of other parties' data by purposely manipulating its own data before executing the protocols. For example, in Protocol 1 Bob wants to know whether $X[i] = 1$ for some i. He may make up a set of numbers with all, but the ith, values being set to 0's (i.e., $Y[i] = 1$ and $Y[j] = 0$ for $j \neq i$). According to Protocol 1, if Bob obtains $\sum_{i=1}^{N} U_x[i] + \sum_{i=1}^{N} U_y[i]$, indicating the total number of counts for both X and Y being 1, then Bob can know that $X[i]$ is 0 if the above result is 0 and $X[i]$ is 1 if the above result is 1. To deal with this problem, we may randomly select a party to hold the frequency counts. For example, let's consider the scenario of three parties. Without loss of generality, we assume Bob is a malicious party. The chance that Bob gets chosen to hold the frequency counts is $\frac{1}{3}$. We consider the following two cases.(Assume that the probability of samples in a sample space are equally likely.)

1. Make a correct guess of both Alice's and Carol's values.
   If Bob is not chosen to hold the frequency counts, he then chooses to randomly guess and the probability for him to make a correct guess is $\frac{1}{4}$. In case Bob is chosen, if the product result is 1 (with the probability of $\frac{1}{4}$), he then concludes that both Alice and Carol have value 1; if the product result is 0 (with the probability of $\frac{3}{4}$), he would have a chance of $\frac{1}{3}$ to make a correct guess. Therefore, we have $\frac{2}{3} * \frac{1}{4} + \frac{1}{3}(\frac{1}{4} + \frac{3}{4} * \frac{1}{3}) \approx 33\%$.

Note that the chance for Bob to make a correct guess, without his data being purposely manipulated, is 25%.

2. Make a correct guess for only one party's (e.g., Alice) value.
If Bob is not chosen to hold the frequency counts, the chance that his guess is correct is $\frac{1}{2}$. In case Bob is chosen, if the product result is 1, he then knows the Alice's value with certainty; if the result is 0, there are two possibilities that need to be considered: (1) if Alice's value is 0, then the chance that his guess is correct is $\frac{2}{3}$; (2) if Alice's value is 1, then the chance that his guess is correct is $\frac{1}{3}$. Therefore, we have $\frac{2}{3} * \frac{1}{2} + \frac{1}{3}(\frac{1}{4} + \frac{3}{4}(\frac{2}{3} * \frac{2}{3} + \frac{1}{3} * \frac{1}{3})) \approx 56\%$. However, if Bob chooses to make a random guess, he then has 50% of chance to be correct.

It can be shown that the ratio for Bob to make a correct guess with/without manipulating his data in case 1 is (n+1)/n and in case 2 is approaching 1 with an exponential rate of n ($\approx 2^{-(n-1)}$), where n is the number of parties. The probability for a malicious party to make a correct guess about other parties' values decreases significantly as the number of parties increases.

How to deal with information disclosure by the inference from the results

Assume the association rule, $DrugInjection \Rightarrow Hepatitis$, is what we get from the collaborative association rule mining, and this rule has 99% confidence level (i.e., $P(Hepatitis|DrugInjection) = 0.99$). Now given a data item $item_1$ with Alice:(Drug-Injection), Alice can figure out Bob's data (i.e., Bob:(Hepatitis) is in $item_1$) with confidence 99% (but not vice versa). Such an inference problem exists whenever the items of the association rule is small and its confidence measure is high. To deal with the information disclosure through inference, we may enforce the parties to randomize their data as in [4] with some probabilities before conducting the association rule mining.

How to deal with the repeat use of protocol

A malicious party (e.g., Bob) may ask to run the protocol multiple times with different set of values at each time by manipulating his $U_y[.]$ value. If other parties respond with honest answers, then this malicious party may have chance to obtain actual values of other parties. To avoid this type of disclosure, constraints must be imposed on the number of repetitions.

5.2 Communication Analysis

There are three sources which contribute to the total bit-wise communication cost for the above protocols: (1) the number of rounds of communication to compute the number product for a single value (denoted by $NumRod$); (2) the maximum number of bits for the values involved in the protocols (denoted by

M); (2) the number of times (N) that the protocols are applied. The total cost can be expressed by $NumRod * M * N$ where $NumRod$ and M are constants for each protocol. Therefore the communication cost is O(N). N is a large number when the number of parties is big since N exponentially increases as the number of parties expands.

6 CONCLUDING REMARKS

In this paper, we consider the problem of privacy-preserving collaborative data mining with inputs of binary data sets. In particular, we study how multiple parties to jointly conduct association rule mining on private data. We provided an efficient association rule mining procedure to carry out such a computation. In order to securely collecting necessary statistical measures from data of multiple parties, we have developed a secure protocol, namely the number product protocol, for multiple-party to jointly conduct their desired computations. We also discussed the malicious model and approached it by distributing the measure of frequency counts to different parties, and suggested the use of the randomization method to reduce the inference of data disclosure.

In our future work, we will extend our method to deal with non-binary data sets. We will also apply our technique to other data mining computations, such as privacy-preserving clustering.

References

1. R. Agrawal, T. Imielinski, and A. Swami. Mining association rules between sets of items in large databases. In P. Buneman and S. Jajodia, editors, *Proceedings of ACM SIGMOD Conference on Management of Data*, pages 207–216, Washington D.C., May 1993.
2. D. Beaver. Commodity-based cryptography (extended abstract). In *Proceedings of the twenty-ninth annual ACM symposium on Theory of computing*, El Paso, TX USA, May 4-6 1997.
3. W. Du and Z. Zhan. Building decision tree classifier on private data. In *Workshop on Privacy, Security, and Data Mining at The 2002 IEEE International Conference on Data Mining (ICDM'02)*, Maebashi City, Japan, December 9 2002.
4. W. Du and Z. Zhan. Using randomized response techniques for privacy-preserving data mining. In *Proceedings of The 9th ACM SIGKDD International Conference on Knowledge Discovery and Data Mining*, Washington, DC, USA, August 24-27 2003.
5. O. Goldreich. Secure multi-party computation (working draft). http://www.wisdom.weizmann.ac.il/home/oded/ public_html/foc.html, 1998.
6. J. Han and M. Kamber. *Data Mining Concepts and Techniques.* Morgan Kaufmann Publishers, 2001.

7. J.Vaidya and C.W.Clifton. Privacy preserving association rule mining in vertically partitioned data, in the 8th ACM SIGKDD International Conference on Knowledge Discovery and Data Mining, July 23-26, 2002, edmonton, alberta, canada.
8. Y. Lindell and B. Pinkas. Privacy preserving data mining. In *Advances in Cryptology - Crypto2000, Lecture Notes in Computer Science*, volume 1880, 2000.
9. R.Agrawal and S.Ramakrishnan. Privacy-preserving data mining. In *ACM SIGMOD*, pages 439–450, 2000.
10. A. C. Yao. Protocols for secure computations. In *In Proceedings of the 23rd Annual IEEE Symposium on Foundations of Computer Science*, 1982.

Impact of Purity Measures on Knowledge Extraction in Decision Trees

Mitja Lenič, Petra Povalej, Peter Kokol

Laboratory for system design, Faculty of Electrical Engineering and Computer Science, University of Maribor, Maribor, Slovenia

Symbolic knowledge representation is crucial for successful knowledge extraction and consequently for successful data mining. Therefore decision trees and association rules are most commonly used symbolic knowledge representations. Often some sorts of purity measures are used to identify relevant knowledge in data. Selection of appropriate purity measure can have important impact on quality of extracted knowledge. In this paper a novel approach for combining purity measures and thereby altering background knowledge of extraction method is presented. An extensive case study on 42 UCI databases using heuristic decision tree induction as knowledge extraction method is also presented.

Introduction

An important step in the successful data mining process is to select appropriate attributes, which may have significant impact on observed phenomena, before collecting the data. This step introduces an important part of knowledge about the problem in database and therefore represents background knowledge of the database. However, the knowledge extraction method used in the data mining process also includes some predefined background knowledge about the method algorithm, which in the entry uses the background knowledge of the database (its attributes, their types and definition domain) and the data collected.

Background knowledge of knowledge extraction method is usually hard coded into the induction algorithm and depends on its target knowledge representation.

Generally knowledge extraction methods can be divided into search space methods that use some sort of non-deterministic algorithm and heuristic methods that use heuristic function to accomplish goal. Heuristic function is therefore an important part of background knowledge of knowledge extraction method. For decision tree and rule induction methods this heuristics are named (im)purity measures. Since induction algorithm for decision trees and rules is commonly fixed, the only tunable parameters that can adjust background knowledge are the selection of purity measure and discretization method. In this paper the main focus is on impact of purity measures on induction method although also different discretization methods are applied. We introduce new hybrid purity measures that change background knowledge of induction method. Hybrid purity measures are composed out of most commonly used single purity measures. In order to demonstrate the effectiveness of newly introduced hybrid purity measures, a comparison to its single components is performed on 42 UCI databases. We also study the impact of boosting to hybrid and commonly used single purity measures as an alternative (second opinion) knowledge. Additionally the effect of pruning is considered. The paper is focused on the use of purity measures on induction of decision trees, however our findings are not limited only to decision trees and can be applied also in other knowledge extraction methods like rule extraction method, discretization method, etc.

Heuristic induction of decision tree

Decision tree is a hierarchical knowledge representation consisting of tests in inner nodes and their consequences in leafs. Tests define specific attributes and values, representing the inherent knowledge hidden in a database. Since tests and thereafter the extracted knowledge depends mainly on a selected purity measure, the relation purity measure – knowledge is very important and we study and analyze it very profoundly in the present paper.

Decision tree induction

Greedy top-down decision tree induction is a commonly used method. Starting with an empty tree and the entire training set the following algorithm is applied until no more splits are possible:

A greedy decision tree induction algorithm:

1. If all training samples at the current node *t* belong to the same class *c*, create a leaf node with a class *c*.
2. Otherwise, for each attribute compute its purity with a respect to the class attribute using a purity measure.
3. Select the attribute A_i with the highest value of purity measure with a respect to the discretization as the test at the current node.
4. Divide the training samples into separate sets, so that within a set all samples have the same value of A_i using selected discretization. Create as many child nodes as there are distinct values of A_i.
5. Label edges between the parent and the child nodes with outcomes of A_i and partition the training samples into the child nodes.
6. A child node is said to be "pure" if all the training samples at the node belong to the same class.
7. Repeat the previous steps on all impure child nodes.

A new sample is thus dropped down the decision tree from the root of the tree to a single leaf node. As a consequence, the instance space is partitioned into mutually exclusive and exhaustive regions, one for each leaf. The number of leafs is therefore a good measure for complexity of the decision tree. A quality of induced decision tree is then tested on unseen testing instances and described with average total and average class accuracy. Since average total accuracy can be very misleading for assessing a reliability of induced decision trees in our experiments we focused mostly on average class accuracy.

Discretization methods

The values of the attributes can be nominal or discrete, but all nominal attributes have to be mapped into discrete space. Thereafter one of the most often addressed issues in building a decision tree is dealing with often addressed issues in building a decision tree is dealing with continuous (numeric) attributes. Continuous attributes must be replaced by discrete attributes, where each of discrete values corresponds to a subinterval of the continuous attribute. Finding lower and upper limits of those subintervals is therefore a task that must be performed before the attribute can be used. Those limits have to be chosen very carefully to preserve all the information contained in the attribute. Consequently the very method of discretization we choose can have an important impact on the quality of induced decision tree.

In this research nodal discretization was used. Thus in every decision tree node all continuous attributes are mapped into discrete values using only the instances that are available at that node. In each node all possible discretization methods are considered and the discretization method with the highest value of purity measure is then selected for the test.

Following methods of nodal discretization are considered:

- Nodal equidistant discretization is the most simple way of splitting the interval into equidistant subintervals. This approach has proved to be surprisingly good when used as nodal discretization. The drawback of this discretization is that it does not consider the nature of the attribute, neither it considers the distribution of training samples.
- Nodal threshold discretization splits the interval into two subintervals by determining a threshold with the help of information gain function and training samples. The drawback of this approach is that it tries to picture the world just as "black or white", since it always splits the interval into two subintervals.
- Nodal dynamic discretization first splits the interval into many subintervals, so that every training sample's value has its own subinterval. In the second step it merges together smaller subintervals that are labeled with the same outcome into larger subintervals. In each of the following steps three subintervals are merged together: two "stronger" subintervals with one "weak" interval, where "weak" interval lies between those two "strong" subintervals. Here "strong" and "weak" applies to number of training samples in the subinterval tree. In comparison to previous two approaches the dynamic discretization returns more 'natural' subintervals, which in most instances results in better and smaller decision trees. For quality assessment of dynamic discretization commonly purity measure is used.

Single purity measures

As described in a previous section purity measures have a central role in the heuristic knowledge extraction methods. In this section we introduce four of the most popular purity measures that were used in our experiments. Before we proceed let us define the terminology.

Let S be the whole training set described with A attributes and C classes. Let V be the number of values of a given attribute respectively. Let T denote available tests in specific inner node. Let n denote the number of training instances, $n_{i.}$ the number of training instances from class C_i, $n_{.j}$ the number of instances with j-th value of a given attribute and n_{ij} the number of instances from class C_i with a j-th value of a given attribute. Let further $p_{ij} = n_{ij}/n_{..}$, $p_{i.} = n_{i.}/n_{..}$, $p_{.j} = n_{.j}/n_{..}$, and $p_{i|j} = n_{ij}/n_{.j}$ denote the probabilities from the training set.

Information gain, information gain ratio

One of the first and most commonly used purity measures, which was already used in Quinlan's ID3 algorithm [1], is the information gain. It is based on Shannon's entropy from information theory [2], which has its origins in thermodynamics and statistical physics. In the latter entropy represents the degree of disorder in a substance or system. Similarly, entropy in information theory measures the uncertainty of a message as an information source. The more information contains the message, the smaller the value of the entropy.

Let E_C, E_T, E_{CT} denote the entropy of class distribution, the entropy of the values of a given test and the entropy of the joint distribution class - test outcome:

$$E_C = -\sum_i p_{i.} \log_2 p_{i.} \qquad E_T = -\sum_j p_{.j} \log_2 p_{.j}$$

$$E_{CT} = -\sum_i \sum_j p_{ij} \log_2 p_{ij}$$

The expected entropy of the class distribution with regard to test T is defined as

$$E_{C|T} = E_{CT} - E_T .$$

When compared to the entropy E_C of the class distribution, the $E_{C|T}$ gives the reduction of the entropy (the gain of information) to be expected when the test T is selected for the inner node. Hence the information gain I_{gain} is defined as

$$I_{gain}(T) = E_C - E_{C|T}.$$

Therefore, in every inner node test that yields the highest value of I_{gain} is selected.

As information gain shows a strong bias towards the multi-valued test, Quinlan introduced the information gain ratio in C4.5, which is defined as

$$I_{gainratio}(T) = \frac{I_{gain}(T)}{E_T}.$$

Dividing the information gain by the entropy of the test outcome distribution strongly reduces the bias towards the multi-valued tests [3].

Gini index

Another well-known purity measure is Gini index that has been also used for tree induction in statistics by Breiman et al. [4] (i.e. CART). It is defined as:

$$Gini(T) = -\sum_j p_{.j} \sum_i p_{i|j}^2 - \sum_i p_{i.}^2.$$

The test yielding the highest value of Gini index is selected for the inner node. It emphasizes equal sized offspring and purity of both children. Breiman et al. also pointed out that Gini index has difficulties when the class attribute has relatively large number of classes.

Chi-Square goodness-of-fit

Different types of chi-square tests [5] are frequently used for significance testing in statistics.

The chi-square goodness-of-fit test is used to test if an observed distribution conforms to any other distribution, such as one based on theory (exp. normal distribution) or one based on some known distribution.

It compares the expected frequency e_{ij} with the observed frequency n_{ij} of instances from class C_i with a j-th value of a given test **[6]**. More specifically,

$$\chi^2(T) = \sum_i \sum_j \frac{(e_{ij} - n_{ij})^2}{e_{ij}} \qquad e_{ij} = \frac{n_{.j} n_{i.}}{n_{..}} .$$

Clearly a larger value of χ^2 indicates that the test is more pure. Just as for information gain and gini index the test with the highest value of χ^2 is selected for the node.

J-measure

J-measure was introduced by Smyth and Goodman [7] as an informatics theoretic means of quantifying the information content of the rule.

The J_j-measure or cross-entropy is appropriate for selecting a single attribute value of a give test T for rule generation and it is defined by the equation

$$J_j(T) = p_{.j} \sum_i p_{i|j} \log \frac{p_{i|j}}{p_{i.}} .$$

Generalization upon all outcomes of the test gives the test purity measure:

$$J(T) = \sum_j J_j(T) .$$

J-measure was also used as a basis for reducing overfitting by prepruning branches during decision tree induction [8].

Hybrid purity measures

Each purity measure uses a different approach for assessing the knowledge gain of a test respecting the outcome and in such way each purity measure introduces its own background knowledge. Therefore selecting the appropriate purity measure for a specific problem is a very demanding task. To combine different aspects of each single purity measure we decided to introduce hybrid purity measures, shortly described in subsequent subsections.

Sum and product of pairs

First new purity measure is constructed based on a sum of pairs of different purity measures:

$$H^{+}(M_1, M_2, T) = M_1(T) + M_2(T)$$

where M_i represents a single purity measure and H represents the hybrid measure.

More specifically, in each node both purity measures are separately calculated for each test and then summed together (i.e. $H^{+}(I_{gainratio}, Gini, T)$ represents a sum of Information gain ratio and Gini). A test with the highest summed value is chosen for the inner node. Thus, the whole decision tree is induced on the basis of a hybrid purity measure H^{+}.

In a similar way a hybrid purity measure based on a product of pairs of different purity measures is defined (i.e. Information gain ratio * Chi square):

$$H^{\bullet}(M_1, M_2, T) = M_1(T) \cdot M_2(T)$$

Linear combination of purity measures

Motivated by the basic idea of hybrid purity measures described above we tried to find a generalized way to combine all single purity measures. Observing the idea of base purity measures the next logical step seemed to be a construction of general linear combination of those.

$$H^L(\mathbf{M},\mathbf{w},T) = \mathbf{w}^T \cdot \mathbf{M}(T) = \sum_i w_i M_i(T),$$

where w_i are randomly generated coefficients.

Each new hybrid purity measure H^L is defined with a vector **w**.

Voting of purity measures

Our next idea was to use multiple purity measures in a single knowledge representation model. In the case of decision trees, each test node can be constructed using different purity measures. Therefore we constructed a new algorithm for classic decision tree induction where in each inner node a greedy search for best test using each purity measure is performed. The final test in the inner node is determined by voting of different purity measures where each purity measure gives one vote for the best test chosen by it. The test with the most votes is than chosen for the inner node. In the case where two or more different tests have the highest vote the final test is determined randomly.

Weighted voting of purity measures

The new induction algorithm based on voting of purity measures (described in previous subsection) was modified with the introduction of weighted voting. A weight is assigned to each purity measure, resulting in a vector **w**. The test in the inner node is determined by voting of different purity measures where each purity measure gives one weighted vote for the best test chosen by it. The test with the most votes is than selected for the inner node.

Boosting

Boosting is a general method for improving the accuracy of any given induction algorithm. It works by running the induction algorithm on the training set multiple times, each time focusing attention on the difficult cases. At every iteration knowledge is extracted as result of the weighted training samples and each sample is then reweighted according to the consistency of extracted knowledge with facts. After the boosting process is finished, the composite knowledge is obtained with its relevance factor for given problem. From the classification point of view a new sample will have a class with the greatest total vote assigned. Boosting has so far proved to be highly accurate on the training set and usually that also stands for the testing set.

From the data mining point of view, extracting alternative aspect from the same data can be viewed as second opinion or as council of expert knowledge bases that expose different aspects of single problem.

In this section we will shortly describe the popular AdaBoost algorithm derived by Freund and Schapire. We used the AdaBoost algorithm to boost the induction algorithms.

Algorithm AdaBoost

As mentioned above, AdaBoost algorithm, introduced by Freund and Schapire, is a method for generating an ensemble of classifiers by successive reweightings of the training samples [9].

The final composite classifier generally performs well on the training samples even when its constituent classifiers are weak. Each classifier (if appropriate method is selected) contains different aspect of knowledge hidden in data. Although boosting in general increases the accuracy, it sometimes leads to deterioration. That can be put down to overfitting or very skewed class distribution across the weight vectors w_t.

AdaBoost algorithm

Given: a set of training samples $i = 1,2,...,N$

Trials: $t = 1,2,...,T$

Initialize: for every sample i initial weight $w_1[i] = 1/N$ ($w_t[i]$...weight for case i in the trial t)

For trial $t = 1,2,...,T$:

- Train classifier C_t from the training samples using the weights w_t.
- Calculate the error rate ε_t of the classifier C_t on the training samples as the sum of the weights $w_t[i]$ for each misclassified case i.
- If $\varepsilon_t = 0$ or $\varepsilon_t \geq ½$, terminate the process
  - otherwise update the weights $w_{t+1}[i]$ as follows:

$$w_{t+1}[i] = \begin{cases} \dfrac{w_t[i]}{2\varepsilon_t}; & \text{...if } C_t \text{ misclassifies sample } i \\ \dfrac{w_t[i]}{2(1-\varepsilon_t)}; & \text{...otherwise.} \end{cases}$$

- To classify a sample x:
  - Choose class k to maximize the sum $\sum \log \frac{1-\varepsilon_t}{\varepsilon_t}$ for every classifier C_t that predicts class k.

Experiment setup

The effectiveness of newly introduced hybrid purity measures has to be evaluated in practical experiment. For that reason we selected 42 databases from UCI repository [10] that were already divided in learning and testing sets. We used the following evaluation criteria:

$$accuracy = \frac{num\ of\ correctly\ classified\ objects}{num.\ of\ all\ objects}$$

$$accuracy_c = \frac{num\ of\ correctly\ classified\ objects\ in\ class\ c}{num.\ of\ all\ objects\ in\ class\ c}$$

$$average\ class\ accuracy = \frac{\sum_c accuracy_c}{num.\ of\ classes}$$

$$size = num.\ of\ leafs$$

In the case of knowledge extracted with boosting, size is calculated as sum of individual knowledge representation sizes. In general there is a connection between generalization abilities and knowledge representation size.

We made four experiments where hybrid purity measures were compared to single ones. First experiment included general comparison between hybrid and single purity measures without pruning and without boosting. In the second experiment pruning was included to reduce overfitting. The third experiment included comparison of boosted hybrid and boosted single purity measures without pruning. In the last experiment boosting and pruning was included.

Results and discussion

In all experiments described before the results of the comparison between single and hybrid purity measures on all 42 was made. First, the decision trees induced on the basis of single and hybrid purity measures were compared according to accuracy and average class accuracy achieved on the test set. We were interested in the number of databases on which specific method (single or hybrid purity measure) achieved exclusively better accuracy / average class accuracy (see Table 1). The results clearly show that hybrid purity measures achieved better results according to accuracy and average class accuracy on more databases than single purity measures. Nevertheless, Pearson Chi-square test of significance was made in order to backup our findings. The significance test showed significant difference in all comparisons except in two experiments: (1) the comparison of accuracy on unpruned single and hybrid purity measures and (2) the comparison of pruned single and hybrid purity measures on the basis of average class accuracy. Nevertheless the results show non negligible difference that speaks in favor of hybrid purity measures in both cases.

	Accuracy		Average class accuracy	
	Single	Hybrid	Single	Hybrid
Unpruned	11	19	10	23
Pruned	3	14	7	14
Boosted	4	26	5	28
Pruned boosted	6	29	7	27

Table 1. Comparison of single and hybrid purity measures based on accuracy and average class accuracy on the test set. Number of databases on which a specific method achieved better accuracy. Grey cells mark significant difference between single and hybrid purity measures.

The size of induced decision trees was also one of the evaluation criteria for assessing the quality of hybrid purity measures. Since smaller decision trees are preferable the single and hybrid purity measures were compared according to the number of databases on which a specific method induced smaller decision tree. The results in the Table 2 show that in most experiments the use of hybrid purity measures resulted in more decision trees with higher number of leaves. The exception is the comparison between pruned boosted single and hybrid purity measures where hybrid purity measures induced smaller decision trees on almost 60% of the databases.

	Size	
	Single	Hybrid
Unpruned	21	10
Pruned	6	8
Boosted	22	17
Pruned boosted	13	25

Table 2. Comparison of single and hybrid purity measures based on the size of induced decision tree(s). Number of databases on which a specific method induced smaller decision tree(s).

The experiments showed that although some decision trees induced on the basis of hybrid purity measures are bigger in size, the test accuracy and average class accuracy is often higher than when single purity measures are used. The size of induced decision tree is certainly important evaluation criteria however it can be overlooked on account of significant improvement in the accuracy.

Conclusion

In this paper an extensive case study on 42 UCI databases using heuristic decision tree induction as knowledge extraction method was presented. A novel approach for combining purity measures and thereby altering background knowledge of the extraction method was introduced.

Experiments with hybrid purity measures showed that their quality differs on different databases. It is clearly visible, that by changing the background knowledge with selection of different purity measure extracted knowledge can express different aspects of a specific problem.

With weighted hybrid purity measures additional tuning of extraction method is possible, by changing influence weights of each purity measure in the hybrid purity measure. That enables smooth changing of background knowledge and might improve quality of extracted knowledge.

References

[1] Quinlan, J.R. Discovering Rules by Induction from Large Collections of Examples, Expert Systems in the Microelectronic Age, Ed. D. Michie, Edinburgh University Press, 1979.

[2] Shannon, Weaver The mathematical theory of communications, Urbana: The University of Illinois Press, 1949.

[3] Konenko, I. On Biases in Estimating Multi-Valued Attributes, Proc. 1st Int. Conf. On Knowledge Discovery and Data Mining, 1034-1040, Montreal, 1995.

[4] Breiman, L.; Friedman, J. H.; Olshen, R. A.; Stone, C. J. Classification and Regression Trees, Wadsworth International Group, Belmond, CA, 1984.

[5] Snedecor, George W.; and Cochran, William G. (1989), Statistical Methods, Eighth Edition, Iowa State University Press.

[6] White, A. P.; Liu W. Z. Bias in information-based measures in decision tree induction, Machine Learning, 1994, 15:321-329.

[7] Smyth, P.; Goodman, R. M. Rule induction using information theory, In: Piatetsky-Schapiro, G.; Frawley, W. J. (Eds.), Knowledge Discovery in Databases, AAAI Press, 1991, 159-176.

[8] Bramer M.A. Using J-pruning to reduce overfitting in classification, In: Knowledge Based Systems, Vol. 15, Issues 5-6, pp. 301-308, July 2002.

[9] Freund, Y.; Schapire, R. E. Experiments with a new boosting algorithm, In Proceedings Thirteenth International Conference on Machine Learning, Morgan Kaufman, San Francisco, 1996, 148-156.

[10] Blake, C.L. & Merz, C.J. (1998). UCI Repository of machine learning databases [http://www.ics.uci.edu/~mlearn/MLRepository.html]. Irvine, CA: University of California, Department of Information and Computer Science.

Multidimensional On-line Mining

Ching-Yao Wang[1], Tzung-Pei Hong[2], and Shian-Shyong Tseng[1]

[1] Institute of Computer and Information Science, National Chiao-Tung University, Taiwan, R.O.C.
[2] Department of Electrical Engineering, National University of Kaohsiung, Taiwan, R.O.C.

Abstract. In the past, incremental mining approaches usually considered getting the newest set of knowledge consistent with the entire set of data inserted so far. Users can not, however, use them to obtain rules or patterns only from their interesting portion of the data. In addition, these approaches only focused on finding frequent patterns in a specified part of a database. That is, although the data records are collected in under certain time, place and category, such contexts (circumstances) have been ignored in conventional mining algorithms. It will cause the lack of patterns or rules to help users solve problems at different aspects and with diverse considerations. In this paper, we thus attempt to extend incremental mining to online decision support under multidimensional context considerations. We first propose the *multidimensional pattern relation* to structurally and systematically retain the additional context information and mining information for each inserted dataset into a database. We then develop an algorithm based on the proposed multidimensional pattern relation to correctly and efficiently fulfill diverse on-line mining requests.

Keywords: data mining, association rule, incremental mining, multidimensional mining, constraint-based mining, data warehouse.

1 Introduction

Knowledge discovery and data mining technology has recently gotten much attention in the field of large databases and data warehouses. It attempts to discovery non-trivial, implicit, previously unknown and potentially useful knowledge from databases or warehouses [2][7][13], and thus aids managers to make correct decisions. Among the various types of databases and mined knowledge, mining association rules from transaction databases is most commonly seen [1][3]. It discovers relationships among items such that the presence of certain items in a transaction tends to imply the presence of certain other items. Since this mining process is rather time-consuming, many approaches have been proposed to reduce the computation time and improve the performance. Some famous mining algorithms are Apriori [3], DHP [21], Partition [24], DIC [5], Sampling [19], GSP [4][25], and among others. These algorithms process data in a batch way and reprocess the entire database whenever either the data stored in a database or the thresholds (i.e. the minimum support or the minimum confidence) set by users are changed. They do not utilize previously mined patterns for later maintenance, and will require considerable computation time when a database is massive in size.

Incremental mining algorithms were thus proposed, which utilized previously mining information (such as large itemsets) to reduce the cost of re-computation [8][9][14][15][18][23][26][27][28]. These algorithms usually consider getting the set of knowledge consistent with the entire set of data inserted so far. Users can not, however, easily use them to obtain rules or patterns only from their interesting portion of data.

In addition, data records are usually collected in blocks in different contexts (circumstances). The context attributes such as time, place and category have been ignored in conventional mining algorithms [12]. If mining algorithms can consider related context information, they could help users solve problems at different aspects and with diverse considerations. Constraint-based and multidimensional mining techniques were thus proposed to achieve this purpose [11][12][13][17][20][22].

In this paper, we attempt to extend the concept of effectively utilizing previously discovered patterns in incremental mining to support online generation of association rules under multidimensional considerations. We first propose the *multidimensional pattern relation* to structurally and systematically store the additional context information and mining information for each inserted dataset. This idea is conceptually similar to the construction of a data warehouse for OLAP [6][16][29]. Both of them preprocess the underlying data in advance, integrate related information, and consequently store the results in a centralized structural repository for later use and analysis. The mining information in a multidimensional pattern relation is, however, unlike the summarized information of *fact attributes* in a data warehouse. They can not be easily aggregated to fulfill users' mining requests. We thus develop an on-line mining algorithm based on the proposed multidimensional pattern relation to correctly and efficiently fulfill different mining requests.

2 Some Related Works

As mentioned above, most mining algorithms process data in a batch way and must reprocess the entire database whenever either the data stored in a database or the thresholds set by users are changed. They do not use previously mining information and may need considerable computation time to get the newest set of rules or patterns [8]. In real-word applications, new records may be inserted and old records may be deleted or modified along with time.

Designing a mining algorithm that can efficiently maintain previously mined patterns or rules as databases grow is thus important.

Recently, some researchers have developed incremental mining algorithms for maintaining association rules and sequential patterns. Examples for associations rules are the FUP algorithm proposed by Cheung et al. [8][9], the adaptive algorithm proposed by Sarda and Srinivas [23], the incremental mining algorithm based on the pre-large concept proposed by Hong et al. [14][15], and the incremental updating technique based on negative borders proposed by Thomas et al. [26]. Examples for sequential patterns are Lin and Lee's FASTUP algorithm for inserted records [18], and Wang et al's maintenance algorithms for deleted records [27] and for modified records [28]. The common idea in these approaches is that previously mining information should be utilized as much as possible to reduce maintenance costs. Intermediate results, such as large itemsets, are kept to save computation time for maintenance, although original databases may still need to be reprocessed in some situations.

The above maintenance algorithms usually consider getting the set of knowledge consistent with the entire set of data inserted so far. Users can not, however, use them to obtain rules or patterns only from their interesting portion of the data. For example, assume the transaction data in a company are collected and analyzed each month. The maintenance algorithms can easily obtain the up-to-date mined knowledge. But if a manager wants to know the item association on data from all the first quarters, the above approaches may not work out.

Venkatesh and his co-workers thus proposed an efficient algorithm [10], called *DEMON*, capable of allowing users to select a temporal subset of the database for mining desirable patterns in a dynamic environment. DEMON assumed that real-life data did not evolve in an arbitrary way. Instead, it assumed that data were inserted or deleted in a block during a fixed time interval such as a month. The database was then divided into several units, each with the data in a time interval. By this way, users could flexibly select the blocks that they were interested in, and DEMON could combine them together for mining interesting patterns or models. Fig. 1 exemplifies a database of DEMON consisting of 12 blocks from 2002/1 to 2002/12.

Similar to incremental mining algorithms, DEMON also kept several models (mined patterns) which were derived from previous requests for later use and analysis. It still needed to reprocess the data in the newly selected blocks in a batch way when the kept models did not satisfy users' new requests. Moreover, it considered only data division in different time intervals and could not fully meet the diverse demands of online decision support under multidimensional considerations. The following scenarios illustrate the considerations.

- Scenario 1: A marketing analyst may want to know what patterns are significant when the minimum support increases from 5% to 10%.

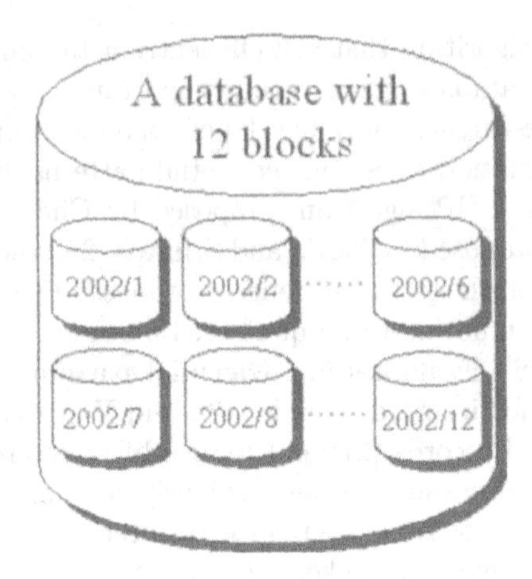

Fig. 1. An example of a database structure in DEMON

- Scenario 2: A decision-maker may have known which product combinations sold in last August were popular, and wants to know which product combinations sold in last September were also popular.
- Scenario 3: A marketing analyst may want to analyze the data collected from the branches in Los Angeles and San Francisco in all the first quarters in the last five years.
- Scenario 4: A decision-maker may have known that people often buy beers and diapers together from a transaction database, and want to further know under what contexts (e.g., place, month, or branch) this pattern is significant or, oppositely, under what contexts this pattern becomes insignificant.
- Scenario 5: A decision-maker may want to know what the mined patterns this year differ from those last year, such as what new patterns appear and what old patterns disappear.

Below, we propose a multidimensional mining framework to efficiently solve the above scenarios.

3 A Multidimensional Mining Framework

The proposed multidimensional mining framework maintains additional context information for diverse decision supports and additional mining information for efficient on-line mining. Assume that data are inserted or deleted in a block during a time interval such as a month. Whenever a new block of data is inserted into the database, significant patterns are mined from this block as the mining information based on an initial minimum support. The mining information along with the corresponding context information will then be stored into the predefined *multidimensional pattern relation*. The details of the multidimensional pattern relation will be described in Section 4. On the other hand, when an old block is deleted from the database, its corresponding context information and mining information are also removed from the multidimensional pattern relation. Fig. 2 shows an example of the multidimensional mining framework. The database consists of the blocks in different branches from 2002/1 to 2002/12. The context information and mining information along with each block of data form a tuple in the corresponding multidimensional pattern relation. When a new block of data from the Los Angeles branch in 2003/1 is inserted, it is first stored in the underlying database. The significant patterns for this block are mined and then stored, with the other context information, in the multidimensional pattern relation.

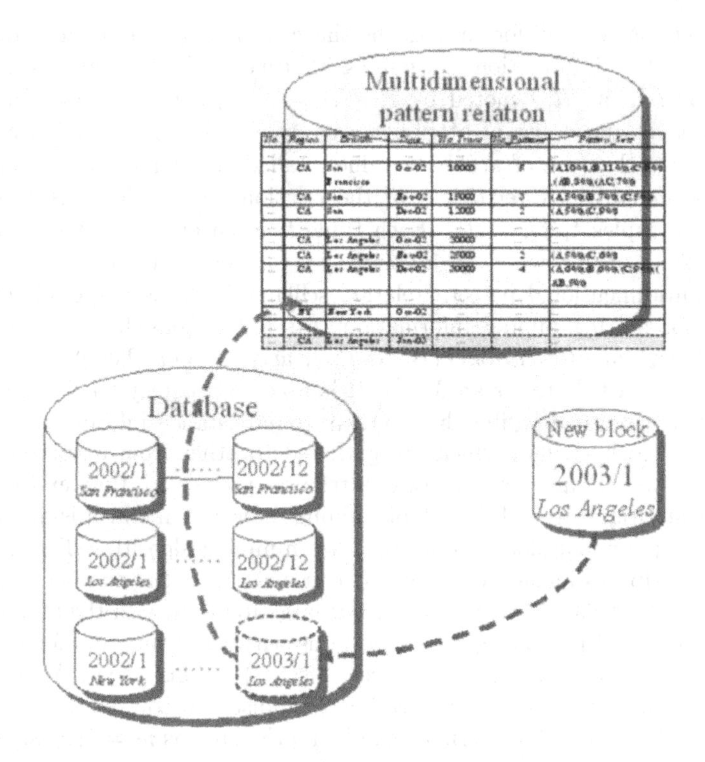

Fig. 2. An example of the multidimensional mining framework

The mining information kept for each block can be used to efficiently support on-line analysis since the overload for responding to a new mining request may be reduced. This idea is conceptually similar to the construction of a data warehouse for OLAP. A data warehouse is an integrated, subject-oriented, and nonvolatile data repository used to store historical aggregated data of an organization for supporting decision-making processes [6][16][29]. Here, a multidimensional pattern relation acts as an integrated, subject-oriented, and nonvolatile knowledge repository for supporting future mining requests. Both of them preprocess the underlying data in advance, integrate related information, and consequently store the results in a centralized structural repository for later use and analysis. The mining information in a multidimensional pattern relation is, however, unlike the summarized information of *fact attributes* in a data warehouse. They can not be easily aggregated to fulfill users' mining requests. A more sophisticated aggregation procedure thus needs to be designed. We will describe the proposed aggregation procedure in Section 5.

4 The Multidimensional Pattern Relation

In this section, we will formally define the *multidimensional pattern relation* to store context information and mining information for later analysis. First, a relation schema R, denoted by $R(A_1, A_2, \cdots, A_n)$, is made up of a relation name R and a list of attributes $A_1, A_2, \cdots, A_n$. Each attribute A_i is associated with a set of attribute values, called the domain of A_i and denoted by $dom(A_i)$. A relation r of the relation schema $R(A_1, A_2, \cdots, A_n)$ is a set of tuples $t_1, t_2, \cdots, t_m$. Each tuple t_i is an ordered list of n values $< v_i1, v_i2, \cdots, v_in >$, where each value v_ij is an element of $dom(A_i)$.

A multidimensional pattern relation schema MPR is a special relation schema for storing mining information. An MPR consists of three types of attributes, $identification(ID)$, *context*, and *content*. There is only one identification attribute for an MPR. It is used to uniquely label the tuples. Context attributes describe the contexts (circumstance information) of an individual block of data which are gathered together from a specific business viewpoint. Examples of context attributes are region, time and branch. Content attributes describe available mining information which is discovered from each individual block of data by a batch mining algorithm. Examples of content attributes include the number of transactions, the number of mined patterns, and the set of previously mined large itemsets with their supports.

The set of all previously mined patterns with their supports for an individual block of data is called a *pattern set* (*ps*) in this paper. Assume the minimum support is s and there are l large itemsets discovered from an individual block of data. A pattern set can be represented as $ps = \{(x_i, s_i) | s_i \geq s$ and $1 \leq i \leq l\}$, where x_i is a large itemset and s_i is its support. The pattern set is thus a principal content attribute for an inserted block of data.

A multidimensional pattern relation schema MPR with n_1 context attributes and n_2 content attributes can be represented as $MPR(ID, CX_1, CX_2, \cdots, CX_{n_1}, CN_1, CN_2, \cdots, CN_{n_2})$, where ID is an identification attribute, CX_i, $1 \leq i \leq n_1$, is a context attribute, and CN_i, $1 \leq i \leq n_2$, is a content attribute. Assume a multidimensional pattern relation *mpr*, which is an instance of the given MPR, includes tuples $\{t_1, t_2, \cdots, t_m\}$. Each tuple t_i $= (id_i, cx_i1, cx_i2, \cdots, cx_{in_1}, cn_i1, cn_i2, \cdots, cn_{in_2})$ in *mpr* indicates that for the block of data under the contexts of $cx_i1, cx_i2, \cdots$, and cx_{in_1} , the mining information contains $cn_i1, cn_i2, \cdots$, and cn_{in_2}.

Example 1: Table 1 shows a multidimensional pattern relation with the initial minimum support set at 5%. *ID* is an identification attribute, *Region*, *Branch* and *Time* are context attributes, and *No_Trans*, *No_Patterns* and *Pattern_Sets* are content attributes. The *Pattern_Sets* attribute records the sets of mined large itemsets from the previous data blocks. For example, the tuple with $ID = 1$ shows that five large itemsets, {(A, 10%), (B, 11%), (C, 9%), (AB, 8%), (AC, 7%)}, are discovered from some 10000 transactions, with the minimum support = 5% and under the contexts of *Region* = **CA**, *Branch* = **San Francisco** and *Time* = **2002/10**. The other tuples have similar meaning.

Table 1. The mined patterns with the minimum support = 5%

ID	Region	Branch	Time	No_Trans	No_Patterns	Pattern_Sets (Itemset, Support)
1	CA	San Francisco	2002/10	10000	5	(A,10%),(B,11%),(C,9%), (AB,8%),(AC,7%)
2	CA	San Francisco	2002/11	15000	3	(A,5%),(B,7%),(C,5%)
3	CA	San Francisco	2002/12	12000	2	(A,5%),(C,9%)
4	CA	Los Angeles	2002/10	20000	3	(A,7%),(C,7%),(AC,5%)
5	CA	Los Angeles	2002/11	25000	2	(A,5%),(C,6%)
6	CA	Los Angeles	2002/12	30000	4	(A,6%),(B,6%),(C,9%), (AB,5%)
7	NY	New York	2002/10	18000	3	(B,8%),(C,7%),(BC,6%)
8	NY	New York	2002/11	18500	2	(B,8%),(C,6%)
9	NY	New York	2002/12	19000	5	(A,5%),(B,9%),(D,6%), (C,8%),(BC,6%)

5 Multidimensional On-line Mining

A multidimensional on-line mining algorithm based on the proposed multidimensional pattern relation is developed to efficiently fulfill diverse mining requests. The proposed algorithm can easily find large itemsets satisfying user-concerned context constraints and minimum supports, as long as the minimum supports are larger or equal to the initial minimum support used in the construction of the multidimensional pattern relation. The proposed algorithm first selects the tuples satisfying the context constraints in the mining request from the multidimensional pattern relation and then generates the candidate itemsets. The possible maximum supports for the candidate itemsets are also calculated. After that, according to the calculated possible maximum supports, the number of candidate itemsets is reduced by two strategies. Finally, for the remaining candidate itemsets, the proposed algorithm utilizes an Apriori-like mining process to find the final large itemsets and then derives the association rules. The details are described below.

5.1 Generation of Candidate Itemsets with Possible Maximum Supports

When processing a new mining request, the proposed algorithm first selects the tuples in the multidimensional pattern relation, with their contexts matching the user's requirements. All the itemsets kept in the field of the pattern sets in the matched tuples form the set of candidate itemsets. If a candidate itemset appears in all the matched tuples, its actual support can be easily calculated from the kept mining information. In this case, the possible maximum support of a candidate itemset is its actual support. On the contrary, if a candidate itemset does not appear in at least one matched tuple, its actual support cannot be obtained unless the underlying blocks of data for the matched tuples in which the candidate itemset does not appear are reprocessed. In this case, the initial minimum support used in the construction of the multidimensional pattern relation is therefore assumed as the possible maximum support of a candidate itemset in these matched tuples. Let A_x be the set of matched tuples in which an itemset x appears in pattern sets and B_x be the set of matched tuples in which x does not appear in pattern sets.The *possible maximum support* s_x^{max} of x can be thus calculated by the following formula:

$$s_x^{max} = \frac{\sum_{t_i \in A_x} t_i.trans * t_i.ps.x.support + \sum_{t_i \in B_x} (t_i.trans * s_{min}) - 1}{\sum_{t_i \in matched\ tuples} t_i.trans}, \quad (1)$$

where t_i is the i-th matched tuple, $t_i.trans$ is the number of transactions kept in t_i, $t_i.ps$ is the pattern set in t_i, $t_i.ps.x.support$ is the support of x in t_i, and s_min is the initial minimum support in the construction of the multidimensional pattern relation. The first term in the numerator straightly sums up the counts of the candidate itemsets appearing in the matched tuples. The second term sums up the possible maximum counts of the candidate itemsets not appearing in the matched tuples.

Example 2: For the multidimensional pattern relation given in Table 1, assume a mining request q is to get the patterns under the contexts of *Region* = **CA** and *Time* = **2002/11~2002/12** and satisfying the minimum support = 5.5%. The matched tuples are shown in Table 2. The set of candidate itemsets is $\{\{A\}, \{B\}, \{C\}, \{AB\}\}$, which is the union of the itemsets appearing in the pattern sets and with their supports larger than 5.5%. The possible maximum supports of these candidate itemsets are then calculated as follows:

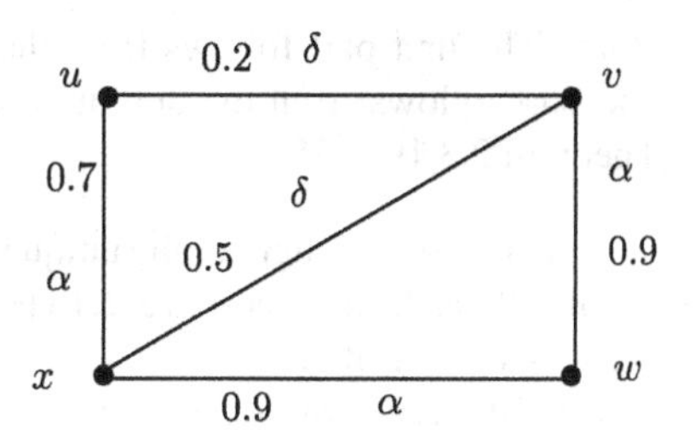

Fig. 3.11 A fuzzy tree with δ^* edge

Also, because F is a subgraph of G,

$$CONN_F(x, y) \leq CONN_{G-xy}(x, y). \quad (3.2)$$

From (3.1) and (3.2), $\mu(xy) < CONN_{G-xy}(x, y)$, which implies that xy is a δ-edge, which is a contradiction. Thus, G contains no β-strong edges.

Conversely, suppose that G is connected and has no β-strong edges. If G has no cycles, then G is a fuzzy tree. Now, assume that G has cycles. Let C be a cycle in G. Then C will contain only α-strong edges and δ-edges. Also, all edges of C cannot be α-strong because otherwise it will contradict the definition of α-strong edges. Thus, there exists at least one δ-edge in C. Then by Theorem 2.3.1, it follows that G is a fuzzy tree. ■

Thus, all strong edges of a fuzzy tree are α-strong and hence Proposition 3.1.5 can be restated as follows.

Theorem 3.2.15 *An edge xy in a fuzzy tree $G = (\sigma, \mu)$ is α-strong if and only if xy is an edge of the spanning tree $F = (\sigma, \nu)$ of G.*

A fuzzy tree can have δ^*-edges as seen from the following example.

Example 3.2.16 Let $G = (\sigma, \mu)$ with $\sigma^* = \{u, v, w, x\}$, $\sigma(s) = 1$ for all $s \in \sigma^*$ and $\mu(uv) = 0.2$, $\mu(xu) = 0.7$, $\mu(vw) = 0.9 = \mu(wx)$, $\mu(vx) = 0.5$. Then G is a fuzzy tree with vw, wx and xu as α-strong and vx and uv as δ-edges. Also, vx is a δ^*-edge because $\mu(vx) > \mu(uv)$, where uv is a weakest edge of G (Fig. 3.11.)

Theorem 3.2.17 *G is a fuzzy tree if and only if there exists a unique α-strong path between any two vertices in G.*

Proof The proof follows from Proposition 3.1.8 and Theorem 3.2.14. ■

G is a fuzzy tree if and only if G has a unique MST and all edges in the MST are α-strong edges. In general, we have the following theorem.

Theorem 3.2.18 *Let T be any spanning tree of a fuzzy graph G. Then T is an MST of G if and only if T contains no δ-edges. Further, an MST T is unique for G if and only if T contains no β-strong edges.*

Proof The first part follows from the definitions of δ edge and MST and the second part follows from the definition of β-strong edge, Theorem 3.2.15 above and Theorem 2.3.19. ■

Note that the strength of the unique x-y path in any MST of G gives $CONN_G(x, y)$ and it follows from Theorem 3.2.18 that there exists strong x-y path between any two vertices x and y of G.

Next the types of edges in fuzzy cycles are discussed. It is observed that there are no δ-edges in a fuzzy cycle G. For, if uv is a δ-edge in G, then it becomes the unique weakest edge of G, which contradicts that G is a fuzzy cycle. Also, a fuzzy cycle cannot have all its edges α-strong because the weakest edges in the fuzzy cycle cannot be α-strong and note that these weakest edges are β-strong edges and all other edges are α-strong. This leads to the following theorem.

Theorem 3.2.19 *Let G be a fuzzy graph such that G^* is a cycle. Then G is a fuzzy cycle if and only if G has at least two β-strong edges.*

Note that in a fuzzy graph G such that G^* is a cycle, w is a fuzzy cutvertex if and only if it is a common vertex of at least two fuzzy bridges and using Theorem 3.2.9 we have the following theorem.

Theorem 3.2.20 *Let G be a fuzzy graph such that G^* is a cycle. If G contains at most one α-strong edge, then G has no fuzzy cutvertices.*

Converse of Theorem 3.2.20 is not true. The condition for the converse to be true is given in the following theorem whose proof is obvious.

Theorem 3.2.21 *If there exists a unique strongest path between any two vertices x, and y in a fuzzy graph G, then it is a strong $x - y$ path.*

Now, we discuss types of edges in a complete fuzzy graph (CFG). In the following results, the number of β-strong edges in a CFG is calculated and the existence of a β-strong path between any two vertices of a CFG is proved. In a complete graph, there are no bridges, but a CFG may contain fuzzy bridges. Hence, we have the following two lemmas.

Lemma 3.2.22 *A complete fuzzy graph has no δ-edges.*

Proof Let G be a complete fuzzy graph. If possible assume that G contains a δ-edge uv (say). Then

$$\mu(uv) < CONN_{G-uv}(u, v).$$

That is, there exists a stronger path P other than the edge uv from u to v in G. Let $\mu(uv) = p$ and the strength of the path P be q. Then $p < q$. Let w be the first vertex in P after u. Then

$$\mu(uw) > p. \tag{3.3}$$

Table 2. The matched tuples in Example 2

ID	Region	Branch	Time	No_Trans	No_Patterns	Pattern_Sets (Itemset, Support)
2	CA	San Francisco	2002/11	15000	3	$(A,5\%),(B,7\%),(C,5\%)$
3	CA	San Francisco	2002/12	12000	2	$(A,5\%),(C,9\%)$
5	CA	Los Angeles	2002/11	25000	2	$(A,5\%),(C,6\%)$
6	CA	Los Angeles	2002/12	30000	4	$(A,6\%),(B,6\%),(C,9\%),$ $(AB,5\%)$

$$s_A^{max} = \frac{15000*5\%+12000*5\%+25000*5\%+30000*6\%}{15000+12000+25000+30000} = 0.0537,$$

$$s_B^{max} = \frac{15000*7\%+12000*5\%-1+25000*5\%-1+30000*6\%}{15000+12000+25000+30000} = 0.0573,$$

$$s_C^{max} = \frac{15000*5\%+12000*9\%+25000*6\%+30000*9\%}{15000+12000+25000+30000} = 0.0735, \text{ and}$$

$$s_{AB}^{max} = \frac{15000*5\%-1+12000*5\%-1+25000*5\%+30000*5\%}{15000+12000+25000+30000} = 0.04996.$$

5.2 Pruning of Candidate Itemsets

The candidate itemsets generated will be reduced by two strategies. The first strategy removes the candidate itemsets whose possible maximum supports are less than the minimum support given in the new mining request. This strategy also uses the *monotone* property of itemsets [3][20] to increase the pruning efficiency. That is, if an itemset x is removed, all the proper supersets of x are also directly removed. The second strategy is used for the candidate itemsets which appear in all the matched tuples. If a candidate itemset appears in all the matched tuples and its possible maximum support is larger than or equal to the minimum support in the mining request, then it is directly put in the set of final large itemsets.

Example 3: Continuing with Examples 2, the first pruning strategy will remove the candidate itemsets $\{A\}$ and $\{AB\}$ since their possible maximum supports are less than 5.5%. The second pruning strategy will put $\{C\}$ in the set of final large itemsets since it appears in all the matched tuples and its possible maximum support is larger than 5.5%. The remaining candidate itemset $\{B\}$ will be further processed.

5.3 Generation of Large Itemsets and Association Rules

Each remaining candidate itemset has enough possible maximum support (larger than or equal to the minimum support given in the new mining request) but does not appear in at least one matched tuple. The proposed algorithm thus has to re-process the underlying blocks of data for these tuples to get their actual supports. The support of a remaining itemset x can easily be calculated by the following formula:

$$s_x = \frac{\sum_{t_i \in A_x} t_i.trans * t_i.ps.x.support + \sum_{t_i \in B_x} t_i.x.count}{\sum_{t_i \in matched\ tuples} t_i.trans}, \qquad (2)$$

where $t_i.x.count$ is the actual support of x obtained by re-processing the blocks of data indicated by t_i and the other terms are the same as before. Like the Apriori mining algorithm, this phase processes the candidate itemsets in a level-wise way. It first processes candidate 1-itemsets. If the actual support of a candidate 1-itemset is less than the minimum support in the mining request, it and its proper supersets will be removed. Otherwise, the 1-itemset is a large itemset for the mining request. This procedure is then repeated for itemsets with more items until all the remaining itemsets have been processed. After the final large itemsets are found, the association rules can then easily be generated from them.

5.4 The Proposed Multidimensional On-line Mining Algorithm

The proposed multidimensional online mining algorithm is described below, where STEPs 1 to 3 are executed for generation of candidate itemsets, STEPs 4 to 6 are executed for pruning of candidate itemsets, and STEPs 7 to 11 are executed for generation of large itemsets and association rules.

The Proposed Multidimensional Online Mining Algorithm:

INPUT: A multidimensional pattern relation based on an initial minimum support s_{min} and a new mining request q with a set of contexts cx_q, a minimum support s_q and a minimum confidence $conf_q$.

OUTPUT: A set of association rules satisfying the mining request q.

STEP 1: Select the set of tuples M which match cx_q.

STEP 2: Set the union of the itemsets appearing in M as the set of candidate itemsets C.

STEP 3: Calculate the possible maximum support s_c^{max} of each candidate itemset c by Eq.(1).

STEP 4: Set $k = 1$, where k is used to keep the number of items in a candidate itemset currently being processed.

STEP 5: Do the following substeps for each c_k in C:

Substep:5-1: If $s_{c_k}^{max} < s_q$, then remove the candidate itemsets including c_k from C. That is, $C = C$- $\{c_h — c_k \subseteq c_h, c_h \in C\}$.

Substep:5-2: If c_k appears in all the tuples in M and $s_{c_k}^{max} \geq s_q$, then put it in the large itemsets. That is, $L = L \cup \{c_k\}$ and $C = C - \{c_k\}$.

STEP 6: Set $k = k + 1$, and repeat STEPs 5 and 6 until all the candidate itemsets are processed.

STEP 7: Set $k = 1$, where k is used to keep the number of items in a candidate itemset currently being processed.

STEP 8: For each c_k in C_k, re-process each underlying block of data D_i for tuple t_i in which c_k does not appear to calculate the actual support of c_k by Eq.(2).

STEP 9: If $s_{c_k} < s_q$, then remove the candidate itemsets including c_k from C, that is $C = C - \{c_h | c_k \subseteq c_h, c_h \in C\}$. Otherwise, $L = L \cup \{c_k\}$.

STEP 10: Set $k = k + 1$, and repeat STEPs 8 and 10 until all the candidate itemsets are processed.

STEP 11: Drive association rules with confidence values larger than or equal to $conf_q$ from the set of large itemsets L.

6 Experiments

Before showing the experimental results, we first describe the experimental environments and the datasets used.

6.1 The Experimental Environments and The Datasets Used

The experiments were implemented in Java on a workstation with dual XEON 2.8GHz processors and 2048MB main memory, running RedHat 9.0 operation system. The datasets were generated by a generator similar to that used in [3]. The parameters listed in Table 3 were considered when generating the datasets.

Table 3. The parameters considered when generating the datasets

Parameter	Description
D	The number of transactions
N	The number of items
L	The number of maximal potentially large itemsets
T	The average size of items in a transaction
I	The average size of items in a maximal potentially large itemset

The generator first generated L maximal potentially large itemsets, each with an average size of I items. The items in a potentially large itemset were randomly chosen from the total N items according to its actual size. The generator then generated D transactions, each with an average size of T items. The items in a transaction were generated according to the L maximal potentially large itemsets in a probabilistic way. The details of the dataset generation process can be referred to in [3].

Two groups of datasets generated in the above way and used in our experiments are listed in Table 4, where the datasets in the same group had the same D, T and I values but different L or N values. Each dataset was treated as a block of data in the database. For example, Group 1 in Table 4 contained ten blocks of data named from $T10I8D10KL^1$ to $T10I8D10KL^{10}$, each of which consisted of 10000 transactions of average 10 items generated according to 200 to 245 maximal potentially large itemsets with an average size of 8 from totally 100 items.

Table 4. The two groups of datasets generated for the experiments

Group	Size	Datasets	D	T	I	L	N
1	10	T10I8D10KL^1 to T10I8D10KL^{10}	10000	10	8	200 to 245	100
2	10	T20I8D100KL^1 to T20I8D100KL^{10}	100000	20	8	400 to 490	200

6.2 The Experimental Results

Each group of datasets in Table 4 was first used to derive its corresponding multidimensional pattern relation. When the initial minimum support was set at 2%, the mining results for the two groups are summarized in Table 5.

Table 5. The summarized mining information for the two groups

Group	Initial minimum support	Average length of maximal large itemsets from a dataset	Average size of large itemsets from a data set
1	2%	11	90061
2	2%	9	121272

The proposed multidimensional on-line mining algorithm and the Apriori algorithms were then run for Groups 1 to 2 along with different minimum supports ranging from 0.022 to 0.04 in the mining requests. The execution times spent by the two algorithms for each group are respectively shown in Fig. 3 and Fig. 4. It is easily seen that the execution time by the proposed algorithm was always much less than that by the Apriori algorithm.

7 Conclusion

In this paper, we have extended the concept of effectively utilizing previously discovered patterns in incremental mining to online decision support under the multidimensional context considerations. By structurally and systematically retain the additional circumstance and mining information in the multidimensional pattern relation, our proposed multidimensional on-line mining

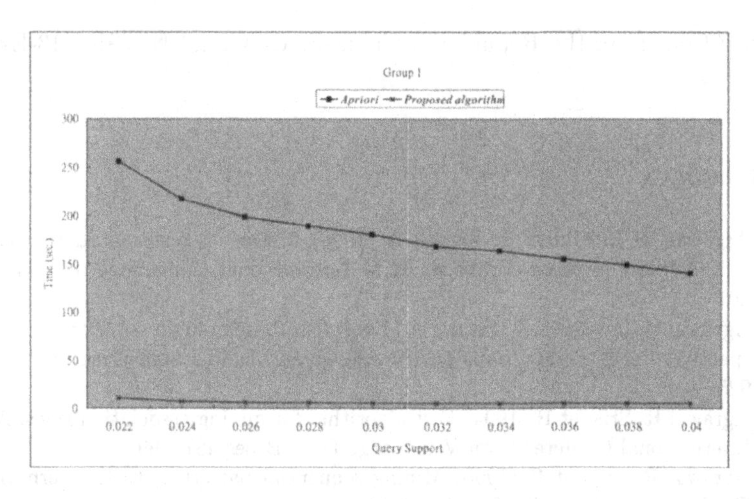

Fig. 3. The execution time spent by the two experimental algorithms for Group 1

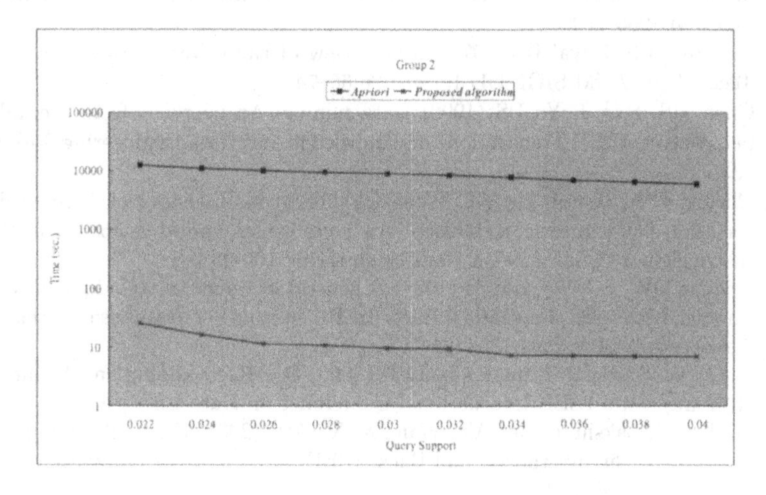

Fig. 4. The execution time spent by the two experimental algorithms for Group 2

algorithm can easily and efficiently derive the association rules satisfying diverse user-concerned constraints. Therefore, the users' on-line mining requests can be rapidly fulfilled.

8 Acknowledgement

This research was supported by MOE Program for Promoting Academic Excellence of Universities under the grant number 89-E-FA04-1-4 and National

Science Council of the Republic of China under Grand No. 91-2213-E-009-007-.

References

1. Agrawal R, Imielinksi T, Swami A (1993) Mining association rules between sets of items in large database, ACM International Conference on SIGMOD 207–216
2. Agrawal R, Imielinksi T, Swami A (1993) Database mining: a performance perspective. IEEE Transactions on Knowledge and Data Engineering 5(6): 914—925
3. Agrawal R, Srikant R (1994) Fast algorithm for mining association rules. ACM International Conference on Very Large Data Bases 487–499
4. Agrawal R, Srikant R (1995) Mining sequential patterns. IEEE International Conference on Data Engineering 3–14
5. Brin S, Motwani R, Ullman JD, Tsur S (1997) Dynamic itemset counting and implication rules for market basket data. ACM International Conference on SIGMOD 255–264
6. Chaudhuri S, Dayal U (1997) An Overview of Data Warehousing and OLAP Technology. ACM SIGMOD Record 26:65–74
7. Chen MS, Han J, Yu PS (1996) Data mining: An overview from a database perspective. IEEE Transactions on Knowledge and Data Engineering 8(6) 866–883
8. Cheung DW, Han J, Ng VT, Wong CY (1996) Maintenance of discovered association rules in large databases: An incremental updating approach. IEEE International Conference on Data Engineering 106–114
9. Cheung DW, Lee SD, Kao B (1997) A general incremental technique for maintaining discovered association rules. In Proceedings of Database Systems for Advanced Applications 185–194
10. Ganti V, Gehrke J, Ramakrishnan R (2000) DEMON: Mining and Monitoring Evolving Data. IEEE International Conference on Data Engineering
11. Grahne G, Lakshmanan L VS, Wang X, Xie MH (2001) On Dual Mining: From Patterns to Circumstances, and Back. IEEE International Conference on Data Engineering 195–204
12. Han J, Lakshmanan LVS, Ng R (1999) Constraint-based, multidimensional data mining. IEEE Computer Magazine 2–6
13. Han J, Kamber M (2001) Data Mining: Concepts and Techniques, Morgan Kaufmann Publishers
14. Hong TP, Wang CY, Tao YH (2000) Incremental data mining based on two support thresholds. International Conference on Knowledge-Based Intelligent Engineering Systems & Allied Technologies
15. Hong TP, Wang CY, Tao YH (2001) A new incremental data mining algorithm using pre-large itemsets. International Journal on Intelligent Data Analysis
16. Immon WH (1996) Building the Data Warehouse, Wiley Computer Publishing
17. Lakshmanan LVS, Ng R, Han J, Pang A (1999) Optimization of Constrained Frequent Set Queries with 2-variable Constraints. ACM International Conference on SIGMOD 157–168

18. Lin MY, Lee SY (1998) Incremental update on sequential patterns in large databases. IEEE International Conference on Tools with Artificial Intelligence 24–31
19. Mannila H, Toivonen H, Verkamo AI (1994) Efficient algorithm for discovering association rules. The AAAI Workshop on Knowledge Discovery in Databases 181–192
20. Ng R, Lakshmanan LVS, Han J, Pang A (1998) Exploratory Mining and Pruning Optimizations of Constrained Associations Rules. ACM International Conference on SIGMOD 13–24
21. Park JS, Chen MS and Yu PS (1997) Using a hash-based method with transaction trimming for mining association rules. IEEE Transactions on Knowledge and Data Engineering 9(5) 812–825
22. Pinto H, Han J, Pei J, Wang K (2001) Multi-dimensional Sequential Pattern Mining. ACM International Conference on Information and Knowledge Management 81–88
23. Sarda NL, Srinivas NV (1998) An adaptive algorithm for incremental mining of association rules. IEEE International Workshop on Database and Expert Systems 240–245
24. Savasere A, Omiecinski E, Navathe S (1995) An efficient algorithm for mining association rules in large database. ACM International Conference on Very Large Data Bases 432–444
25. Srikant R, Agrawal R (1995) Mining sequential patterns: generalizations and performance improvements. ACM International Conference on Knowledge Discovery and Data Mining 269–274
26. Thomas S, Bodagala S, Alsabti K, Ranka S (1997) An efficient algorithm for the incremental update of association rules in large databases. ACM International Conference on Knowledge Discovery and Data Mining 263–266
27. Wang CY, Hong TP, Tseng SS (2001) Maintenance of sequential patterns for record deletion. IEEE International Conference on Data Mining 536–541
28. Wang CY, Hong TP, Tseng SS (2002) Maintenance of sequential patterns for record modification using pre-large sequences. IEEE International Conference on Data Mining 693–696
29. Widom J (1995) Research Problems in Data Warehousing. ACM International Conference on Information and Knowledge Management

Quotient Space Based Cluster Analysis[1]

[1,3]Ling Zhang and [2,3]Bo Zhang

[1]Artificial Intelligence Institute, Anhui University, Anhui, China
[2]Computer Science & Technology Department, Tsinghua University, Beijing, China
[3]State Key Lab of Intelligent Technology & Systems, Tsinghua University

Abstract: In the paper, the clustering is investigated under the concept of granular computing, i.e., the framework of quotient space theory. In principle, there are mainly two kinds of similarity measurement used in cluster analysis: one for measuring the similarity among objects (data, points); the other for measuring the similarity between objects and clusters (sets of objects). Therefore, there are mainly two categories of clustering corresponding to the two measurements. Furthermore, the fuzzy clustering is gained when the fuzzy similarity measurement is used. From the granular computing point of view, all these categories of clustering can be represented by a hierarchical structure in quotient spaces. From the hierarchical structures, several new characteristics of clustering can be obtained. It may provide a new way for further investigating clustering.

Keywords: Cluster analysis, granular computing, quotient space theory, hierarchical structure, fuzzy clustering

1. Introduction

In machine learning, there exist two basic problems: classification and clustering. In classification, there are many well-known theories and approaches such as SVM, neural networks, etc. In cluster analysis, there are many successful approaches as well. For example, partitioning method [1][2], density-based [3][4], k-means [5], k-nearest neighborhood [4], neural networks [6], etc. In despite of the existence of different clustering approaches, the aim of the clustering is to group the objects (or data, points) in a space into

[1] Supported by Chinese National Nature Science Foundation (Grant No. 60135010)

clusters such that objects within a cluster are similar to each other and objects in different clusters have a high degree of dissimilarity.

From granular computing viewpoint, the objects within a cluster can be regarded as an equivalence class. Then, a clustering of objects in space X corresponds to an equivalence relationship defined on the space. From the quotient space theory [7][8], it's known that it corresponds to constructing a quotient space of X. In each clustering algorithm, it's needed to define a specific metric to measure the similarity (or dissimilarity) among objects. So far various metrics have been adopted such as Euclidean distance, Manhattan distances, inner product, fuzzy membership function, etc. No matter what kind of measurement is used, in principle, there are basically two kinds: one for measuring the similarity among objects (or data, points), the other for measuring the similarity between objects and clusters (sets of objects).

From the above viewpoint, some clustering methods can be restated below.

Partitioning clustering: Given a universe X, a similarity function s(x, y) (≥0), and a threshold s. X is assumed to be partitioned into subsets $V_1, V_2, ..., V_k$ satisfying (1) $V_1, V_2, ..., V_k$ is a partition of X, (2) $\forall x, y \in X,\ if\ \ s(x,y) \geq s,$ then x and y belong to the same cluster.

k-means method: Given a set V in a distance space, and an integer k. V is grouped into k subsets $V_1, V_2, ..., V_k$. For each subset V_i, $x \in V_i \Leftrightarrow s(x, a_i) = \min\{s(x, a_j), j = 1, 2, ..., k\}$, where a_i is the center of set V_i. In the method, the similarity between object x and a cluster represented by a_i is used as well.

Density-based clustering CURD [9]: Clustering using references and density (CURD) is a revised version of CURE [10][11]. It is a bottom-up hierarchical clustering algorithm. First, taking each object (point) as a clustering center, then the most similar points are gradually grouped into clusters until k clusters are obtained. In this algorithm, the similarity function is described by two variables, so it is a multivariate clustering. Its clustering process will be stated in session 3.

Furthermore, the fuzzy clustering is discussed based on the fuzzy equivalence relation.

From the granular computing point of view, all these categories of clustering can be represented by a hierarchical structure in quotient spaces. From the hierarchical structures, several new characteristics of clustering can be obtained. It may provide a new way for further investigating clustering.

2. Univariate Cluster Analysis

A clustering with univariate measure is called a univariate clustering. Taking univariate partitioning clustering as an example, we discuss its similarity function below

.

Definition 2.1. Given a universe X, a similarity function s(x, y), and a threshold s. X is assumed to be partitioned into several subsets and satisfies

(1) The sufficient condition: $\forall x, y \in X,\ \ if\ \ s(x,y) \geq s,$ then x and y belong to the same cluster.

(2) The necessary condition: if x and y belong to the same cluster, there must exist $x = x_1, x_2, \ldots, x_n = y$ such that $s(x_i, x_{i+1}) \geq s$.

(3) X is partitioned into the union of mutual disjoint subsets, $X = S_1 \cup S_2 \ldots \cup S_k,\ \ S_i \cap S_j = \varnothing, i \neq j$.

A clustering with s(x, y) as its similarity function is called a univariate partitioning clustering.

Given a similarity function s(x, y) and a threshold s, then we have a clustering denoted by C(s), $0 \leq s \leq S$. When X is finite, S is finite either. Obviously, C(s) is a partition of X and is equivalent to an equivalence relation on X. An equivalence relation given on X corresponds to a given quotient space of X from quotient space theory [7][8]. Thus, the quotient space theory can be used to deal with clustering. Some concepts and properties of quotient spaces are shown below.

Definition 2.2. Given a universe X. R is an equivalence relation on X. Let $[x] = \{y | yRx,\ \ y \in X\},\ \ [X]_R = \{[x] | x \in X\}$, where $[X]_R$ is called a quotient space of X corresponding to R, or denoted by [X] for simplicity, xRy denotes that x and y are equivalent.

R denotes all equivalence relations on X. Define a relation "<" on **R** as $R_1 R_2 \in \mathbf{R},\ R_2 < R_1 \Leftrightarrow x, y \in X,\ \ if\ \ xR_1 y\ \ then\ \ xR_2 y$.

Proposition 2.1. **R** composes a semi-order lattice under relation "<".

We will show below that C(s), $0 \le s \le S$, composes a hierarchical structure.

Proposition 2.2. Since C(s) is a clustering on X, C(s) is a quotient space of X and corresponds to an equivalence relation R(s). Given $0 \le s_1 < s_2 \le S$, R(s_1) and R(s_2) are equivalence relations corresponding to C(s_1) and C(s_2), respectively. Obviously, if xR(s_2)y, from definition 2.1, there exist $x = x_1, x_2, ..., x_n = y$ such that $s(x_i, x_{i+1}) \ge s_2$. Since $s(x_i, x_{i+1}) \ge s_2 \ge s_1$, xR($s_1$)y, i.e., R($s_2$)<R($s_1$). Thus, C($s_2$) is a quotient space of C(s_1). $\{C(s),\ 0 \le s \le S\}$ is a hierarchical structure on X.

From the hierarchical structure, we get the following interesting geometrical view of clustering. If [X] is obtained from X by some clustering, then [X] is a quotient space of X. [X] is coarser and has less information than X. Therefore, in [X] some properties of X may lose. For example, if X is a connected set in a distance space, [X] may not necessarily be a connected set under its corresponding quotient topology. But only in some coarser sense, the properties can be preserved.

Definition 2.3. Assume X is a distance space. X is called as d-connected (d>0), if X can't be partitioned into the union of two subsets satisfying the condition: $X = A \cup B,\ \bar{D}(A, d/2) \cap \bar{D}(B, d/2) = \varnothing$, where $\bar{D}$ is a closure of D,

$D(A,d) = \{y | \exists x \in A,\ d(x,y) \le d/2\}$, d(x, y) the distance between x and y.

From the definition, it's known that an d-connected set X means that a pair of points in X is regarded as connected if their distance is less than and equals to d. In clustering, the similarity function s can be defined as the reciprocal of distance d. From granular computing point of view, the large the d, the bigger the d-connected set and the coarser the quotient space C(s) obtained from clustering.

3. Multivariate Clustering

A clustering with multivariate measure is called as a multivariate clustering. We discuss the CURD clustering with two-variable measurement [9].

Definition 3.1. Given a set V in a distance space X, a radius r, and a density threshold t. For $\forall p \in V$, ϕ(p, r) is a supper-sphere with p as its center and r as its radius. Density D(p, r) is the number of points within ϕ(p, r) in V. The point p is called a reference point with respect to (r, t), when D(p, r)>t.

Definition 3.2. Given r and t. If the distance d(p, q) between reference points p and q is less than and equals to 2r, p and q are called a pair of neighboring reference points.

Definition 3.3. If p is a reference point, the supper-sphere ϕ(p, r) is called a representative region of p.

CURD clustering algorithm:

Given r and t. For a set V, find all its reference points and the connected components $V_1, V_2, ..., V_k$ consisted of the representative regions of the reference points. Thus, $V = V_1 \cup V_2 \cup ... \cup V_k \cup \{V_0\}$, $V_0 = V/(V_1 \cup ... \cup V_k)$, {$V_0$} denotes a set of clusters, where each point in V_0 is regarded as a cluster. Components $V_1, V_2, ..., V_k$ are the clusters grouped by CURD.

CURD is a clustering with two variables r and t. By using the propositions presented in session 2, we have the following constructive theorem of CURD.

Proposition 3.1. Given r, t, and a set V of X. We have a clustering C(r, t) from V by CURD. Fixed r, $\{C(r,t),\ 0 \le t \le k\}$ is a hierarchical structure of X. Similarly, fixed t, $\{C(r,t),\ 0 \le r \le b\}$ is a hierarchical structure as well.

Proof: Fixed r, assume $t_1<t_2$. If point p is a reference point with respect to (r, t_2), it is much more a reference point under (r, t_1). Assume that x and y are equivalent in C(r, t_2), i.e., x and y belong to the same connected component in C(r, t_2). There exist $x = x_1, x_2, ..., x_n = y$ and reference points $p_1, p_2, ..., p_n$ such that

$$x_i \in (\phi(p_i, r) \cap \phi(p_{i+1}, r)),\ i = 1,2,...,n-1,\ x \in \phi(p_1, r), y \in \phi(p_n, r) \quad (3.1)$$

Since $t_1<t_2$, $p_1, p_2, ..., p_n$ are reference points with respect to (r, t_1). From form (3.1), x and y belong to the same component in C(r, t_1), i.e., x and y are equivalent in C(r, t_1) either. Thus, C(r, t_1) is a

quotient space of $C(r, t_2)$. Finally, $\{C(r,t),\ 0 \le t \le k\}$ is a hierarchical structure on X, where k is the cardinality of set V.

Fixed t, given $r_1 < r_2$. The reference points with respect to (r_1,t) must be the reference points with respect to (r_2,t). The volume of the representative regions of the latter is not less than the former. And the volume of the connected components of the latter is not less than the former. Thus, given x and y, if they are equivalent in $C(r_1, t)$, then they are equivalent in $C(r_2, t)$ as well. $C(r_2, t)$ is a quotient space of $C(r_1, t)$. #

Letting s=1/r, we have C(s, t) instead of C(r, t) in order that the smaller the variable s the coarser the space is and we normalize variable s such that s∈[0, 1].

Definition 3.4. Assume that $B[a_1,a_2;b_1,b_2]$ is a rectangle in a Euclidean plane. Define a semi-order "<": $x=(x_1,x_2)<y=(y_1,y_2) \Leftrightarrow x_1 \le y_1$ *and* $x_2 \le y_2$. We have a set of semi-order rectangles

Theorem 3.1. Assume that $C[0,1;1,k]=\{C(s,t)|0 \le s \le 1, 1 \le t \le k\}$ is a set of clusters obtained from CURD clustering and is a set V of a distance space X. By transforming $T: C[0,1;1,k] \to B[0,1;1,k]$, $T(C(s,t))=(s,t)$, then C[0,1; 1,k] and semi-order rectangle B[0,1; 1,k] are homomorphism. If $(s_1, t_1)<(s_2, t_2)$ in B[0,1; 1,k], then $C(s_1, t_1)$ is a quotient space of $C(s_2, t_2)$ in B[0,1; 1,k].

$\{C(s,t)|0 \le s \le 1, 1 \le t \le k\}$ is a hierarchical structure with two variables.

4. Fuzzy Cluster Analysis

We first introduce some basic concepts of fuzzy set.

Definition 4.1. X is a universe. A fuzzy set A on X is defined as: ∀x∈X, given $\mu_A \in [0,1]$, μ_A is called a membership of x with respect to A. Map $\mu_A : x \to [0,1], x \to \mu_A(x)$ is called a membership function. Let T(X) be a set of all fuzzy subsets on X. T(X) is a function space consisted of functions $\mu : X \to [0,1]$.

Definition 4.2. $R \in T(X \times X)$ is called a fuzzy equivalence relation on X, if

(1) $\forall x \in X$, $R(x, x)=1$
(2) $\forall x, y \in X$, $R(x, y)=R(y, x)$
(3) $\forall x, y, z \in X$, $R(x, z) \geq \sup_y (\min(R(x, y), R(y, z))$

Note that if $R(x, y) \in \{0,1\}$, then R is a common equivalence relation.

Proposition 4.1. Assume R is a fuzzy equivalence relation. Define $\forall x, y \in X, x \sim y \Leftrightarrow R(x,y)=1$. Then "~" is a common equivalence relation and let [X] be the corresponding quotient space.

Theorem 4.1. R is a fuzzy equivalence relation and [X] is a quotient space defined in proposition 4.1. Define $\forall a, b \in [X],\ d(a,b)=1-R(x,y), \forall x \in a, y \in b$ (4.1)

Then d(.,.) is a distance function on [X] and called a distance function with respect to R.

Proposition 4.2. R is a fuzzy equivalence relation on X. Let $R_\lambda = \{(x,y) | R(x,y) \geq \lambda\}$, $0 \leq \lambda \leq 1$. R_λ is a common equivalence relation and called as a cut relation of R.

Definition 4.3. For a normalized distance space (*X*, *d*), i.e., $\forall a,b \in [X], d(a,b) \leq 1$, if any triangle formed by connecting any three points on X not in a straight line is an equicrural triangle and its crus is the longest side of the triangle, the distance is called as an equicrural distance.

Proposition 4.3. If d is a normalized distance with respect to fuzzy equivalence relation R, then d is equicrural distance.

Theorem 4.2. Assume [X] is a quotient space of X. Given a normalized equicrural distance function d(.,.) on [X]. Letting $\forall x, y \in X, R(x,y)=1-d(x,y)$, R(x, y) is a fuzzy equivalence relation on X.

Theorem 4.3. Assume that $\{X(\lambda) | 0 \leq \lambda \leq 1\}$ is a hierarchical structure on X. There exists a fuzzy equivalence relation R on X such that $X(\lambda)$ is a quotient space with respect to R_λ, where R_λ is the cut relation of R, $\lambda \in [0,1]$.

Theorem 4.4. The following three statements are equivalent, i.e.,
(1) Given a fuzzy equivalence relation on X,
(2) Given a normalized equicrural distance on some quotient space of X,
(3) Given a hierarchical structure on X.

Definition 4.4. Cut relation $R(\lambda)$ of R is a common equivalence relation and corresponds to a clustering on X, where R is a fuzzy equivalence relation on X. All possible clusters with respect to different R(λ), λ∈[0,1] are called a cluster set of R.

From theorem 4.3, it's known that the structure induced from fuzzy clustering based on fuzzy equivalence relation R and the structure induced from common partitioning clustering are the same. No matter what kind of similarity functions is used, all possible clusters obtained from a partitioning clustering must compose a hierarchical structure on X. The cluster set obtained from a fuzzy clustering and that obtained from a normalized equicrural distance are equivalent. In other word, a fuzzy clustering and a clustering based on distance function are equivalent.

Proposition 4.4. Let d(.,.) be a normalized equicrural distance on X, X be a distance space. We have a hierarchical structure $\{C(\lambda)|0 \le \lambda \le 1\}$ based on a partitioning clustering using λ=(1-d(.,.)) as its similarity function. In the structure, C(0) is the coarsest quotient space. ∀x, y, if x and y are equivalent ⇔ x and y belong to the same (1-λ)-connected component. C(λ) is a quotient space consisted of the (1-λ)-connected components in X.

From granular computing point of view, the larger the value (1-λ) the bigger the (1-λ)-connected set, and the coarser the corresponding quotient space. Therefore, different quotient space C(λ) consists of the connected components with different granularity.

Now we consider the multivariate fuzzy clustering. Under the same symbols used in theorem 3.1, we have

Theorem 4.5. There exist k normalized equicrural distance functions $d_1(.,.),...,d_k(.,.)$ on X such that x and y are equivalent on $C(s,t) \Leftrightarrow d_t(x,y) \le 1-s$.

Proof: Fixed t, from proposition 3.1, $\{C(s,t), 0 \le s \le 1\}$ is a hierarchical structure. From theorem 4.4, the structure can be represented by a normalized equicrural distance $d_t(.,.)$. Letting t be 1,2,…,k, we have $d_1(.,.),...,d_k(.,.)$.#

Theorem 4.6. Assume $\{d_{t0}, d_{t1}, ..., d_{tk}\}$ is a set of distance functions obtained from theorem 4.5. Then, $\forall x, y \in X, t_i < t_j$ *have* $d_{ti}(x,y) \le d_{tj}(x,y)$.

Proof: Given a hierarchical structure $\{C(s,t_j), 0 \le s \le 1\}$ with respect to t_j. For a pair of points x and y

$$d_{tj}(x,y)=\begin{cases}1-\inf\{s|x \ \ and \ \ y \ \ are \ \ non-equivalent \ \ in \ \ C(s,t_j)\} \\ 0, \ x \ \ and \ \ y \ \ are \ \ equivalent \ \ in \ \ any \ \ C(s,t_j)\end{cases}$$

Assume that $d_{tj}(x,y)=s^*$, i.e., x and y are not equivalent in $C(s^*,t_j)$ but are equivalent in C(s, t_j), s<s*. From proposition 3.1, fixed s<s*, $\{C(s,t_j),0\le t\le k\}$ is a hierarchical structure in X. From t_i<t_j, when s<s*, $C(s,t_j)$ is a quotient space of $C(s,t_i)$. Then, if x and y are equivalent in $C(s,t_j)$, they must be equivalent in $C(s,t_i)$. So $d_{ti}(x,y)\le 1-s^*\le d_{tj}(x,y)$. #

Theorem 4.6 shows that distance $d_t(.,.)$ is a monotonic function of t. The larger the distance the higher the resolution is. That is, the larger the t the finer the grain-size of clusters is. In other word, a hierarchical structure with two variables can be described by a set of monotonic distance functions. The structure can also be represented by a set of monotonic fuzzy equivalence relations based on theorem 4.4. Among the three representations, the hierarchical structure representation is essential, since the distance function and fuzzy equivalence relation representations are not unique. Therefore, the latter two representations have some redundancy.

Conclusions

In the paper, we discuss the clustering from granular computing. Generally, there are two kinds of similarity information: the similarity among objects, and the similarity between objects and clusters (sets of objects). Based on the quotient space theory, all possible clusters obtained from a univariate clustering algorithm compose a hierarchical structure {C(d)} on X. {C(d)} and line-segments are homomorphism. For example, in clustering in order to optimized parameter d, f(C(d)) may be used as an objective function. Similarly, all possible clusters obtained from a multivariable (two variables) clustering algorithm compose a hierarchical structure with two variables {C(d, t)}. {C(d, t)} are homomorphism with a set of semi-order rectangles on a plan. The optimization of the clustering can also be regarded as that of a function with two variables. Different clustering just corresponds to the connected components with different grain-size. The multi-granular view of clustering may provide a new way for its further investigation.

References

[1] MacQueen J, Some methods for classification and analysis of multivariate observations. Proc. 5th Berkley Symp. Math. Statist. Prob., 1967,1:281-297.

[2] C. J. Veenman et al, A maximum variance cluster algorithm, IEEE Trans. on PAMI, vol24, no9, Sept. 2002, 1273-1280.

[3] Jain A. K, Dubes R. C, Algorithms for clustering, Englewood, Cliffs N. J, Prentice Hall.1988.

[4] Jain A. K, Murry M. N, Flynn P. J, Data clustering: A survey, ACM Comput. Surve.,1999, 31: 264-323.

[5] Ester M, Kriegel HP, Sander J, Xu X.X, Density based algorithm for discovering clusters in large spatial databases with noise. In: Simoudis E, Han JW, Fayyad UM, eds., Proceedings of the 2nd International Conference on Knowledge Discovery and Data Mining, Portland : AAAI Press, 1996:226-231.

[6] J-H Wang, et al, Two-stage clustering via neural networks, IEEE Trans. on Neural Networks, vol14, no3, May 2003, 606-615.

[7] B. Zhang, L. Zhang, Theory and Application of Problem Solving, North-Holland Elsevier Science Publishers B.V. 1992

[8] Zhang, L. and Zhang, Bo. The quotient space theory of problem solving, Proceedings of International Conference on Rough Sets, Fuzzy Set, Data Mining and Granular Computing, LNCS 2639, 11-15, 2003.

[9] Ma, Shuai, et al, A clustering based on reference points and density (Chinese), Chinese Journal of Software, vol14, no6, 2003, 1089-1095.

[10] Guha S, Rastogi R, Shim K, CURE: An efficient clustering algorithm for large databases. In: Hass LM, Tiwary A, eds. Proceedings of the ACM SIGMOD International Conference on Management of Data Seattle: ACM Press, 1998:73-84.

[11] Agrawal R, Gehrke J, Gunopolos D, Raghavan P, Automatic subspace clustering of high dimensional data for data mining application. In: Hass L M, Tiwary A, eds. Proceedings of the ACM SIGMOD International Conference on Management of Data Seattle: ACM Press, 1998:94-105.

Part III

Novel Applications

Research Issues in Web Structural Delta Mining

Qiankun Zhao[1], Sourav S. Bhowmick[1], and Sanjay Madria[2]

[1] School of Computer Engineer, Nanyang Technological University, Singapore.
qkzhao@pmail.ntu.edu.sg, assourav@ntu.edu.sg
[2] Department of Computer Science, University of Missouri-Rolla, USA.
madrias@umr.edu

Summary. Web structure mining has been a well-researched area during recent years. Based on the observation that data on the web may change at any time in any way, some incremental data mining algorithms have been proposed to update the mining results with the corresponding changes. However, none of the existing web structure mining techniques is able to extract useful and hidden knowledge from the *sequence of historical web structural changes*. While the knowledge from snapshot is important and interesting, the knowledge behind the corresponding changes may be more critical and informative in some applications. In this paper, we propose a novel research area of web structure mining called **web structural delta mining**. The distinct feature of our research is that our mining object is the sequence of historical changes of web structure (also called **web structural deltas**). For web structural delta mining, we aim to extract useful, interesting, and novel web structures and knowledge considering their historical, dynamic, and temporal properties. We propose three major issues of web structural delta mining, *identifying useful and interesting structures, discovering associations from structural deltas,* and *structural change pattern based classifier*. Moreover, we present a list of potential applications where the web structural delta mining results can be used.

1 Introduction

With the progress of World Wide Web (WWW) technologies, more and more data are now available online for web users. It can be observed that web data covers a wide spectrum of fields from governmental data via entertaining data, commercial data, etc. to research data. At the same time, more and more data stored in traditional repositories are emigrating to the web. According to most predictions in 1999, the majority of human data will be available on the web in ten years [9]. The availability of huge amount of web data does not imply that users can get whatever they want more easily. On the contrary, the massive amount of data on the web has already overwhelmed our abilities to find the desired information. It has been observed that 99% of the data reachable on the web is useless to 99% of the users [10]. However, the huge and diverse properties of web data do imply that there should be useful knowledge hidden behind web data, which cannot be easily interpreted by human intuition.

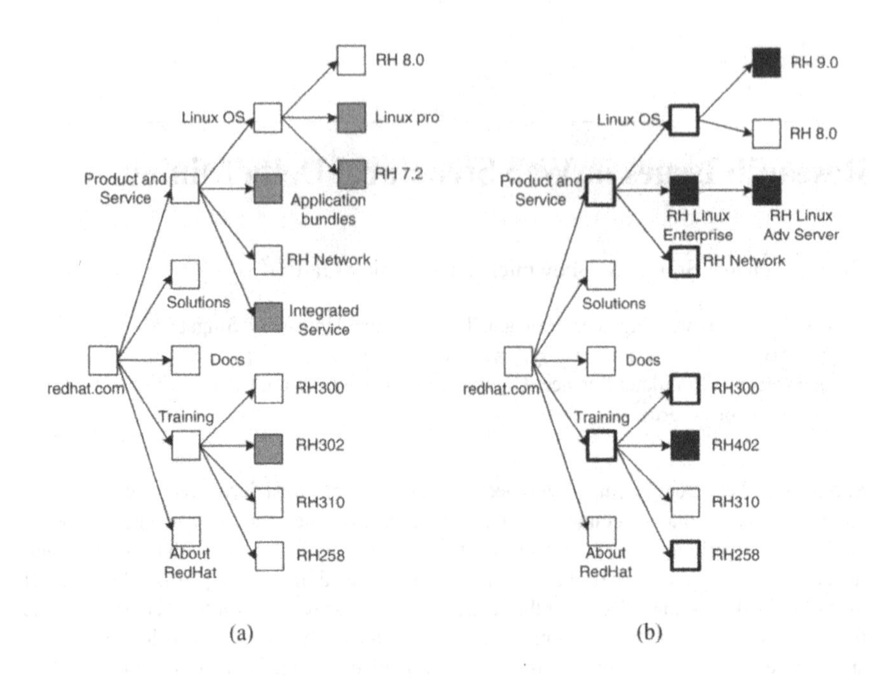

Fig. 1. Two Versions of the Web Site Structure of redhat.com

Under such circumstance, web mining was initially introduced to automatically discover and extract useful but hidden knowledge from web data and services [8]. Web mining was defined as a converging research area from several research communities, such as database, information retrieval, machine learning, and natural language processing [13]. The objects of web mining are web resources, which can be web documents, web log files, structure of web sites, and structure of web documents themselves. Recently, many research efforts have been directed to web mining and web mining is now widely used in different areas [13]. Search engines, such as *Google*, use web mining technique to rank query results according to the importance and relevance of these pages [18]. Web mining is also expected to create structural summarizations for web pages [17], which can make search engines work more efficiently as well. Moreover, web mining is used to classify web pages [12], identify web communities [4], etc.

One of the key features of web data is that it may change at any time in any way. New data are inserted to the web; obsolete data are deleted while some others are modified. Corresponding to the types of web data, changes can be classified into three categories: *changes of web content*, *changes of web structure*, and *changes of web usage*. Due to the autonomous nature of the web, these changes may occur without notifying the users. We believe that the dynamic and autonomous properties of web data pose both challenges and opportunities to the web mining community.

Let us elaborate on this further. The knowledge and information mined from obsolete data may not be valid and useful any more with the changes of web data. Let us take one of the classic web structure mining algorithms *HITS* [12] for example. With the evolution of pages on the web, more and more web pages are created and linked to some of the existing web pages; some outdated web documents are deleted along with corresponding links while others hyperlinks may be updated due to changes of web content. Consequently, the set of *authoritative* and *hub pages* [12] computed at time t_1 may change at time t_2. That is some of the previously *authoritative* pages may not be authoritative any more. Similar cases may happen to *hub* pages. Thus, the mining results of the *HITS* algorithm may not be accurate and valid any more with the changes of web data.

With the dynamic nature of web data, there is an opportunity to get novel, more useful and informative knowledge from their historical changes, which cannot be discovered using traditional web mining techniques on snapshot data. For example, suppose we have two versions of the *redhat.com* web site structure as shown in Figure 1. In this figure, we use the grey boxes to represent pages deleted from the previous version, black boxes to denote newly inserted pages and bolded boxes to represent updated pages. From the changes of web structure in the two versions, it can be observed that when the information of products changes such as new products are added or outdated products are deleted, the information on training will change accordingly. For instance, in Figure 1(b), when a new product *RH 9.0* is added, a new training corresponding to this product *RH402* is inserted. In this example, it can be inferred that the information of *Product and Service* and *Training* may be associated. Such inference can be verified by examining the historical web structural deltas. Then knowledge about associations of structure changes can be extracted by applying association rule mining techniques to the historical changes of web structural data. The extracted knowledge can be rules such as changes of substructure *A* imply changes of substructure *B* within certain *time window* with certain *support* and *confidence*. Besides the association rules, interesting substructures, and enhanced classifiers can also be extracted from the historical web structural changes. From the above example, we argue that knowledge extracted from historical structural deltas is more informative compared to the mining results from snapshot data.

Based on the above observations, in this paper, we propose a novel approach to extract hidden knowledge from the changes of the historical web structural data (also known as web structural delta). Firstly, this approach is expected to be more efficient since the mining object is the sequence of deltas that is generally much smaller than the original sequence of structural data in terms of size. Secondly, novel knowledge that cannot be extracted before can be discovered by incorporating the dynamic property of web data and temporal attributes. The intuition behind is that while the knowledge from snapshot is important and interesting, the knowledge behind the corresponding changes may be more critical and informative. Such knowledge can be extracted by using different data mining techniques such as association rule mining, sequential pattern mining, classification, and clustering [10]. In this paper, we focus on exploring research issues for mining knowledge from the historical changes of web structural data.

The organization of this paper is as following. In Section 2, we present a list of related works. It includes web structure mining techniques and change detection systems for web data. The formal definition of web structural delta mining is presented in Section 3. In addition, different research issues are discussed in this section. The list of applications where the web structural delta mining results can be used is presented in Section 4. Finally, the last section concludes the paper.

2 Related Work

Our proposed web structural delta mining research is largely influenced by two research communities. The web mining community has looked at developing novel algorithms to mine snapshots of web data. The database community has focused on detecting, representing, and querying changes to the web data. We review some of these technologies here.

2.1 Web Structure Mining

Over the last few years, web structure mining has attracted a great deal of attention in the web mining community. Web structure mining was initially inspired by the study of social network and citation analysis [13]. Web structure mining was defined to generate structural summary about web sites and web pages. It includes the study of hyperlinked structure of the web [18], categorizing web pages into *authoritative pages* and *hub pages* [12], and generating community information with respect to the similarity and relations between different web pages and web sites [4]. We give a brief review of these techniques that includes two classic web structure mining algorithms and some algorithms for identifying web communities.

PageRank Algorithm:

One of the algorithms that analyze the hyperlink structure of the web is *PageRank* [18]. *PageRank* is an algorithm developed in Stanford University and now employed by the web search engine *Google*. PageRank is used to rank the search results of the search engine according to the corresponding *PageRank* values, which is calculated based on the structure information. In essence, *PageRank* interprets a link from page A to page B as a vote, by page A, for page B. However, *PageRank* looks at more than the sheer volume of votes, or links a page receives; it also analyzes the page that casts the vote. Votes cast by pages that are themselves "important" weigh more heavily and help to make other pages "important". Important and high quality web sites in the search results receive high ranks.

HITS Algorithm:

Most research efforts for classifying web pages try to categorize web pages into two classes: *authoritative* pages and *hub* pages. The idea of *authoritative* and *hub* pages was initialized by Kleinberg in 1998 [12]. The *authoritative* page is a web page that is linked to by most web pages that belong to this special topic, while a *hub* page is a web page that links to a group of authoritative web pages. One of the basic algorithms to detect *authoritative* and *hub* pages is named *HITS* [12]. The main idea is to classify web pages into *authoritative* and *hub* pages based on the *in-degree* and *out-degree* of corresponding pages after they have been mapped into a graph structure. This approach is purely based on hyperlinks. Moreover, this algorithm focuses on only web pages belonging to a specific topic.

Web Community Algorithms:

Cyber community [4] is one of the applications based on the analysis of similarity and relationship between web sites or web pages. The main idea of web community algorithms is to construct a community of web pages or web sites that share a common interest. Web community algorithms are clustering algorithms. The goal is to maximize the similarity within individual web communities and minimize the similarity between different communities. The measures of similarity and relationship between web sites or web pages are based on not only hyperlinks but also some content information within web pages or web sites.

2.2 Change Detection for Web Data

Considering the dynamic and autonomous properties of web data, recently many efforts have been directed into the research of change detection for web data [7, 16, 15, 6, 19]. According to format of web documents, web data change detection techniques can be classified into two categories. One is for HTML document, which is the dominant of current web data. Another is for XML document, which is expected to be the dominant of web data in the near future. We briefly review some of the web data change detection techniques now.

Change Detection for HTML Document:

Currently, most of the existing web documents are in HTML format, which is designed for the displaying purpose. An HTML document consists of markup tags and content data, where the markup tags are used to manipulate the representation of the content data. The changes of HTML documents can be changes of the HTML markup tags or the content data. The changes can be *sub page* level or *page* level. The AT&T Internet Difference Engine (AIDE) [7] was proposed to find and display changes to web pages. It can detect changes of *insertion* and *deletion*. WebCQ [16] is a system for monitoring and delivering web information. It provides personalized services for notifying and displaying changes and summarizations of corresponding interested web pages. SCD algorithm [15] is a sub page level change detection algorithm, which detects semantic changes of hierarchical structured data contents in any two HTML documents.

Change Detection for XML Document:

Recently, XML documents are becoming more and more popular to store and exchange data in the web. Different techniques of detecting changes for XML documents have been proposed [6, 19]. For instance, XyDiff [6] is used to detect changes of ordered XML documents. It supports three types of changes: *insertion, deletion,* and *updating*. X-Diff [19] is used to detect changes of unordered XML documents. It takes the XML documents as unordered tree, which makes the change detection process more difficult. In this case, two trees are equivalent if they are isomorphic, which means that they are identical except for the orders among siblings. The X-Diff algorithm can also identify three types of changes: *insertion, deletion,* and *updating*.

3 Web Structure Delta Mining

Based on recent research work in web structure mining, besides the validity of mining results and the hidden knowledge behind historical changes that we mentioned earlier, we observed that there are two other important issues have not been addressed by the web mining research community.

- The first observation is that existing web structure mining algorithms focus only on the *in degree* and *out degree* of web pages [12]. They do not consider the *global structural property* of web documents. Global properties such as the hierarchy structure, location of the web page among the whole web site and relations among ancestor and descendant pages are not considered. However, such information is important to understand the overall structure of a web site. For example, each part of the web structure is corresponding to a underlining concept and its instance. Consequently, the hierarchy structure represents the relations among the concepts. Moreover, based on the location of each concept in the hierarchy structure, the focus of a web site can be extracted with the assumption that the focus of the web site should be easy to be accessed.
- The second observation is that most web structure mining techniques ignored the structural information within individual web documents. Even if there are some algorithms designed to extract the structure of web documents [2], these data mining techniques are used to extract intra-structural information but not to mine knowledge behind the intra-structural information. With the increasing popularity of XML documents, this issue is becoming more and more important because XML documents carry more structural information compared to its HTML counterpart.

Web structural delta mining is to address the above limitations. As shown in Figure 2, it bridges the two popular research areas, web structure mining and change detection of web data. In this section, we will first give a formal definition of web structural delta mining from our point of view. Next, we elaborate on the major research issues of web structural delta mining.

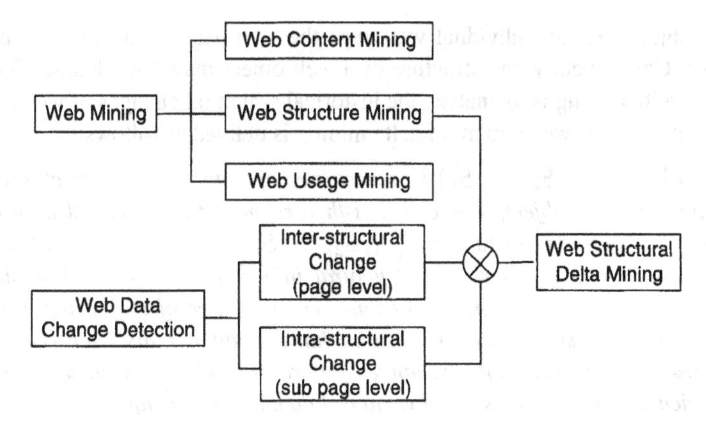

Fig. 2. Web Structural Delta Mining

3.1 Problem Statement

The goal of web structural delta mining is to extract any kind of interesting and useful information from the historical web structural changes. As the object of web structural delta mining can be structures of web sites, structures of a group of linked web page and even structures within individual web page, we introduce the term *web object* to define such objects.

Definition 1. *Let* $O=\{w_1, w_2, \cdots, w_n\}$ *be a set of web pages. O is a* ***web object*** *if it satisfies any one of the following constraints: 1) n=1; 2) For any* $1 \leq i \leq n$, w_i *links to or is linked by at least one of the pages from* $\{w_1, \cdots, w_{i-1}, w_{i+1}, \cdots, w_n\}$.

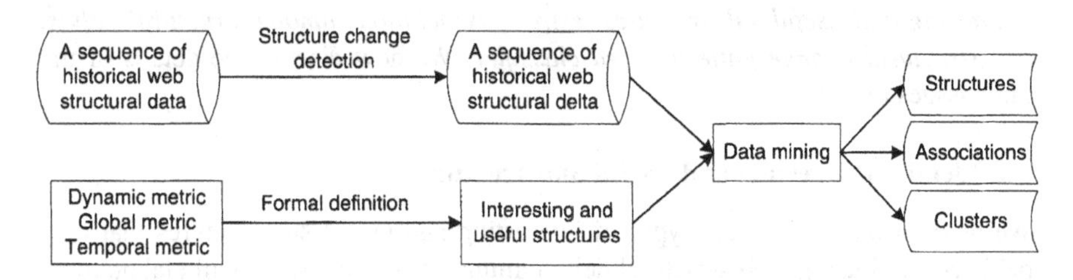

Fig. 3. Architecture of Web Structural Delta Mining

From the definition, we can see that a *web object* can be either an individual web page or a group of linked web pages. Thus, the structure of a web object O refers to the intra-structure within the web page if web object O includes only one web page, otherwise it refers to the inter-structure among web pages in this web object. The web object is defined in such a way that each web object corresponds to an instance of a semantic concept. With respect to the dynamic property of web data, we observed that some web *pages* or *links* might be *inserted into* or *deleted from* the web object and for individual web page the web page itself may also change over time. Consequently, the structure of a web object may also change. Our web structural delta mining is to analyze the historical structural changes of web objects. In our point of view, web structural delta mining is defined as follows.

Definition 2. *Let $\langle S_1, S_2, \cdots, S_n\rangle$ be a sequence of historical web structural information about a web object, where S_i is i-th version of the structural information about the web object O at time t_i. $\langle S_1, S_2, \cdots, S_n\rangle$ are in the order of time sequence. Assume that this series of structural information records all versions of structural information for a period of time. The objective of* ***web structural delta mining*** *is to extract structures with certain changes patterns, discover associations among structures in terms of their changes patterns, and classify structures based on the historical change patterns using various data mining techniques.*

Our definition of web structural delta mining is different from existing web structure mining definitions. In our definition, we incorporate the temporal (by taking different versions of web structures as a sequence), dynamic (by detecting the changes between different versions of web structures), and hierarchical property of web structural data (by taking into account the hierarchy structures of web sites) as shown in Figure 3. The basic idea is as follows. First, given a sequence of historical web structural data, by using the modification of some existing web data change detection systems, the sequence of historical web structural deltas can be extracted. On the other hand, based on the *dynamic metric, global metric* and *temporal metric*, different types of interesting structures can be defined based on their historical change patterns. Based on these definitions, the desired structures can be extracted from the sequence of historical web structural delta by using some data mining techniques. Besides interesting substructures, other knowledge such as association among structural deltas, structural change pattern based classifiers can also be discovered. According to the definition, web structural delta mining includes three issues: *identify interesting and useful substructures, extract associations among structural deltas,* and *structural change pattern based classifier.* We now discuss the details of the three issues in turn.

3.2 Identify Interesting and Useful Substructures

We now introduce different types of interesting and useful substructures that may be discovered using web structural delta mining. In Section 4, we will elaborate on the applications of these structures. The interesting substructures include *frequently changing structure, frozen structure, surprising structure, imbalanced structure, periodic dynamic structure, increasing dynamic structure*, and *decreasing dynamic structure*. Due to the lack of space, we will elaborate only on the first four of them.

Frequently Changing Structure:

Given a sequence of historical web structural data about a web object, we may observe that different substructures change at different frequency with different significance. Here *frequently changing structure* refers to substructures that change more frequently and significantly compared with other substructures [22]. Let us take the inter-structure of the *www.redhat.com* in Figure 1 as an example. In this figure, we observed that some of the substructures have changed between the two versions while others did not. The substructures rooted at nodes of *Product and Service* and *Training* changed more frequently compared to other substructures. In order to identify the frequently changing structure, we proposed two dynamic metrics *node dynamic* and *version dynamic* in [22]. The *node dynamic* measures the significance of the structural changes. The *version dynamic* measures the frequency of the structural changes against the history. Based on the dynamic metrics, the frequently changing structure can be defined as structures whose *version dynamic* and *node dynamic* are no less than the predefined thresholds. Two algorithms for discovering frequently changing structures from historical web structural delta have been proposed in [22].

Frozen Structure:

Frozen structures are the inverse to the frequently changing structures, as these structures seldom change or never change. To identify such kind of structures, we introduce another type of useful structure named *frozen structure*. Frozen structure, from the words themselves, refers to those structures that are relatively stable and seldom change in the history. Similarly, based on the dynamic metric, *frozen structure* can be defined as structures whose values of *node dynamic* and *version dynamic* do not exceed certain predefined thresholds.

Surprising Structure:

Based on the historical dynamic property of certain structures, the corresponding evolutionary patterns can be extracted. However, for some structures, the changes may not always be consistent with the historical patterns. In this case, a metric can be proposed to measure the surprisingness of the changes. If for certain number of changes, the surprisingness exceeds certain threshold, then, the structure is defined as a *surprising structure*. Any structure whose change behavior deviate from the knowledge we learned from the history is a surprising structure. For example, a frozen structure that suddenly changed very significantly and frequently may be a surprising structure; a frequently changing structure suddenly stopped to change is also a surprising structure. The surprising structure may be caused by some abnormal behaviors, mistakes, or fraud actions.

Imbalanced Structure:

Besides the dynamic property of web structure, there is another property, *global property*, which has not been considered by the web structure mining community. If we represent the World Wide Web as a tree structure, here *global property* of the structure refers to the depth and cardinality of a node, which was introduced in the imbalance structure research initiated in the context of hypertext structure analysis by

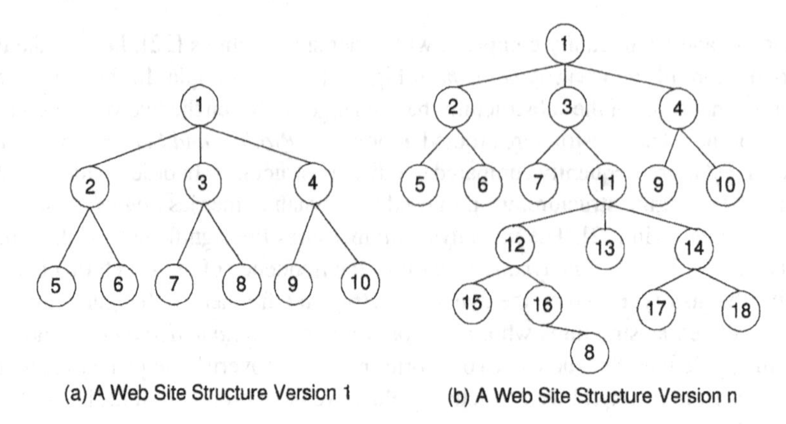

Fig. 4. Two Versions of a Web Site Structure

Botafogo et al [3]. In their work, two imbalance metrics, *depth imbalance* and *child imbalance*, were used to measure the global property of a structure. The imbalanced metric is based on the assumption that each node in the structure carries the same amount of information and the links from a node are further development of the node. Based on the assumption, an ideal structure of information should be a balanced tree. However, a good structure may not be definitely balanced since some structures are designed to be imbalanced by purpose.

In our study, we assume that the first version of any structure is well designed even if there are some imbalanced structures. Our concern is that as web data is autonomous, some balanced structure may become imbalanced due to changes to web data. We argue that sometimes such kind of changes is undesirable. For example, suppose Figure 4 (a) is the first version of a web site structure. Figure 4 (b) depicts the modified web site structural after a sequence of change operations over a period of time. From the two versions, we can observe the following changes. The depth of node *8* changed from 2 to 5; the number of descendant nodes of node *3* increases dramatically while the numbers of descendant nodes of its two siblings did not change. One of the consequences of these changes is that the cost of accessing node *8* from the root node in Figure 4 (b) is more expensive than the cost in Figure 4 (a). The depth of a node reflects the cost to access this particular node from the root node. Nodes that are very deep inside the tree are unlikely to be visited by the majority of the users. Another consequence is that node *3* is developed with more information in Figure 4 (b) than in Figure 4 (a). The number of descendant nodes indicates the importance of the node. With such information, the values of the imbalance metrics can be calculated and the imbalance structures can be identified. Based on the imbalance structures, web designers can check whether such imbalance structures are desirable or not. Consequently, a web structure of higher quality can be maintained. Besides identifying the imbalanced structures, we want to further analyze the changes to find out what kind of changes may have the potential to cause imbalance of a structure.

With such knowledge, the undesirable imbalance structures can be avoided by taking appropriate actions in advance.

3.3 Extract Associations among Structural Deltas

Another important research issue of web structural delta mining is to discover association of structural deltas [5]. The basic idea is to extract correlation among the occurrences of different changes of different web structures. It is similar to the traditional association rule mining in some senses if we treat each possible substructure as an item in the transaction database. However, our structure association rule mining is more complex due to following reasons. Firstly, not every combination of two length (*i-1*) items in the database can be a candidate length *i* item since if they cannot form a complete structure it will be meaningless. Therefore, a more complex candidate item generation method is desirable compared to the Apriori in [1]. Secondly, for each item in the database the types of changes can be different such as *insertion*, *deletion*, and *update*, while in traditional transaction data there is no such attribute. This makes the mining process more complex [1]. There are two types of association rules among the structural deltas.

Structural Delta Association Rule Mining:

Considering the changes of web structural data, it can be observed that some of the substructures changed simultaneously more often than others. The goal of structural delta association rule mining is to discover the sets of substructures that change together with certain *support* and *confidence* [5]. An example of the structural delta association rule is $S_1 \rightarrow S_2$, where S_1 and S_2 are two different substructures. For instance, based on a sequence of historical versions of the structure in Figure 1, we may discover that the substructures rooted at nodes *Product and Service* and *Training* are strongly associated with respect to the changes. Whenever the structure rooted at node *Product and Service* changes, the structure rooted at node *Training* also changes with certain confidence in the history. However, besides the presence of the substructures, the types of changes and the significance of the changes can be different for different substructures. Consequently, more informative and useful structural delta association rules can be extracted by using multiple dimensional association rule mining techniques.

Semantic Delta Association Rule Mining:

Besides the structural delta association rule mining, in which only the structural information is considered, semantic association rules can also be extracted from historical structural deltas if we incorporate the meta data such as summaries or keywords of web pages in the web site into the association rules. In this case, we want to extract knowledge such as which semantic objects are associated in terms of their change histories. An example of the semantic delta association rule can be that more and more new clients are added under product A while some of the clients for product B are keeping deleted in product client list XML document. Such rules may imply the relation between the two products. In this case, they may be substitutable. The reason for such changes is that products A are becoming more and more popular than products B. If products in both categories are inserted or deleted simultaneously, then it may indicate the complement relation of the two products.

3.4 Structure Change Pattern Based Classifier

Besides the above-mentioned two issues, it is also interesting to cluster semi-structured data based on their historical structural change patterns. Recently, a structural rule based classification algorithm has been proposed for semi-structured data [21]. In their approach, the discriminatory structures are extracted by using the TreeMiner [20], and it has been proved to be very efficient. The structure change pattern based classifier is proposed to classify web structures according to the corresponding change behaviors in the history. The intuition behind structure change pattern based classifier is that records from the same class are expected to share the same evolutionary pattern is the case in the biology evolution history. The basic idea is to relate the presence of particular structure change patterns in the records to the likelihood of belonging to a particular class. The key issue is to extract the discriminatory structure change pattern from the data. We claim that the historical structure change pattern is more informative in evolutionary datasets such as biological data than other features extracted from snapshot data. The structure change pattern based classifier is expected to improve existing classification techniques by incorporating the historical change patterns.

4 Applications

Based on the issues we discussed above, the output of our web structural delta mining should include interesting substructures, associations among changes of substructures, different clusters of substructures with respect to the evolutionary patterns. According to the way they are defined, such structural knowledge can be widely used for many different applications such as multi-frequency crawler, efficient XML indexing, semantic meaning extraction, enhanced wrapper, focused change detection system, efficient caching for wireless applications, etc. Note that by no means we claim that the list is exhaustive. Only the first three are discussed in more detail.

4.1 Multi-frequency Crawler

As we know, for traditional web crawler, the crawling strategy is to crawl all the desired web pages at a uniform frequency even though some web pages may never change. This strategy makes the crawling process more time consuming and inefficient. Concerning the historical changes, a multi-frequency crawler that crawls different part of the web site at different frequency can be constructed.

Based on the frequency of the structures changed in the history (such as *frozen structure*, *frequently changing structure*, etc), together with the corresponding frequency of content changes, web documents can be categorized into different classes. For example, for those documents whose structures are frozen, they can be separated into two classes. One class consists of documents whose content change frequently, the other includes these documents whose content hardly change. Based on the frequency of the structural changes and content changes, different crawling strategy can be adapted.

Similarly, for documents whose structures change frequently, they can be clustered into different groups by integrating the structural change frequency and corresponding content change frequency in the history. For documents in the same cluster, they are crawled at the same frequency. By using such strategy, the multi-frequency crawler should be more efficient than traditional crawlers, and the overhead of network communication can be relieved as well.

4.2 Efficient XML Indexing

Another application of the structural delta mining results is for XML indexing. As we know that, the major issue of XML indexing is to identify the ancestor and descendant relationship quickly. To do this, different numbering schemes have been proposed [14, 11]. Li and Moon proposed a numbering scheme in XISS (XML Indexing and Storage System) [14]. The XISS numbering scheme uses an *extended preorder* and a *size*. The *extended preorder* will allow additional nodes to be inserted without reordering and the *size* determines the possible number of descendants.

More recently, XR-Tree [11] was proposed for indexing XML data for efficient structural joins. Compared with the XR-tree, XISS numbering scheme is more flexible and can deal with dynamic updates of XML data more efficiently. Since extra spaces can be reserved in the *extended preorder* to accommodate future insertion, global reordering is not necessary until all the reserved spaces are consumed. However, Li and Moon did not highlight how much extra space to allocate. Allocating too little extra space will leads to the ineffectiveness of the numbering scheme and allocating too much extra space may eventually leads to very large numbers being assigned to nodes for very large XML document. In the XISS approach, the gaps are equally allocated, while in practice different part of the document change at different frequency with different significance.

Based on our mining results from the historical structural changes, the numbering scheme can be improved by allocating the gaps in a more intelligent manner. For the parts of structure that change frequently and significantly, larger gaps are allocated while for frozen structures only very tiny gaps are left. By using this strategy, the numbering scheme should be more efficient in terms of both index maintenance and space allocation.

4.3 Semantic Meaning Extraction

Based on the values of dynamic metrics, some semantic meaning can be extracted from the interesting structures with related domain knowledge. For example, suppose that based on the historical structural information we find out that the substructure rooted at node *Training* is a frequently changing structure with more and more subtrees inserted with similar labels as shown in Figure 1. Then, it can be inferred that the service *Training* is becoming more and more popular or profitable.

Similarly, semantic meaning can also be extracted from other types of structures and associations among structural changes etc. The basic idea is to incorporate some meta data (such as labels of the edges and nodes, types of changes, etc.) into the interesting structures to get the semantic implications. The semantic meaning extracted from the structural delta mining results can be widely used in e-commerce for tasks such as monitoring and predicting the competitors' strategies.

5 Conclusion

In this paper, we present a novel area of web mining named web structural delta mining. Besides the formal definition, three major issues in web structural delta mining have been identified. They are *identifying interesting and useful structures, discovering association of web structural deltas,* and *structure change pattern based classifier*. Different applications, which can be enhanced by using the web structural delta mining results, are also discussed. Currently, we are studying some of the issues.

References

1. Rakesh Agrawal, Tomasz Imieliski and Arun Swami. Mining association rules between sets of items in large databases. In *Proceedings of ACM SIGMOD International Conference on Management of Data*, pages 207–216, 1993.
2. Arasu Arvind and Hector Garcia-Molina . Extracting structured data from web pages. In *The 2003 ACM SIGMOD International Conference on Management of Data*, pages 337–348, 2003.
3. Rodrigo A. Botafogo, Ehud Rivlin and Ben Shneiderman. Structural analysis of hypertexts: identifying hierarchies and useful metrics. *ACM Transactions on Information Systems (TOIS)*, 10(2):142–180, 1992.
4. Soumen Chakrabarti, Byron E. Dom, S. Ravi Kumar, Prabhakar Raghavan, Sridhar Rajagopalan, Andrew Tomkins, David Gibson and Jon Kleinberg. Mining the Web's link structure. *Computer*, 32(8):60–67, 1999.
5. Ling Chen, Sourav S. Bhowmick and Liang Tien Chia. Web structural delta association rule mining. In *Pacific-Asia Conference on Knowledge Discovery and Data Mining*, pages 452–457, 2003.
6. Gregory Cobena, Serge Abiteboul and Amelie Marian. Detecting changes in XML documents. In *18th IEEE International Conference on Data Engineering*, pages 41–52, 2002.
7. Fred Douglis, Thomas Ball, Yih-Farn Chen and Eleftherios Koutsofios. The AT&T internet difference engine: Tracking and viewing changes on the web. *World Wide Web*, 1(1):27–44, 1998.
8. Oren Etzioni. The world-wide web: Quagmire or gold mine? *Communications of the ACM*, 39(11):65–68, 1996.

9. Minos N. Garofalakis, Rajeev Rastogi, S. Seshadri and Kyuseok Shim. Data mining and the web: past, present and future. In *2nd ACM International Workshop on Web Information and Data Management (WIDM)*, pages 43–47, 1999.
10. Jiawei Han and Micheline Kamber. *Data Mining Concepts and Techniques*. Morgan Kanufmann, 2000.
11. Haifeng Jiang, Hongjun Lu, Wei Wang, and Beng Chin Ooi. XR-Tree: Indexing xml data for efficient structural joins. In *IEEE International Conference on Data Engineering*, pages 253–264, 2003.
12. Jon Kleinberg. Authoritative sources in a hyperlinked environment. *Journal of the ACM*, 46(5):604–632, 1998.
13. Raymond Kosala and Hendrik Blockeel. Web mining research: A survey. *Newsletter of the ACM Special Interest Group on Knowledge Discovery and Data Mining*, 2, 2000.
14. Quanzhong Li and Bongki Moon. Indexing and querying XML data for regular path expressions. In *Proceedings of the 27th International Conference on Very Large Data Bases*, pages 361–370, 2001.
15. Seung-Jin Lim and Yiu-Kai Ng . An automated change detection algorithm for HTML documents based on semantic hierarchies. In *17th IEEE International Conference on Data Engineering*, pages 303–312. IEEE Computer Society, 2001.
16. Ling Liu, Calton Pu, and Wei Tang. WebCQ-detecting and delivering information changes on the Web. In *9th International Conference on Information and Knowledge Management*, pages 512–519, 2000.
17. Sanjay Madria, Sourav S. Bhowmick, Wee keong Ng, and Ee Peng. Lim. Research issues in Web data mining. In *Data Warehousing and Knowledge Discovery*, pages 303–312, 1999.
18. Page Lawrence, Brin Sergey, Motwani Rajeev and Winograd Terry. The PageRank citation ranking: Bringing order to the Web. Technical report, Stanford Digital Library Technologies Project, 1998.
19. Yuan Wang, David J. DeWitt and Jin-Yi Cai. X-diff: An effective change detection algorithm for XML documents. In *19th IEEE International Conference on Data Engineering*, 2003.
20. Mohammed J. Zaki. Efficiently mining frequent trees in a forest. In *Proceedings of the eighth ACM SIGKDD international conference on Knowledge discovery and data mining*, pages 71–80, 2002.
21. Mohammed J. Zaki and Charu Aggarwal. XRules: An effective structural classifier for XML data. In *Proceedings of the ninth ACM SIGKDD international conference on Knowledge discovery and data mining*, pages 316–325, 2003.
22. Qiankun Zhao, Sourav S. Bhowmick, Mukesh Mohania and Yahiko Kambayashi. Mining frequently changing structure from historical web structural deltas. Technical report, CAIS, NTU, Singapore, 2004.

Workflow Reduction for Reachable-path Rediscovery in Workflow Mining

Kwang-Hoon Kim[1] and Clarence A. Ellis[2]

[1] Collaboration Technology Research Lab.
Department of Computer Science
KYONGGI UNIVERSITY
San 94-6 Yiuidong Youngtongku Suwonsi Kyonggido, South Korea, 442-760
kwang@kyonggi.ac.kr

[2] Collaboration Technology Research Group
Department of Computer Science
UNIVERSITY OF COLORADO AT BOULDER
Campus Box 0430, Boulder, Colorado, USA, 80309-0430
skip@cs.colorado.edu

Summary. This paper[3] newly defines a workflow reduction mechanism that formally and automatically reduces an original workflow process to a minimal set of activities, which is called minimal-workflow model in this paper. It also describes about the implications of the minimal-workflow model on workflow mining that is a newly emerging research issue for rediscovering and reengineering workflow models from workflow logs containing workflow enactment and audit information gathered being executed on workflow engine. In principle, the minimal-workflow model is reduced from the original workflow process by analyzing dependencies among its activities. Its main purpose is to minimize discrepancies between the modeled workflow process and the enacted workflow process as it is actually being executed. That is, we can get a complete set of activity firing sequences (all reachable-paths from the start to the end activity on a workflow process) on buildtime. Besides, we can discover from workflow logs that which path out of all reachable paths a workcase (instance of workflow process) has actually followed through on runtime. These are very important information gain acquiring the runtime statistical significance and knowledge for redesigning and reengineering the workflow process. The minimal-workflow model presented in this paper is used to be a decision tree induction technique for mining and discovering a reachable-path of workcase from workflow logs. In a consequence, workflow mining methodologies and systems are rapidly growing and coping with a wide diversity of domains in terms of their applications and working environments. So, the literature needs various, advanced, and specialized workflow mining techniques and architectures that are used for finally feed-backing their analysis results to the redesign and reengineering phase of the existing workflow and business

[3] This research was supported by the University Research Program (Grant No. C1-2003-03-URP-0005) from the Institute of Information Technology Assessment, Ministry of Information and Communications, Republic of Korea.

process models. We strongly believe that this work might be one of those impeccable attempts and pioneering contributions for improving and advancing the workflow mining technology.

1 Introduction

In recent, workflow (business process) and its related technologies have been constantly deployed and so gradually hot-issued in the IT arena. This atmosphere booming workflows and business processes modeling and reengineering is becoming a catalyst for triggering emergence of the concept of workflow mining that collects data at runtime in order to support workflow design and analysis for redesigning and reengineering workflows and business processes. Especially real workflow models going with e-Commerce, ERP (Enterprise Resource Planning), and CRM (Customer Relationship Management) are getting larger and more complex in their behavioral structures. These large-scaling movements trigger another big changes in workflow administration (the responsibility of redesign) and monitoring (the responsibility of rediscovery) functionality that has been featured and embedded in the workflow build-time and run-time functionality of the traditional workflow systems. The more a workflow system's architecture is distributed, the more its administrative functions ought to be extended and play important roles for improving the integrity of the workflow process models and systems. At the same time, the more transactional applications are involved in workflow models, the more workflow monitoring features are closely related with statistic-oriented workflow monitoring information. Meanwhile, in the traditional workflow systems, the workflow monitoring features generate status-oriented workflow monitoring information.

In other words, there have been prevalent research and development trends in the workflow literature - workflow mining techniques and systems that collect runtime data into workflow logs, and filter out information and knowledge from the workflow logs gathered by the administration and monitoring features of workflow management system. The workflow mining techniques and systems tend to have completely distributed architectures to support very large-scale workflow applications based upon object-oriented and internet-based infrastructures. Their key targets have been transformed from the passive, centralized, human-oriented, and small/medium scale workflow process models and systems to the active (object), distributed (architecture), system-oriented (transaction), and large-scale (application) workflow process models and systems. So, in order for WfMSs to slot in workflow mining features, it is necessary for their administration and monitoring features to be extended to gathering and analyzing statistical or workload status information of the workflow architectural components dispersed on the distributed environment, and performing feedback the analyzed results to the redesigning and reengineering phase of workflow process models. The advanced workflow management systems, such as transactional workflows and very large-scale workflows, need some additional administration and monitoring features that are used for redesigning and rediscovering workflow procedures and applications, which are tightly related with the workflow mining functionality.

Therefore, the workflow mining functionality has been mainly looking for efficient redesigning and reengineering approaches because of not only that workflow procedures are becoming massively large-scaled and complicated, but also, nevertheless, that their life-cycle and recycling (reengineering) periods are becoming swiftly shorter. That is, BPx, such as business process redesign, reengineering, and restructuring, needs to be done frequently and even dynamically as well. In order to perform the BPx efficiently and effectively, we should consider its enactment audit and history information logged and collected from the runtime and diagnosis phases of the workflow and business process management system. To collect the information, many decisions so need to be made in terms of many perspectives, such as control and data flows, work assignment policies and algorithms, logging and diagnosis information, workcase firing sequences (reachable paths), and the number of workcases performed and managed by various users and administrators. These decisions are done based upon the information and knowledge collected in not only the runtime-diagnosis phases, but also the (re)design phase. Also they are further complicated by the requirements such as flexibility, changeability, multi-versioning, speed, grading of actors' behavior, and so forth.

In this paper, our emphasis is placed on the rediscovery problem [14] that investigates whether it is possible to rediscover the workflow process by merely looking at its logs. In a little more detailed idea, this paper gives a way to efficiently rediscover the discrepancy between the original workflow process model as it is built and the enacted workflow process (which is called from now workcase) as it is actually executed. The discrepancy, as you can easily imagine, is caused by the alternative paths exhibited on the model. The number of alternative paths on the model will effect on the degree of the discrepancy. At this moment, we would make an issue that is called 'reachable-path rediscovery problem'. For example, after rediscovering a workcase from workflow logs, we need to know along which reachable-path the workcase has followed. This might be very useful knowledge for workflow administrators and designers to redesign and re-estimate the original workflow process model after being elapsed a specific amount of period. As a matter of facts, this paper proposes a feasible approach (the concept of minimal-workflow model) to handle out the reachable-path rediscovery problem. The minimal-workflow model is automatically reduced from its original workflow process model by analyzing control dependencies existing among activities. It also shows how the solution is applicable to either the forward selection approach or the backward elimination approach in serving as a decision tree induction algorithm for generating a complete set of activity firing sequences - all reachable-paths from the start to the end activity on a workflow process model - that are very important patterns because of that they become criteria for acquiring the runtime information just like how many workcases of the workflow process model have been enacted along with each reachable-path, respectively. This statistical runtime information should be very effective and valuable knowledge for redesigning and reengineering the workflow model as simply introduced just before.

The remainder of this paper is organized as follows. The next section briefly describes about what the workflow process model (Information Control Net) conceptually does mean, and formalizes the problem addressed in this paper. Section 3 works out a way to define a workflow mining framework in a formal manner. And based on the framework, several algorithms are presented that rediscovers reachable-paths of a large class of workcases, and describes about the implications of the minimal-workflow model as a workflow mining methodology and technique. Finally, the paper finalizes with an overview of related work and some conclusions.

2 Preliminaries

This section briefly introduces the information control net used in the paper. First, it describes the basic concept of the information control net. Next formally defines the information control net and its implications. And the problem scope of this paper is defined at the end of this section.

2.1 Information Control Net

The original Information Control Net was developed in order to describe and analyze information flow within offices. It has been used within actual as well as hypothetical automated offices to yield a comprehensive description of activities, to test the underlying office description for certain flaws and inconsistencies, to quantify certain aspects of office information flow, and to suggest possible office restructuring permutations. The ICN model defines an office as a set of related procedures. Each procedure consists of a set of activities connected by temporal orderings called procedure constraints. In order for an activity to be accomplished, it may need information from repositories, such as files, forms, and some data structures.

An ICN captures these notations of workflow procedures, activities, precedence, and repositories. A workflow procedure is a predefined set of work steps, called activities, and a partial ordering of these activities. Activities can be related to each other by conjunctive logic (after activity A, do activities B and C) or by disjunctive logic (after activity A, do activity B or C) with predicates attached. An activity is either a compound activity containing another procedure, or a basic unit of work called an elementary activity. An elementary activity can be executed in one of three modes: manual, automatic, or hybrid. Typically one or more participants are associated with each activity via roles. A role is a named designator for one or more participants which conveniently acts as the basis for partitioning of work skills, access controls, execution controls, and authority / responsibility. An actor is a person, program, or entity that can fulfill roles to execute, to be responsible for, or to be associated in some way with activities and procedures.

As an example, consider an order procedure of a large technology corporation. This workflow procedure shown in Fig. 1, which we have modeled in detail elsewhere [10, 11], consists of several steps from order evaluation to final contract storing activities, such as:

- Order evaluation (α_1) (sales manager evaluates orders on database)
- Letter of regret (α_2) (secretary writes a letter of regret for not delivering the ordered one)
- Billing (α_3) (bill clerk makes a bill)
- Shipping (α_4) (inventory clerk makes the ordered one in ready to be shipped)
- Archive (α_5) (order administrator stores the record on database).

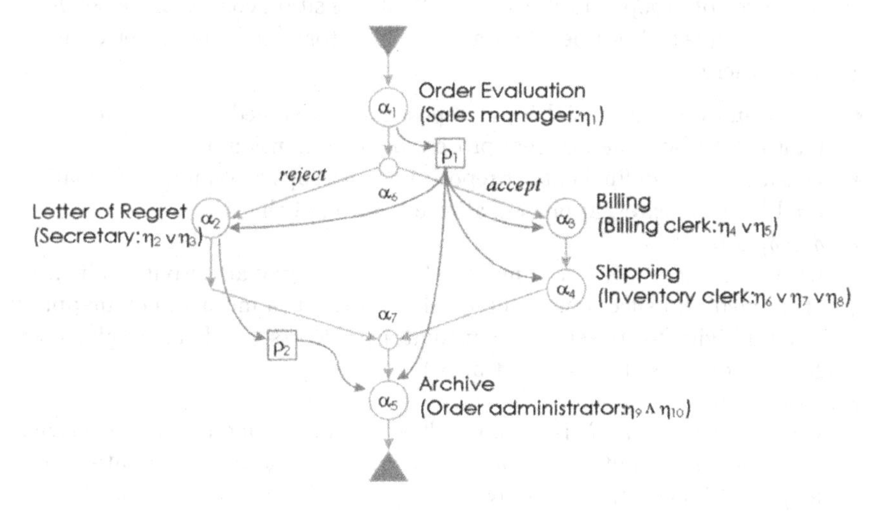

Fig. 1. A Workflow Procedure Model for Order Processing

Notice that there is an implied ordering because each of the activities 1 through 5 needs to be done sequentially in order. In parenthesis, a role is identified as the performer of each activity. This is distinguishable from the actual person (the actor) who does the work. Typically, an activity is executed by a person utilizing a computer workstation. Computer programs that automatically perform activities, or provide automated assistance within hybrid activities are called scripts. Note that some activities such as preliminary assessment are performed by multiple roles, and some activities such as archive could potentially be performed in automatic mode where the computer is the actor. Some activities are obviously compound activities; the role specified to perform this activity is in fact a group.

Most workflow management products also provide a graphical workflow modeling tool to design, analyze, and evolve the workflow procedure specifications. In the graphical notation of ICN, circles represent activities, arcs (which may have transition conditions such as reject and accept) represent the precedence partial order, hollow dot represent or-split and or-join respectively, and small boxes represent relevant data. A workcase is the locus of control for a particular execution of a procedure. It designates a workcase that represents one instance of the workflow procedure, or one customer who is considering for ordering. These workflow modeling tools support alternative and parallel processing and decision points. Thus, note that there are dangling arcs in Fig. 1 representing decisions paths to possibly reject or accept the order.

2.2 Formal Notation of the Information Control Net

A basic ICN is 7-tuple $\Gamma = (\delta, \gamma, \varepsilon, \pi, \kappa, I, O)$ over a set of A activities (including a set of group activities), a set T of transition conditions, a set R of repositories, a set P of roles, and a set C of actors (including a set of actor groups), where

- I is a finite set of initial input repositories, assumed to be loaded with information by some external process before execution of the ICN;
- O is a finite set of final output repositories, perhaps containing information used by some external process after execution of the ICN;
- $\delta = \delta_i \cup \delta_o$
  where, $\delta_o : A \longrightarrow \wp(A)$ is a multi-valued mapping of an activity to its sets of (immediate) successors, and $\delta_i : A \longrightarrow \wp(A)$ is a multi-valued mapping of an activity to its sets of (immediate) predecessors; (For any given set S, $\wp(S)$ denotes the power set of S.)
- $\gamma = \gamma_i \cup \gamma_o$
  where $\gamma_o : A \longrightarrow \wp(R)$ is a multi-valued mapping (function) of an activity to its set of output repositories, and $\gamma_i : A \longrightarrow \wp(R)$ is a multi-valued mapping (function) of an activity to its set of input repositories;
- $\varepsilon = \varepsilon_a \cup \varepsilon_p$
  where $\varepsilon_p : A \longrightarrow P$ is a single-valued mapping of an activity to one of the roles, and $\varepsilon_a : P \longrightarrow \wp(A)$ is a multi-valued mapping of a role to its sets of associated activities;
- $\pi = \pi_p \cup \pi_c$
  where, $\pi_c : P \longrightarrow \wp(C)$ is a multi-valued mapping of a role to its sets of associated actors, and $\pi_p : C \longrightarrow \wp(P)$ is a multi-valued mapping of an actor to its sets of associated roles;
- $\kappa = \kappa_i \cup \kappa_o$
  where κ_i : sets of control-transition conditions, T, on each arc, $(\delta_i(\alpha), \alpha), \alpha \in A$; and κ_o : sets of control-transition conditions, T, on each arc, $(\alpha, \delta_o(\alpha)), \alpha \in A$; where the set $T = \{\mathit{default}, \mathit{or}(\mathit{conditions}), \mathit{and}(\mathit{conditions})\}$.

The following Table 1 is to represent the order processing workflow procedure model illustrated in Fig. 1 by using the formal notation of ICN.

2.3 Implication of the Information Control Net

Given a formal definition, the execution of an ICN can be interpreted as follows: For any activity α, in general,

$$\delta(\alpha) = \{ \\ \{\beta_{11}, \beta_{12}, \ldots, \beta_{1m(1)}\}, \\ \{\beta_{21}, \beta_{22}, \ldots, \beta_{2m(2)}\}, \\ \ldots, \\ \{\beta_{n1}, \beta_{n2}, \ldots, \beta_{nm(n)}\} \\ \}$$

means that upon completion of activity α, a transition that simultaneously initiates all of the activities β_{i1} through $\beta_{im(i)}$ occurs. Only one value of $i(1 \leq i \leq n)$ is selected as the result of a decision made within activity α. (Note that if $n = 1$, then no decision is needed and α is not a decision node.) In general, if $m(i) = 1$ for all i, then no parallel processing is initiated by completion of α. In ICN diagrams, the former, that an activity has a parallel transition, is represented by a solid dot, and the latter, that an activity has a decision (or selective) transition, is represented by hollow dot.

For any activity α, $\varepsilon_p(\alpha) = \{\varpi_1, \varpi_2, \ldots, \varpi_n\}$, where n is the number of roles, $\varpi \in P$, involved in the activity, means that an activity α is performed by one of the roles; $\varepsilon_a(\varpi) = \{\alpha_1, \alpha_2, \ldots, \alpha_m\}$, where m is the number of activities performed by the role, means that a role ϖ is associated with several activities in a procedure.

In terms of the control-transition conditions, there are three kinds of conditions: *default* control-transition conditions, *OR* control-transition conditions, and *AND* control-transition conditions (including selective control-transition conditions). These conditions must be specified by one of the following so that the flow of control can be accomplished by the enactment part:

- by the workflow modelers at the modeling part of workflow management system,
- by APIs on the invoked programs,
- by the statistical or temporal strategies that the system can automatically control,
- by the system.

The execution of an ICN commences by a single λ transition. We always assume without loss of generality that there is a single starting node:

$$\exists \alpha_1 \in A \ni \{\{\lambda\}\} \in \delta_i(\alpha_1).$$

Table 1. Formal Notation of the Order Processing Workflow Model

$\Gamma = (\delta, \gamma, \varepsilon, \pi, \kappa, I, O)$ over A, R, P, C, T	*TheOrderProcedureinICN*
$A = \{\alpha_1, \alpha_2, \alpha_3, \alpha_4, \alpha_5, \alpha_6, \alpha_7, \alpha_I, \alpha_F\}$	*Activities*
$R = \{\rho_1, \rho_2\}$	*Repositories*
$P = \{\varpi_1, \varpi_2, \varpi_3, \varpi_4, \varpi_5\}$	*Roles*
$C = \{\eta_1, \eta_2, \eta_3, \eta_4, \eta_5, \eta_6, \eta_7, \eta_8, \eta_9, \eta_{10}\}$	*Actors*
$T = \{d(default), or_1('reject'), or_2('accept')\}$	*TransitionConditions*
$I = \{\emptyset\}$	*InitialInputRepositories*
$O = \{\emptyset\}$	*FinalOutputRepositories*

$\delta = \delta_i \cup \delta_o$	$\delta_i(\alpha_I) = \{\emptyset\};$	$\delta_o(\alpha_I) = \{\alpha_1\};$
	$\delta_i(\alpha_1) = \{\alpha_I\};$	$\delta_o(\alpha_1) = \{\alpha_6\};$
	$\delta_i(\alpha_2) = \{\alpha_6\};$	$\delta_o(\alpha_2) = \{\alpha_7\};$
	$\delta_i(\alpha_3) = \{\alpha_6\};$	$\delta_o(\alpha_3) = \{\alpha_4\};$
	$\delta_i(\alpha_4) = \{\alpha_3\};$	$\delta_o(\alpha_4) = \{\alpha_7\};$
	$\delta_i(\alpha_5) = \{\alpha_7\};$	$\delta_o(\alpha_5) = \{\alpha_F\};$
	$\delta_i(\alpha_6) = \{\alpha_1\};$	$\delta_o(\alpha_6) = \{\{\alpha_2\}, \{\alpha_3\}\};$
	$\delta_i(\alpha_7) = \{\{\alpha_2\}, \{\alpha_4\}\};$	$\delta_o(\alpha_7) = \{\alpha_5\};$
	$\delta_i(\alpha_F) = \{\alpha_5\};$	$\delta_o(\alpha_F) = \{\emptyset\};$
$\gamma = \gamma_i \cup \gamma_o$	$\gamma_i(\alpha_I) = \{\emptyset\};$	$\gamma_o(\alpha_I) = \{\emptyset\};$
	$\gamma_i(\alpha_1) = \{\emptyset\};$	$\gamma_o(\alpha_1) = \{\rho_1\};$
	$\gamma_i(\alpha_2) = \{\rho_1\};$	$\gamma_o(\alpha_2) = \{\rho_2\};$
	$\gamma_i(\alpha_3) = \{\rho_1\};$	$\gamma_o(\alpha_3) = \{\emptyset\};$
	$\gamma_i(\alpha_4) = \{\rho_1\};$	$\gamma_o(\alpha_4) = \{\emptyset\};$
	$\gamma_i(\alpha_5) = \{\rho_1, \rho_2\};$	$\gamma_o(\alpha_5) = \{\emptyset\};$
	$\gamma_i(\alpha_6) = \{\emptyset\};$	$\gamma_o(\alpha_6) = \{\emptyset\};$
	$\gamma_i(\alpha_7) = \{\emptyset\};$	$\gamma_o(\alpha_7) = \{\emptyset\};$
	$\gamma_i(\alpha_F) = \{\emptyset\};$	$\gamma_o(\alpha_1) = \{\emptyset\};$
$\varepsilon = \varepsilon_a \cup \varepsilon_p$	$\varepsilon_p(\alpha_I) = \{\emptyset\};$	$\varepsilon_a(\varpi_1) = \{\alpha_1\};$
	$\varepsilon_p(\alpha_1) = \{\varpi_1\};$	$\varepsilon_a(\varpi_2) = \{\alpha_2\};$
	$\varepsilon_p(\alpha_2) = \{\varpi_2\};$	$\varepsilon_a(\varpi_3) = \{\alpha_3\};$
	$\varepsilon_p(\alpha_3) = \{\varpi_3\};$	$\varepsilon_a(\varpi_4) = \{\alpha_4\};$
	$\varepsilon_p(\alpha_4) = \{\varpi_4\};$	$\varepsilon_a(\varpi_5) = \{\alpha_5\};$
	$\varepsilon_p(\alpha_5) = \{\varpi_5\};$	
	$\varepsilon_p(\alpha_6) = \{\emptyset\};$	
	$\varepsilon_p(\alpha_7) = \{\emptyset\};$	
	$\varepsilon_p(\alpha_F) = \{\emptyset\};$	
$\pi = \pi_p \cup \pi_c$	$\pi_p(\eta_1) = \{\varpi_1\};$	$\pi_c(\varpi_1) = \{\eta_1\};$
	$\pi_p(\eta_2) = \{\varpi_2\};$	$\pi_c(\varpi_2) = \{\eta_2, \eta_3\};$
	$\pi_p(\eta_3) = \{\varpi_2\};$	$\pi_c(\varpi_3) = \{\eta_4, \eta_5\};$
	$\pi_p(\eta_4) = \{\varpi_3\};$	$\pi_c(\varpi_4) = \{\eta_6, \eta_7, \eta_8\};$
	$\pi_p(\eta_5) = \{\varpi_3\};$	$\pi_c(\varpi_5) = \{\eta_9, \eta_{10}\};$
	$\pi_p(\eta_6) = \{\varpi_4\};$	
	$\pi_p(\eta_7) = \{\varpi_4\};$	
	$\pi_p(\eta_8) = \{\varpi_4\};$	
	$\pi_a(\eta_9) = \{\varpi_5\};$	
	$\pi_p(\eta_{10}) = \{\varpi_5\};$	
$\kappa = \kappa_i \cup \kappa_o$	$\kappa_i(\alpha_I) = \{\emptyset\};$	$\kappa_o(\alpha_I) = \{d\};$
	$\kappa_i(\alpha_1) = \{d\};$	$\kappa_o(\alpha_1) = \{d\};$
	$\kappa_i(\alpha_2) = \{or_1\};$	$\kappa_o(\alpha_2) = \{d\};$
	$\kappa_i(\alpha_3) = \{or_2\};$	$\kappa_o(\alpha_3) = \{d\};$
	$\kappa_i(\alpha_4) = \{d\};$	$\kappa_o(\alpha_4) = \{d\};$
	$\kappa_i(\alpha_5) = \{d\};$	$\kappa_o(\alpha_5) = \{d\};$
	$\kappa_i(\alpha_6) = \{d\};$	$\kappa_o(\alpha_6) = \{or_1, or_2\};$
	$\kappa_i(\alpha_7) = \{d\};$	$\kappa_o(\alpha_7) = \{d\};$
	$\kappa_i(\alpha_F) = \{d\};$	$\kappa_o(\alpha_F) = \{\emptyset\};$

At the commencement, it is assumed that all repositories in the set $I \subseteq R$ have been initialized with data by the external system. The execution is terminated by any one λ output transition. The set of output repositories is data holders that may be used after termination by the external system.

2.4 Problem Scope: The Reachable-path Rediscovery Problem

As stated in the previous section, the scope of problem is related with the rediscovery problem that has stated in [14] as a matter of workflow mining techniques. It investigates whether it is possible to rediscover the workflow process by merely looking at its logs. In this paper, we would try to restate the problem so as for the minimal-workflow model to be well applied as a solution. In a little more detailed idea, it gives a way to efficiently rediscover the discrepancy between the original workflow process model as it is built and the enacted workflow process as it is actually executed. The discrepancy, as you can easily see in Fig. 2, is caused by the alternative paths exhibited on the model. The number of alternative paths on the model will effect on the degree of the discrepancy.

For example, after rediscovering a workcase from workflow logs, we need to know along which reachable-path the workcase has followed. In the figure, the workcase rediscovered from the workflow log belongs to one of the three reachable-paths. For simple workflow processes, this reachable-path determination might be quite easy to handle. However, for massively large and complicated workflow process models this is much more difficult and time-consuming work. At the same time, this needs to be done efficiently as the number of workcases becomes quite large. That's the reason why this paper tries to restate the problem as '***reachable-path rediscovery problem***'.

Therefore, we look for efficient mining approaches to solve the reachable-path rediscovery problem. The minimal-workflow model proposed in this paper is one of those approaches. The reachable-path rediscovery problem might be very important and effective issue in order to perform the redesigning and reengineering of workflow process models. In other words, in the reengineering phase, the critical path, which is the reachable-path having the largest number of workcases, might be the most effective one out of all possible reachable-paths. For example, assume that one hundred thousands workcases have been enacted out of the workflow process model shown in the left-most side of the figure. If 90% of the workcases were associated with the first and the second reachable-paths, then those two reachable-paths become the critical path of the model. So, we can reengineer the original workflow process model by separating the critical path from the third reachable-path. The key point in this problem is how fast and how much efficiently we can rediscover the reachable-paths of all of the workcases. In the next section, we suggest a workflow mining framework using the minimal-workflow model as a solution of the problem.

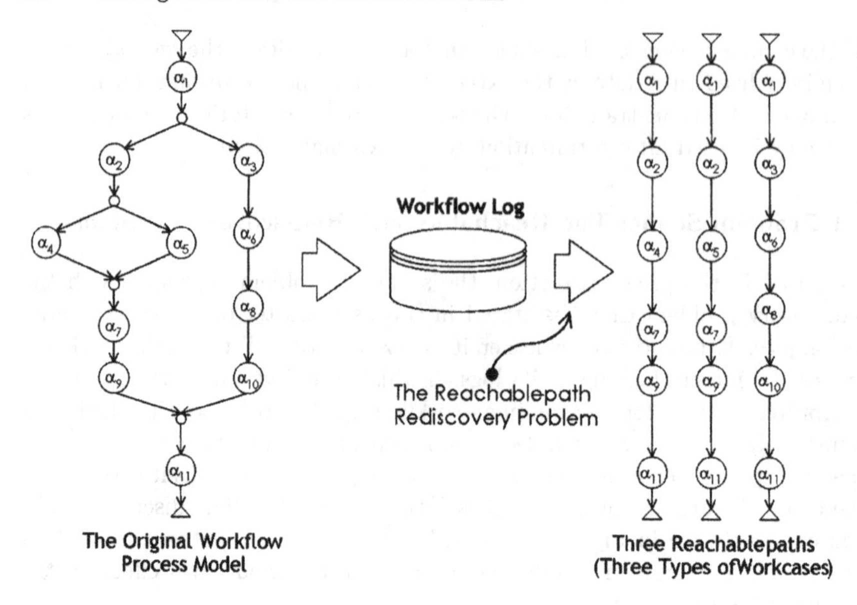

Fig. 2. The reachable-path rediscovery problem: Along which reachable-path does a workcase follow?

3 Workflow Reduction Mechanism

In this section, we give the complete description of a systematic and formal approach to construct a minimal-workflow model from a workflow procedure modeled by the information control net. The minimal-workflow model will play very important role in solving the reachable-path rediscovery problem. That is, it is used as a decision tree induction algorithm in the workflow mining system that has been particularly designed and implemented only for solving the problem. The on-going research project has been carried out through the CTRL research group in Kyonggi University. The construction procedure of the minimal-workflow model and its usages put in the workflow mining framework. The following subsections devote to details and formality of the framework.

3.1 Workflow Reduction Framework

This paper's main contribution is to propose a framework that consists of a series of formal approaches from analyzing workflow dependencies to generating minimal-workflow and algorithms, which is finally used for solving the reachable-path rediscovery problem. The framework, as illustrated in Fig. 3, has three kinds of model transformations. The first transformation is from ICN to workflow dependent net performed through the workflow dependency analysis method, second is from the workflow dependent net to minimal-workflow, and the third one is done from ICN to a set of reachable-paths. These transformations are performed in the modeling time, and finally used for rediscovering a reachable-path of a workcase from workflow log. Based upon these transformations, there might be two possible approaches for solving the reachable-path rediscovery problem as followings:

- Naive approach using the third transformation only (from ICN to a set of reachable-paths)
- Sophisticated approach using the minimal-workflow as a decision tree induction mechanism

As you can easily imagine, the naive approach is straightforwardly done by a simple algorithm generating a set of reachable-paths from an information control net. But, we hardly expect the higher efficiency in mining and rediscovering reachable-paths from workflow logs mounted up by enactments of large-scale and complicated workflow process models. So, this paper is working on the sophisticated approach, which is called workflow mining framework shown in Fig. 3. This approach at first steps out from analyzing control dependencies among activities in a workflow procedure.

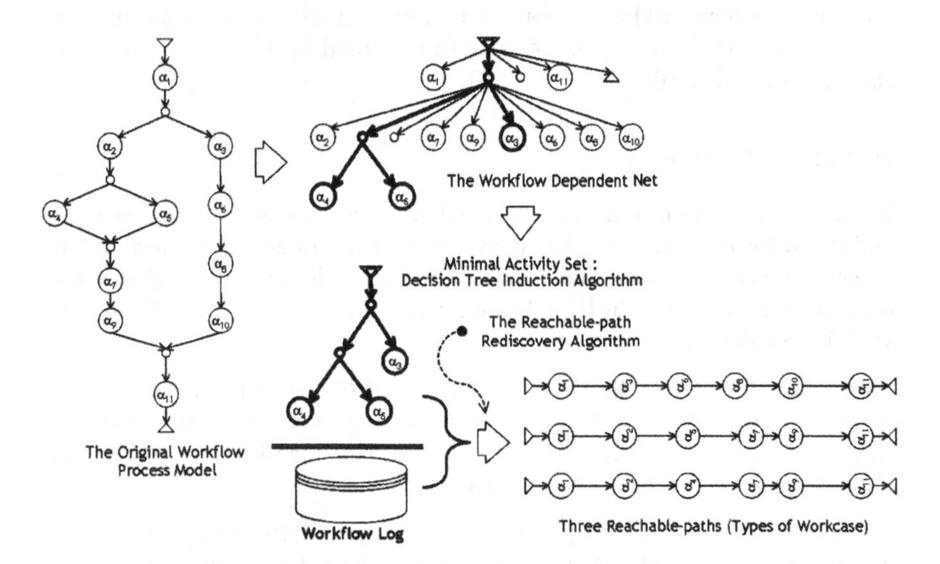

Fig. 3. The Workflow Reduction Framework

Based upon these control dependencies, the framework generates a workflow dependent net by applying the concepts of walk and dominance operations [1]. On the workflow dependent net, it is possible to filter out a minimal-workflow model by using the concepts of immediate backward dominator and immediate forward dominator, which are defined in the next subsection. Finally, the minimal-workflow model serves as a decision tree induction algorithm (generating reduced activity set or minimal activity set) in order to decide along which reachable-path a workcase follows (the reachable-path rediscovery problem). Conclusively, we surely expect the much higher efficiency from the sophisticated approach, because it dramatically reduces the size of activity set that has to be used for rediscovering a reachable-path of a workfcase. At the same time, it is able to filter out a huge amount of enacted activities' data by preprocessing audit data logged on workflow logs.

3.2 Workflow Dependent Net

This subsection describes the formal definitions of workflow dependent net, and conceives an algorithm that constructs a workflow dependent net from the formal specification of an ICN. The workflow dependent net is mainly concerned about the control-flow perspective in a workflow procedure. In particular, it is used to model the effect of conditional branches (*or-split* and *or-join*) and parallel branches (*and-split* and *and-join*) on the behavior of workflow procedure. The workflow dependent net is constructed out of the set δ (control flow part) in the formal notation of ICN model. (Note that the notations and their meanings are same to that used in the ICN definition, if they are not redefined.)

Primitive Operations

From the information control net, the following two operations, such as ***walk*** and ***dominance***, work out. The operations are used for generating a workflow dependent net from an information control net. The basic concepts of the operations were defined in [1], and we revised them so as to fit well into the workflow model.

Definition 1 *A **walk**, W, in an ICN. In an ICN, $\Gamma = (\delta, \gamma, \varepsilon, \pi, \kappa, I, O)$, a **walk** W is a sequence of activities, $\alpha_1, \alpha_2, \ldots, \alpha_n$, such that $n \geq 0$ and $\alpha_{i+1} \in \delta_o(\alpha_i)$ for $i = 1, 2, \ldots, n-1$. The length of a **walk** $W = \alpha_1\alpha_2\ldots\alpha_n$, denoted $|W|$, is the number **n** of activity occurrences in W.*

Note that a *walk* of length zero has no any activity occurrence - such a *walk* is called empty. A nonempty *walk* whose first activity is u and whose last activity is v is called a $u-v$ *walk*. If $W = w_1w_2\ldots w_m$ and $X = x_1x_2\ldots x_n$ are *walks* such that either W is empty, X is empty, or w_m is adjacent to x_1, then the concatenation of W and X, denoted WX, is the *walk* $w_1w_2\ldots w_mx_1x_2\ldots x_n$.

Definition 2 ***Dominance*** *in an ICN. When an ICN, $\Gamma = (\delta, \gamma, \varepsilon, \pi, \kappa, I, O)$, satisfies all of the following conditions: (a) Γ contains two distinguished activities: the initial activity α_I satisfying that $\delta_i(\alpha_I)$ is $\{\emptyset\}$, and the final activity*

α_F satisfying that $\delta_o(\alpha_F)$ is $\{\emptyset\}$; (b) every activity of Γ occurs on somewhere of the $\alpha_I - \alpha_F$ ***walk****, we can define the following types of* ***dominance****:*

- *Let Γ be an ICN. An activity $u \in A$* ***forward-dominates*** *an activity $v \in A$ iff every $v - \alpha_F$ walk in Γ contains u; u* ***properly forward-dominates*** *v iff $u \neq v$ and u forward-dominates v.*
- *Let Γ be an ICN. An activity $u \in A$* ***strongly forward-dominates*** *an activity $v \in A$ iff u forward-dominates v and there is an integer $k \geq 1$ such that every walk in Γ beginning with v and of length k contains u.*
- *Let Γ be an ICN. The* ***immediate forward-dominator*** *of an activity $\alpha \in (A - \{\alpha_F\})$, denoted ifd($\alpha$), is the activity that is the first properly forward-dominator of α to occur on every $\alpha - \alpha_F$ walk in Γ.*
- *Let Γ be an ICN. An activity $u \in A$* ***backward-dominates*** *an activity $v \in A$ iff every $v - \alpha_I$ walk in Γ contains u; u* ***properly backward dominates*** *v iff $u \neq v$ and u backward dominates v.*
- *Let Γ be an ICN. The* ***immediate backward dominator*** *of an activity $\alpha \in (A - \{\alpha_I\})$, denoted ibd($\alpha$), is the activity that is the first proper backward dominator of α to occur on every $\alpha - \alpha_I$ walk in Γ.*

Workflow Dependent Net

Based upon the primitive operations (*walk* and *dominance*), it is possible to generate a set of workflow control dependent relationships being embedded on a workflow model. These dependency relationships are filtered out through the following dependency operations being formally defined in the following theorem. We can automatically construct a workflow dependent net of the workflow model by an algorithm using the dependency operations.

Definition 3 ***Control Dependency*** *in an ICN. Let Γ be an information control net, and let $u, v \in A$.*

- *An activity u is* ***control-dependent*** *on an activity v iff the activity u is a strongly forward-dominator of the activity v.*
- *An activity u is* ***strongly control-dependent*** *on an activity v iff the activity u is not only control-dependent on but also an immediate forward-dominator of the activity v.*
- *The function,* ***fd(v, u)****, returns a set of forward dominators of v in the walk $v - u$.*
- *The function,* ***ifd(v, u)****, returns a set of immediate forward-dominators of the activity v in the walk $v - u$.*

Note that the activities that are strongly control-dependent on the control-transition conditions of an or-split/and-split control construct are those between the fork activity and the join activity of the or-split/and-split control construct. Additionally, the activi-ties, that are strongly control-dependent on the branch condition of a loop-control construct, are the control-transition conditions themselves and the activities in the inside of the loop. But, the loop should not be concerned in this algorithm because it is a special construct of or-split/and-split control construct. The following Fig. 4 is the algorithm to construct a formal workflow dependent net from a workflow procedure in ICN.

Definition 4 ***Workflow Dependent Net*** *of an ICN. A workflow dependent net is formally defined as $\Omega = (\varphi, \xi, I, O)$ over a set A of activities and a set T of transition conditions, where*

- $\varphi = \varphi_i \cup \varphi_o$
  where $\varphi_o : A \longrightarrow \wp(A)$ is a multi-valued mapping of an activity to a set of activities that is ***control-dependent*** *or* ***strongly or multiply control-dependent*** *on the activity, and $\varphi_i : A \longrightarrow \wp(A)$ is a single-valued mapping of an activity to that the activity is* ***control-dependent*** *or* ***strongly or multiply control-dependent****;*
- $\xi = \xi_i \cup \xi_o$
  where ξ_i: a set of control transition conditions, $\tau \in T$, on each arc, $(\varphi_i(\alpha), \alpha)$; and ξ_o: a set of control transition conditions, $\tau \in T$, on each arc, $(\alpha, \varphi_o(\alpha))$, where $\alpha \in A$;
- *I is a finite set of initial input repositories;*
- *O is a finite set of final output repositories;*

In mapping the workflow dependent diagram into its formal definition, a circle represents an activity node, a solid arrow represents a control-dependency between two associated activities, and a control-transition condition is positioned on the solid arrow. Additionally, the workflow dependent net is extensible to accommodate the concept of compound workflow models such as subworkflows, nested workflows, and chained workflows.

Next, how can we build a workflow dependent net from an information control net? We need an algorithm for this. By using those operations such as walk,dominance and control-dependency, we construct the algorithm.

The Workflow Dependent Net Construction Algorithm

A sketch of the algorithm is given as the following:

Input An ICN, $\Gamma = (\delta, \gamma, \varepsilon, \pi, \kappa, I, O)$;
Output A Workflow Dependent Net, $\Omega = (\varphi, \xi)$;
Initialize $T \leftarrow \{\alpha_I\}$; /* u, T are global */
PROCEDURE(In $s \leftarrow \delta_o(\alpha_I)$, **In** $f \leftarrow \{\alpha_F\}$**)** /* Recursive procedure */
BEGIN
$v \leftarrow s; \varphi_i(v) \leftarrow \delta_i(v); \varphi_o(\delta_i(v)) \leftarrow v$; $T \leftarrow T \cup \{v\}$;
$O \leftarrow \delta_o(v)$;
WHILE ($\exists u \in O$ is not equal to f) **DO**
FOR ($\forall u \in O$ and $u \notin T$) **DO**
IF (u is a strongly forward-dominator of v?)

```
          Then do
             IF (u is a multiply forward-dominator of v?)
                Then do
                   Call PROCEDURE(In s ← u, In f ← {'and-join'});
                   φo(δi(v)) ← u; φi(u) ← δi(v); T ← T ∪ {u};
end
                Else do
                   φo(δi(v)) ← u; φi(u) ← v; T ← T ∪ {u}; end
             END IF
          Else do
             Call PROCEDURE(In s ← u, In f ← {'or-join'});
             φo(δi(v)) ← u; φi(u) ← δi(v); T ← T ∪ {u}; end
       END IF
     END FOR
     Replace O To δo(u); /* O ← δo(u); */
   END WHILE
END PROCEDURE
```

The time complexity of the workflow dependent net construction algorithm is $O(n^2)$, where n is the number of activities in an ICN. The functions for deciding both the strongly forward-domination relation and the multiply forward-domination relation between two activities can be computed in $O(n)$, and the recursive procedure itself can be computed in $O(n)$. Therefore, the overall time complexity is $O(n^2)$.

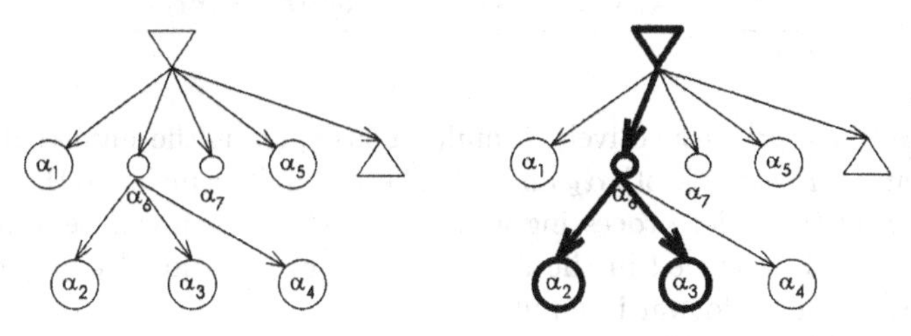

Fig. 4. The Workflow Dependent Net of the Order Processing Workflow Model

The graphical representation of the workflow dependent net for the order processing workflow procedure is presented in Fig. 4. As shown in the figure, there are two nets. The net on the left-hand side represents the workflow dependent net constructed by the algorithm, and the right-hand side one gives marks (bold arrows) on the edges that have some special dominate-relationships, such as ***ifd*** (immediate forward dominator) and ***ibd*** (immediate backward dominator) relationship, with conjunctive or disjunctive logics. For example, α_2 and α_3 (activity nodes) are the immediate forward dominators (*ifd*) of α_6 (disjunctive node - or-split) with respect to transition-conditions,

Table 2. Formal Specification for the Workflow Dependent Net of the Order Processing Workflow Model

$\Omega = (\varphi, \xi, I, O)$ over A, T		*TheOrderProcedureinWDN*
$A = \{\alpha_1, \alpha_2, \alpha_3, \alpha_4, \alpha_5, \alpha_6, \alpha_7, \alpha_I, \alpha_F\}$		*Activities*
$T = \{d(default), or_1('reject'), or_2('accept')\}$		*TransitionConditions*
$I = \{\emptyset\}$		*InitialInputRepositories*
$O = \{\emptyset\}$		*FinalOutputRepositories*
$\varphi = \varphi_i \cup \varphi_o$	$\varphi_i(\alpha_I) = \{\emptyset\}$;	$\varphi_o(\alpha_I) = \{\alpha_1, \alpha_5, \alpha_6, \alpha_7, \alpha_F\}$;
	$\varphi_i(\alpha_1) = \{\alpha_I\}$;	$\varphi_o(\alpha_1) = \{\emptyset\}$;
	$\varphi_i(\alpha_2) = \{\alpha_6\}$;	$\varphi_o(\alpha_2) = \{\emptyset\}$;
	$\varphi_i(\alpha_3) = \{\alpha_6\}$;	$\varphi_o(\alpha_3) = \{\emptyset\}$;
	$\varphi_i(\alpha_4) = \{\alpha_6\}$;	$\varphi_o(\alpha_4) = \{\emptyset\}$;
	$\varphi_i(\alpha_5) = \{\alpha_I\}$;	$\varphi_o(\alpha_5) = \{\emptyset\}$;
	$\varphi_i(\alpha_6) = \{\alpha_I\}$;	$\varphi_o(\alpha_6) = \{\alpha_2, \alpha_3, \alpha_4\}$;
	$\varphi_i(\alpha_7) = \{\alpha_I\}$;	$\varphi_o(\alpha_7) = \{\emptyset\}$;
	$\varphi_i(\alpha_F) = \{\alpha_I\}$;	$\varphi_o(\alpha_F) = \{\emptyset\}$;
$\xi = \xi_i \cup \xi_o$	$\xi_i(\alpha_I) = \{\emptyset\}$;	$\xi_o(\alpha_I) = \{d\}$;
	$\xi_i(\alpha_1) = \{d\}$;	$\xi_o(\alpha_1) = \{\emptyset\}$;
	$\xi_i(\alpha_2) = \{or_1\}$;	$\xi_o(\alpha_2) = \{\emptyset\}$;
	$\xi_i(\alpha_3) = \{or_2\}$;	$\xi_o(\alpha_3) = \{\emptyset\}$;
	$\xi_i(\alpha_4) = \{or_2\}$;	$\xi_o(\alpha_4) = \{\emptyset\}$;
	$\xi_i(\alpha_5) = \{d\}$;	$\xi_o(\alpha_5) = \{\emptyset\}$;
	$\xi_i(\alpha_6) = \{d\}$;	$\xi_o(\alpha_6) = \{or_1, or_2\}$;
	$\xi_i(\alpha_7) = \{d\}$;	$\xi_o(\alpha_7) = \{\emptyset\}$;
	$\xi_i(\alpha_F) = \{d\}$;	$\xi_o(\alpha_F) = \{\emptyset\}$;

'reject' and *'accept'*, respectively. Simultaneously, α_6 is the immediate backward dominator (*ibd*) of both α_2 and α_3. These activities make up the minimal activity set of the order processing workflow model, which will be a minimal-workflow model described in the next subsection. Additionally, α_1, α_5, α_6, α_7 and α_F strongly forward dominate α_I.

Note that the control-transition conditions are not represented in Fig. 4, because they are not directly related with constructing the workflow dependent net from a ICN. It means that special domination-relationships between activities are the necessary-and-satisfactory knowledge for constructing the workflow dependent net model from an ICN model. However, they are becoming essential criteria for finally reducing the workflow dependent net into the minimal workflow model. In the next subsection, we are going to precisely describe about how to use them in the reduction algorithm. The formal specification of the workflow dependent net shown in the Fig. 4 is presented in Table 2.

3.3 The Minimal Workflow Model

In this subsection, we formally define the minimal workflow net model, and derive an algorithm constructing the minimal workflow net from a workflow dependent net. In order to construct the minimal workflow net model, it is necessary to extend the domination-relationship operations (such as *ifd* and *ibd*) so as to incorporate the concept of dependency type. In the workflow dependent net, we treat every node in a net as a unified type - activity. However, in the minimal workflow net model, it is necessary for the nodes to be classified into activity-type with the immediate backward domination, conjunctive-logic-type (*and-split*), and disjunctive-logic-type (*or-split*), which are playing an important role in composing the minimal workflow model. Based upon these operations and classes, the types of dependency are defined as following:

Definition 5 ***Types of Dependency*** *in a workflow dependent net model (WDN). Let Ω be a WDN, $\Omega = (\varphi, \vartheta, I, O)$ over a set of activities, A, and a set of transition-conditions, T. We can define the following* ***types of dependency****:*

- *There exists an* ***ibd-type dependency*** *between two activities, v and u, where $v \in \varphi_i(u) \wedge u \in \varphi_o(v) \wedge v \neq u$ in WDN, which is denoted as* ***ibdtd****(v) and gives an activity, u, that is the immediate backward dominator of v occurring on every $v - \alpha_I$ walk in an ICN.*
- *There exists an* ***conjunctive-type dependency*** *between two activities, v and u, where $v \in \varphi_i(u) \wedge u \in \varphi_o(v) \wedge v \neq u$ in WDN, which is denoted as* ***ctd****(v) and gives an activity, u, that is a 'and-split' activity.*
- *There exists an* ***disjunctive-type dependency*** *between two activities, v and u, where $v \in \varphi_i(u) \wedge u \in \varphi_o(v) \wedge v \neq u$ in WDN, which is denoted as* ***dtd****(v) and gives an activity, u, that is a 'or-split' activity.*

Definition 6 ***Minimal Workflow Net*** *of a workflow dependent net model (WDN). Let M be a MWN, a minimal-workflow net, that is formally defined as $M = (\chi, \vartheta, I, O)$ over a set of activities, A, and a set of transition-conditions, T, where*

- $\chi = \chi_i \cup \chi_o$
  where, $\chi_o : A \longrightarrow \wp(A)$ is a multi-valued mapping of an activity to an another set of activities, each member of which has one of the types of dependency, such as ibd-type, conjunctive-type, or disjunctive-type dependency, and $\chi_i : A \longrightarrow \wp(A)$ is a single-valued mapping of an activity to an another activity that is one of the members in $\{\alpha_I, or - split, and - split;$
- $\vartheta = \vartheta_i \cup \vartheta_o$
  where, ϑ_i: a set of control transition conditions, $\tau \in T$, on each arc, $(\varphi_i(\alpha), \alpha)$; and ϑ_o: a set of control transition conditions, $\tau \in T$, on each arc, $(\alpha, \varphi_o(\alpha))$, where $\alpha \in A$;
- *I is a finite set of initial input repositories;*

- *O is a finite set of final output repositories;*

In mapping a minimal-workflow diagram into its formal definition, solid directed edge coming into a node correspond to χ_i, and solid directed edge going out of a node correspond to χ_o. A minimal-workflow net is formally constructed from a workflow dependent net through the following algorithm. In the algorithm, we need the concepts of the dependency operations, such as *ibdtd*(α), *ctd*(α), and *dtd*(α), newly extended in this section.

The Minimal Workflow Net Construction Algorithm

A sketch of the algorithm is given as the following:

```
Input A Workflow Dependent Net, Ω = (φ, ξ, I, O);
Output A Minimal Workflow Net, M = (χ, ϑ, I, O);
Initialize T ← {∅}; /* T is global */
PROCEDURE(In s ← {α_I}) /* Recursive Procedure */
BEGIN
  v ← s; χ_i(v) ← φ_i(v); χ_o(v) ← {∅}; T ← T ∪ {v};
  O ← φ_o(v);
  FOR (∀u ∈ O) DO
    SWITCH (What type of dependency between v and u is?) DO
        Case 'ibd-type dependency':
            χ_o(v) ← u;          χ_i(u) ← v;
            ϑ_o(v) ← ξ_o(v);     ϑ_i(u) ← ξ_i(u);
            T ← T ∪ {u};
            break;
        Case 'conjunctive-type dependency':
            χ_o(v) ← u;          χ_i(u) ← v;
            ϑ_o(v) ← ξ_o(v);     ϑ_i(u) ← ξ_i(u);
            T ← T ∪ {u};
            Call PROCEDURE(In s ← u);
            break;
        Case 'disjunctive-type dependency':
            χ_o(v) ← u;          χ_i(u) ← v;
            ϑ_o(v) ← ξ_o(v);     ϑ_i(u) ← ξ_i(u);
            T ← T ∪ {u};
            Call PROCEDURE(In s ← u);
            break;
        Default:
            T ← T ∪ {u};
            break;
    END SWITCH
  END FOR
  IF ((x, y ∈ χ_o(v)) ∧ (x ≠ y) ∧ (ϑ_i(x) = ϑ_i(y)) ∧ (x = ibdtp(v)) DO
    Then do
```

Eliminate x (*the ibd-type dependency*)
from the minimal workflow net;
end;
END IF
END PROCEDURE

The minimal workflow net construction algorithm for a workflow dependent net can be computed in $O(n)$, where n is the number of activities in the set, A. Because the statements in the recursive procedure are executed n times that is exactly same to the number of activities, and the time needed for deciding the immediate backward dominator (ibd-type dependency) is $O(1)$. Therefore, the time complexity of the algorithm is $O(n)$.

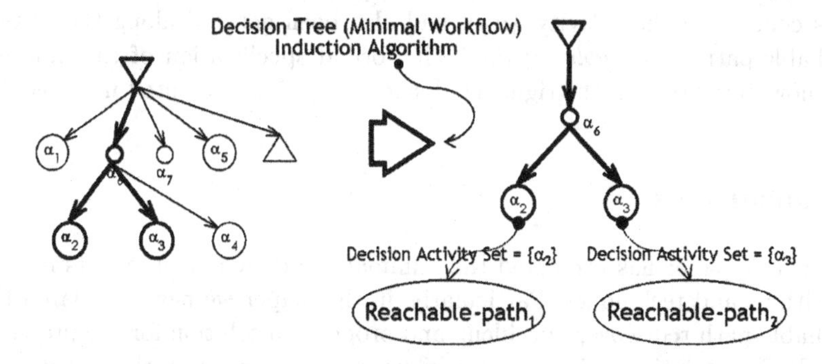

Fig. 5. The Minimal Workflow Net of the Order Processing Workflow Model

Table 3. Formal Specification for the Minimal Workflow Net of the Order Processing Workflow Model

$M = (\chi, \vartheta, I, O)$ over A, T		*TheOrderProcedureinWDN*
$A = \{\alpha_1, \alpha_2, \alpha_3, \alpha_6\}$		*Activities*
$T = \{d(default), or_1('reject'), or_2('accept')\}$		*TransitionConditions*
$I = \{\emptyset\}$		*InitialInputRepositories*
$O = \{\emptyset\}$		*FinalOutputRepositories*
$\chi = \chi_i \cup \chi_o$	$\chi_i(\alpha_1) = \{\emptyset\};$	$\chi_o(\alpha_1) = \{\alpha_6\};$
	$\chi_i(\alpha_2) = \{\alpha_6\};$	$\chi_o(\alpha_2) = \{\emptyset\};$
	$\chi_i(\alpha_3) = \{\alpha_6\};$	$\chi_o(\alpha_3) = \{\emptyset\};$
	$\chi_i(\alpha_6) = \{\alpha_1\};$	$\chi_o(\alpha_6) = \{\alpha_2, \alpha_3\};$
$\vartheta = \vartheta_i \cup \vartheta_o$	$\vartheta_i(\alpha_1) = \{\emptyset\};$	$\vartheta_o(\alpha_1) = \{d\};$
	$\vartheta_i(\alpha_2) = \{or_1\};$	$\vartheta_o(\alpha_2) = \{\emptyset\};$
	$\vartheta_i(\alpha_3) = \{or_2\};$	$\vartheta_o(\alpha_3) = \{\emptyset\};$
	$\vartheta_i(\alpha_6) = \{d\};$	$\vartheta_o(\alpha_6) = \{or_1, or_2\};$

In Fig. 5, the right-hand side is to represent a minimal workflow net extracted from the workflow dependent net (the left-hand side) of the order processing workflow model by the minimal workflow (decision tree) induction algorithm. As a result, the minimal workflow net, which is finally generating the minimal activity set consisting of decision activity sets, will play a very important role in workflow mining mechanism as a reachable-path decision tree induction technique. Based on this minimal activity set, we can easily and efficiently solve the reachable-path rediscovery problem. For example, the minimal activity set for the order processing workflow is $\{\alpha_2, \alpha_3\}$. And the decision activity sets for the $reachable - path_1$ is $\{\alpha_2\}$, and the decision activity set for $reachable - path_2$ is $\{\alpha_3\}$. That is, we can easily see that, from workflow logs, all workcases containing the activity, (α_2), had been followed along the first reachable-path, $(reachable - path_1)$, at the same time, all workcases containing the activity, (α_3), had also been enacted along the second reachable-path, $(reachable - path_2)$.The formal specification of the minimal workflow net shown in the right-hand side of Fig. 5 is presented in Table 3.

4 Conclusions

So far, this paper has presented the minimal-workflow model and its related algorithms and techniques. Particularly, in this paper we newly declared the reachable-path rediscovery problem, and proposed a solution for the problem, as well. The solution is just the workflow mining framework that eventually gives us higher-level of efficiency in rediscovering reachable-paths from workflow logs. In order to construct the model from an information control net, as you've seen, three times of model transformations have to be done, such as the workflow dependent net algorithm, the minimal-workflow net algorithm, and the reachable-path generation algorithm. (Note that the third algorithm has not presented in this paper yet, because it can be straightforwardly done through a simple manner.) These three transformations should be done in modeling or building time. And, the final result of the transformations spawns a minimal activity set of the workflow process model that will be working out with a reachable-path decision algorithm. But, in the algorithms generating the minimal-workflow model we did not take into account the concept of loop construct in workflow model. We would leave it to one of the future works of this paper.

Conclusively, this minimal-workflow model is a sort of decision tree induction technique in workflow mining systems preprocessing data on workflow logs. In recent, the literature needs various, advanced, and specialized workflow mining techniques and architectures that are used for finally feedbacking their analysis results to the redesign and reengineering phase of the existing workflow and business process models. we strongly believe that this work might be one of those impeccable attempts and pioneering contributions for improving and advancing the workflow mining technology, because the minimal-workflow model should be a very useful technique in redesigning and reengineering very large-scale and complicated workflows. The model is going to be applied to the workflow mining system that will be integrated with the workflow management system (e-Chautauqua WfMS) that our research group have been developing through an on-going project.

Notes and Comments. *This research was supported by the University Research Program (URP) (Grant No. C1-2003-03-URP-0005) from the Institute of Information Technology Assessment, Ministry of Information and Communications, Republic of Korea.*

References

1. Andy Podgurski and Lori A. Clarke (1990), "A Formal Model of Program Dependencies and Its Implications for Software Testing, Debugging, and Maintenance", IEEE Trans. on SE, Vol. 16, No. 9
2. Clarence A. Ellis, Gary J. Nutt (1980), "Office Information Systems and Computer Science", ACM Computing Surveys, Vol. 12, No. 1
3. Clarence A. Ellis, Gary J. Nutt (1993), "The Modeling and Analysis of Coordination Systems", University of Colorado/Dept. of Computer Science Technical Report, CU-CS-639-93
4. Clarence A. Ellis (1983), "Formal and Informal Models of Office Activity", Proceedings of the 1983 Would Computer Congress, Paris, France
5. Clarence A. Ellis, Keddara K and Rozenberg G. (1995), "Dynamic Change within Workflow Sys-tems", Proceedings of the ACM SIGOIS Conference on Organizational Computing Systems, Milpitas, CA.
6. C.A. Ellis and J. Wainer (1994), "Goal-based Models of Collaboration", Journal of Collaborative Computing, Vol. 1, No. 1
7. Diimitrios Georgakopoulos, Mark Hornick (1995), "An Overview of Workflow Management: From Process Modeling to Workflow Automation Infrastructure", Distributed and Parallel Data-bases, 3, pp. 115-153
8. Gustavo Alonso and Hans-Joerg Schek (1996), "Research Issues in Large Workflow Management Systems", Proceedings of NSF Workshop on Workflow and Process Automation in Infor-mation Systems: State-of-the-Art and Future Directions
9. James H. Bair (1990), "Contrasting Workflow Models: Getting to the Roots of Three Vendors", Proceedings of International CSCW Conference
10. Kwang-Hoon Kim (1996), "Practical Experience on Workflow: Hiring Process Automation by FlowMark", IBM Internship Report, IBM/ISSC Boulder Colorado
11. Kwang-Hoon Kim and Su-Ki Paik (1996), "Practical Experiences and Requirements on Workflow", Lecture Notes Asian '96 Post-Conference Workshop: Coordination Technology for Col-laborative Applications, The 2nd Asian Computer Science Conference, Singapore
12. Kwang-Hoon Kim and Clarence A. Ellis (1997), "A Framework for Workflow Architectures", University of Colorado/Department of Computer Science, Technical Reports, CU-CS-847-97

13. Stefan Jablonski and Christoph Bussler (1996), "Workflow Management: Modeling Concepts, Architectures and Implementation", International Thomson Computer Press
14. W.M.P. van der Aslst, et el (2002), "Workflow Mining: Which processes can be rediscovered?", Technical Report, Department of Technology Management, Eindhoven University of Technology

Principal Component-based Anomaly Detection Scheme*

Mei-Ling Shyu

Department of Electrical and Computer Engineering
University of Miami
Coral Gables, FL, USA
shyu@miami.edu

Shu-Ching Chen

School of Computer Science
Florida International University
Miami, FL, USA
chens@cs.fiu.edu

Kanoksri Sarinnapakorn

Department of Electrical and Computer Engineering
University of Miami
Coral Gables, FL, USA
ksarin@miami.edu

LiWu Chang

Center for High Assurance Computer Systems
Naval Research Laboratory
Washington, DC, USA
lchang@itd.nrl.navy.mil

* The preliminary version of the current draft was published in [23].

Abstract. *In this chapter, a novel anomaly detection scheme that uses a robust principal component classifier (PCC) to handle computer network security problems is proposed. An intrusion predictive model is constructed from the major and minor principal components of the normal instances, where the difference of an anomaly from the normal instance is the distance in the principal component space. The screening of outliers prior to the principal component analysis adds the resistance property to the classifier which makes the method applicable to both the supervised and unsupervised training data. Several experiments using the KDD Cup 1999 data were conducted and the experimental results demonstrated that our proposed PCC method is superior to the k-nearest neighbor (KNN) method, density-based local outliers (LOF) approach, and the outlier detection algorithm based on the Canberra metric.*

Keywords: *Anomaly detection, principal component classifier, data mining, intrusion detection, outliers, principal component analysis.*

1 Introduction

A rapid technological progress has brought a new era to the way people communicate. With the merging of computers and communications, we now find ourselves highly depend on the digital communication networks in everyday life. For example, people can look for information from everywhere on the Internet and check e-mails or messages at any place, either from a desktop personal computer, a laptop, or even a mobile phone. While we treasure the ease and convenience of being connected, it is also recognized that an intrusion of malicious or unauthorized users from one place can cause severe damages to wide areas. This introduces a serious issue in computer network security. Heady et al. [9] defined an intrusion as "any set of actions that attempt to compromise the integrity, confidentiality or availability of information resources." The identification of such a set of malicious actions is called intrusion detection problem that has received great interest from the researchers.

The existing intrusion detection methods fall in two major categories: *signature recognition* and *anomaly detection* [11,18,19]. For signature recognition techniques, signatures of the known attacks are stored and the monitored events are matched against the signatures. When there is a match, the techniques signal an intrusion. An obvious limitation of these techniques is that they cannot detect new attacks whose signatures are unknown. In contrast, anomaly detection algorithms build a model from the normal training data and detect the deviation from the normal model in the new piece of test data, where a large departure from the normal model is likely to be anomalous. The advantage of the anomaly detection algorithms is that they can detect new types of intrusions [4] with the trade-off of a high false alarm rate. This is because the previously unseen, yet legitimate, system behaviors may also be recognized as anomalies [5,19].

Various intrusion detection techniques in the anomaly detection category that have been proposed in the literature include the robust support vector machines [10] in machine learning and the statistical-based methods. An extensive review of a number of approaches to novelty detection was given in [20,21]. Statistical-based anomaly detection techniques use statistical properties of the normal activities to build a norm profile and employ statistical tests to determine whether the observed activities deviate significantly from the norm profile. A multivariate normal distribution is usually assumed, which can be a drawback. A technique based on a chi-square statistic that has a low false alarm and a high detection rate was presented in [24]. The Canberra technique that is a multivariate statistical based technique was developed in [6]. Though this method does not suffer from the normality assumption of the data, their experiments showed that the technique performed very well only in the case where all the attacks were placed together. Ye et al. [23] proposed a multivariate quality control technique based on Hotelling's T test that detects both counterrelationship anomalies and mean-shift anomalies. The authors showed that their proposed technique detected all intrusions for a small data set and 92% of the intrusions for a large data set, both with no false alarms.

Many anomaly detection techniques employ the outlier detection concept. A detection technique that finds the outliers by studying the behavior of the projections from the data set was discussed in [1]. Another approach called the local outlier factor (LOF) approach was proposed in [3]. LOF is the degree of being an outlier that is assigned to each object, where the degree depends on how isolated the object is with respect to the surrounding neighborhood. Lazarevic et al. [17] proposed several detection schemes for detecting network intrusions. A comparative study of these schemes on DARPA 1998 data set indicated that the most promising technique was the LOF approach [17].

In this chapter, we propose a novel anomaly detection scheme, called principal component classifier (PCC), based on the principal components, outlier detection, and the assumption that the attacks appear as outliers to the normal data. Our proposed principal component-based approach has some advantages. First, unlike many statistical-based intrusion detection methods that assume a normal distribution or resort to the use of the central limit theorem by requiring the number of features to be greater than 30 [24,25], PCC does not have any distributional assumption. Secondly, it is typical for the data of this type of problem to be high dimensional, where dimension reduction is needed. In PCC, robust principal component analysis (PCA) is applied to reduce the dimensionality to arrive at a simple classifier which involves two functions of some principal components. Since only a few parameters of the principal components need to be retained for future detection, the benefit is that the statistics can be computed in little time during the detection stage. Being an outlier detection method, our proposed PCC scheme can find itself in many applications other than intrusion detection, e.g., fault detection, sensor detection, statistical process control, distributed sensor network, etc. Our experimental results show that the method has a good detection rate with a low false alarm, and it outperforms the k-nearest neighbor (KNN) method, the LOF approach, and the Canberra metric.

This chapter is organized as follows. Section 2 presents a brief overview of some statistical data mining tools used in our work. This includes the concept of PCA, distance function, and outlier detection. The proposed PCC scheme is described in Section 3. Section 4 gives the details of the experiments followed by the results and the discussions. Conclusions are presented in Section 5.

2 Statistical Data Mining

2.1 Principal Component Analysis

As data involved in the intrusion detection problems are of large scale, a data mining tool that will come in handy is the principal component analysis (PCA). PCA is often used to reduce the dimension of the data for easy exploration and further analysis. It is concerned with explaining the variance-covariance structure of a set of variables through a few new variables which are functions of the original variables. Principal components are particular linear combinations of the p random variables X_1, X_2, …, X_p with three important properties. First, the principal components are uncorrelated. Second, the first principal component has the highest variance, the second principal component has the second highest variance, and so on. Third, the total variation in all the principal components combined is equal to the total variation in the original variables X_1, X_2, …, X_p. They are easily obtained from an eigenanalysis of the covariance matrix or the correlation matrix of X_1, X_2, …, X_p [12,13,14].

Principal components can be obtained from the covariance matrix or from the correlation matrix. However, they are usually not the same and not simple functions of the others. When some variables are in a much bigger magnitude than others, they will receive heavy weights in the leading principal components. For this reason, if the variables are measured on scales with widely different ranges or if the units of measurement are not commensurate, it is better to perform PCA on the correlation matrix.

Let **R** be a p x p sample correlation matrix computed from n observations on each of p random variables, $X_1, X_2, \ldots, X_p$. If $(\lambda_1, \mathbf{e}_1), (\lambda_2, \mathbf{e}_2), \ldots, (\lambda_p, \mathbf{e}_p)$ are the p eigenvalue-eigenvector pairs of **R**, $\lambda_1 \geq \lambda_2 \geq \ldots \geq \lambda_p \geq 0$, then the ith sample principal component of an observation vector $\mathbf{x} = (x_1, x_2, \ldots, x_p)'$ is

$$y_i = \mathbf{e}_i'\mathbf{z} = e_{i1}z_1 + e_{i2}z_2 + \ldots + e_{ip}z_p, \quad i = 1,2,\ldots,p \tag{1}$$

where

$\mathbf{e}_i = (e_{i1}, e_{i2}, \ldots, e_{ip})'$ is the ith eigenvector

and

$\mathbf{z} = (z_1, z_2, \ldots, z_p)'$ is the vector of standardized observations defined as

$$z_k = \frac{x_k - \bar{x}_k}{\sqrt{s_{kk}}}, \quad k = 1,2,\ldots,p$$

where $\bar{x}_k$ and s_{kk} are the sample mean and the sample variance of the variable X_k.

The ith principal component has sample variance λ_i and the sample covariance of any pair of principal components is 0. In addition, the total sample variance in all the principal components is the total sample variance in all standardized variables $Z_1, Z_2, \ldots, Z_p$, i.e.,

$$\lambda_1 + \lambda_2 + \ldots + \lambda_p = p \tag{2}$$

This means that all of the variation in the original data is accounted for by the principal components.

2.2 Distance Functions

The most familiar distance function is the Euclidean distance that is frequently used as a measure of similarity in the nearest neighbor method. Let $\mathbf{x} = (x_1, x_2, \ldots, x_p)'$ and $\mathbf{y} = (y_1, y_2, \ldots, y_p)'$ be two p-dimensional observations. The Euclidean distance between **x** and **y** is defined as

$$d(\mathbf{x}, \mathbf{y}) = \sqrt{(\mathbf{x} - \mathbf{y})'(\mathbf{x} - \mathbf{y})} \tag{3}$$

In Euclidean distance, each feature contributes equally to the calculation of the distance. However, when the features have very different variability or different features are measured on different scales, the effect of the features with large scales of measurement or high variability would dominate others that have smaller scales or less variability. Therefore, this distance measure is undesirable in many applications.

An alternative distance function is the well-known Mahalanobis distance, where a measure of variability can be incorporated into the distance metric directly. Let **S** be the sample covariance matrix, the Mahalanobis distance is defined as follows.

$$d^2(\mathbf{x},\mathbf{y}) = (\mathbf{x}-\mathbf{y})'\mathbf{S}^{-1}(\mathbf{x}-\mathbf{y}) \tag{4}$$

Another distance measure that has been used in the anomaly detection problem is the Canberra metric. It is defined for nonnegative variables only.

$$d(\mathbf{x},\mathbf{y}) = \sum_{i=1}^{p} \frac{|x_i - y_i|}{(x_i - y_i)} \tag{5}$$

2.3 Outlier Detection

In most cases, the data sets usually contain one or a few unusual observations. When an observation is different from the majority of the data or is sufficiently unlikely under the assumed probability model of the data, it is considered an outlier. For the data on a single feature, unusual observations are those that are either very large or very small relative to the others. If the normal distribution is assumed, any observation whose standardized value is large in an absolute value is often identified as an outlier. However, the situation becomes complicated with many features. In high dimensions, there can be outliers that do not appear as outlying observations when considering each dimension separately and therefore will not be detected from the univariate criterion. Thus, a multivariate approach that considers all the features together needs to be used.

Let $\mathbf{X}_1, \mathbf{X}_2, \ldots, \mathbf{X}_n$ be a random sample from a multivariate distribution.

$$\mathbf{X}_j = (X_{j1}, X_{j2}, \ldots, X_{jp})', \quad j = 1,2,\ldots,n$$

The procedure commonly used to detect multivariate outliers is to measure the distance of each observation from the center of the data. If the distribution of $\mathbf{X}_1, \mathbf{X}_2, \ldots, \mathbf{X}_n$ is multivariate normal, then for a future observation $\mathbf{X}$ from the same distribution, the statistic T^2 based on the Mahalanobis distance

$$T^2 = \frac{n}{n+1}(\mathbf{X}-\overline{\mathbf{X}})'\mathbf{S}^{-1}(\mathbf{X}-\overline{\mathbf{X}}) \tag{6}$$

is distributed as $\frac{(n-1)p}{n-p}F_{p,n-p}$, where

$$\overline{\mathbf{X}} = \frac{1}{n}\sum_{j=1}^{n}\mathbf{X}_j, \quad \mathbf{S} = \frac{1}{n-1}\sum_{j=1}^{n}(\mathbf{X}_j - \overline{\mathbf{X}})(\mathbf{X}_j - \overline{\mathbf{X}})' \tag{7}$$

and $F_{p,n-p}$ denotes a random variable with an F-distribution with p and n-p degrees of freedom [13]. A large value of T^2 indicates a large deviation of the observation $\mathbf{X}$ from the center of the population and the F-statistic can be used to test for an outlier.

Instead of the Mahalanobis distance, other distance functions such as Euclidean distance and Canberra metric can be applied. Any observation that has the distance larger than a threshold value is considered an outlier. The threshold is typically determined from the empirical distribution of the distance. This is because the distributions of these distances are hard to derive even under the normality assumption.

PCA has long been used for multivariate outlier detection. Consider the sample principal components, $y_1, y_2, \ldots, y_p$, of an observation $\mathbf{x}$. The sum of the squares of the standardized principal component scores,

$$\sum_{i=1}^{p}\frac{y_i^2}{\lambda_i} = \frac{y_1^2}{\lambda_1}+\frac{y_2^2}{\lambda_2}+\ldots+\frac{y_p^2}{\lambda_p} \tag{8}$$

is equivalent to the Mahalanobis distance of the observation $\mathbf{x}$ from the mean of the sample [12].

It is customary to examine individual principal components or some functions of the principal components for outliers. Graphical exploratory methods such as bivariate plotting of a pair of principal components were recommended in [7]. There are also several formal tests, e.g., the tests based on the first few components [8]. Since the sample principal components are uncorrelated, under the normal assumption and assuming the sample size is large, it follows that

$$\sum_{i=1}^{q}\frac{y_i^2}{\lambda_i} = \frac{y_1^2}{\lambda_1}+\frac{y_2^2}{\lambda_2}+\ldots+\frac{y_q^2}{\lambda_q}, \quad q \le p \tag{9}$$

has a chi-square distribution with the degrees of freedom q. For this to be true, it must also be assumed that all the eigenvalues are distinct and positive, i.e., $\lambda_1 > \lambda_2 > \ldots > \lambda_p > 0$. Given a significance level α, the outlier detection criterion is then

Observation **x** is an outlier if $\sum_{i=1}^{q} \frac{y_i^2}{\lambda_i} > \chi_q^2(\alpha)$

where $\chi_q^2(\alpha)$ is the upper α percentage point of the chi-square distribution with the degrees of freedom q. The value of α indicates the error or false alarm probability in classifying a normal observation as an outlier.

The first few principal components have large variances and explain the largest cumulative proportion of the total sample variance. These major components tend to be strongly related to the features that have relatively large variances and covariances. Consequently, the observations that are outliers with respect to the first few components usually correspond to outliers on one or more of the original variables. On the other hand, the last few principal components represent the linear functions of the original variables with the minimal variance. These components are sensitive to the observations that are inconsistent with the correlation structure of the data but are not outliers with respect to the original variables [12]. The large values of the observations on the minor components will reflect multivariate outliers that are not detectable using the criterion based on the large values of the original variables. In addition, the values of some functions of the last r components, e.g., $\sum_{i=p-r+1}^{p} \frac{y_i^2}{\lambda_i}$ and $\max_{p-r+1 \leq i \leq p} \left| \frac{y_i}{\sqrt{\lambda_i}} \right|$, can also be examined. They are useful in determining how much of the variation in the observation **x** is distributed over these latter components. When the last few components contain most of the variation in an observation, it is an indication that this observation is an outlier with respect to the correlation structure.

3 Principal Component Classifier

In our proposed principal component classifier (PCC) anomaly detection scheme, we assume that the anomalies are qualitatively different from the normal instances. That is, a large deviation from the established normal patterns can be flagged as attacks. No attempt is made to distinguish different types of attacks. To establish a detection algorithm, we perform PCA on the correlation matrix of the normal group. The correlation matrix is used because each feature is measured in different scales. Many researchers have applied PCA to the intrusion detection problem [2]. Mostly, PCA is used as a data reduction technique, not an outlier detection tool. Thus, it is our interest to study the use of PCA in identifying attacks or outliers in the anomaly detection problem. Although the graphical methods can identify multivariate outliers effectively, particularly when working on principal components, they may not be practical for real-time detection applications. Applying an existing formal test also presents a difficulty since the data need to follow some assumptions in order for the tests to be valid. For example, the data must have a multivariate normal distribution. Therefore, we aim at developing a novel anomaly detection scheme based on the principal components that is computationally fast and easy to employ without imposing too many restrictions on the data. PCA is a mathematical technique. It does not require data to have a certain distribution in order to apply the method. However the method will not be useful if the features are not correlated.

Due to the fact that the outliers can bring large increases in variances, covariances and correlations, it is important that the training data are free of outliers before they are used to determine the detection criterion. The relative magnitude of these measures of variation and covariation has a significant impact on the principal component solution, particularly for the first few components. Therefore, it is of value to begin a PCA with a robust estimator of the correlation matrix. One simple method to obtain a robust estimator is multivariate trimming. First, we use the Mahalanobis metric to identify the 100γ% extreme observations that are to be trimmed. Beginning with the conventional estimators $\bar{\mathbf{x}}$ and $\mathbf{S}$, the distance $d_i^2 = (\mathbf{x}_i - \bar{\mathbf{x}})'\mathbf{S}^{-1}(\mathbf{x}_i - \bar{\mathbf{x}})$ for each observation $\mathbf{x}_i$ (i=1,2,...,n) is calculated. For a given γ, the observations corresponding to the $\gamma * n$ largest values of $\{d_i^2, i = 1,2,\ldots,n\}$ are removed. In our experiments, γ is set to 0.005. New trimmed estimators $\bar{\mathbf{x}}$ and $\mathbf{S}$ of the mean and the covariance matrix are computed from the remaining observations. A robust estimator of the correlation matrix is obtained using the elements of $\mathbf{S}$. The trimming process can be repeated to ensure that the estimators $\bar{\mathbf{x}}$ and $\mathbf{S}$ are resistant to the outliers. As long as the number of observations remaining after trimming exceeds p (the dimension of the vector $\bar{\mathbf{x}}$), the estimator $\mathbf{S}$ determined by the multivariate trimming will be positive definite [12].

This robust procedure incidentally makes the PCC well suited for unsupervised anomaly detection. We cannot expect that the training data will always consist of only normal instances. Some suspicious data or intrusions may be buried in the data set. However, in order for the anomaly detection to work, we assume that the number of normal instances has to be much larger than the number of anomalies. Therefore, with the trimming procedure as described above, anomalies would be captured and removed from the training data set. The proposed PCC comprises two functions of principal component scores.

1. The first function is from the major components $\sum_{i=1}^{q} \frac{y_i^2}{\lambda_i}$. This function, which is the one that has been used in the literature, is to detect extreme observations with large values on some original features. The number of major components is determined from the amount of the variation in the training data that is accounted for by these components. Based on our experiments, we suggest using q major components that can explain about 50 percents of the total variation in the standardized features.
2. The second function is from the minor components $\sum_{i=p-r+1}^{p} \frac{y_i^2}{\lambda_i}$. Different from other existing approaches, we propose the use of this function in addition to the first function to help detect the observations that do not conform to the normal correlation structure. When the original features are uncorrelated, each principal component from the correlation matrix has an eigenvalue equal to 1. On the other hand, when some features are genuinely correlated, the eigenvalues of some minor components will be zero. Hence, the r minor components used in PCC are those components whose variances or eigenvalues are less than 0.20 which would indicate some linear dependencies among the features. The use of value 0.20 for eigenvalues can also be justified from the concept of the coefficient of determination in regression analysis.

A clear advantage of the PCC method over others is that it provides the information concerning the nature of the outliers whether they are extreme values or they do not have the same correlation structure as the normal instances. In the PCC approach, to classify an observation **x**, the principal component scores of **x** for which the class is to be determined is first calculated. Let c_1 and c_2 be the outlier thresholds such that the classifier would produce a specified false alarm rate, **x** is classified as follows.

Classify **x** as a normal instance if $\sum_{i=1}^{q} \frac{y_i^2}{\lambda_i} \le c_1$ and $\sum_{i=p-r+1}^{p} \frac{y_i^2}{\lambda_i} \le c_2$

Classify **x** as an attack if $\sum_{i=1}^{q} \frac{y_i^2}{\lambda_i} > c_1$ or $\sum_{i=p-r+1}^{p} \frac{y_i^2}{\lambda_i} > c_2$

Assuming the data are distributed as multivariate normal, the false alarm rate of this classifier is defined as

$$\alpha = \alpha_1 + \alpha_2 - \alpha_1 \alpha_2 \qquad (10)$$

where $\alpha_1 = P\left(\sum_{i=1}^{q}\frac{y_i^2}{\lambda_i} > c_1 \middle| \mathbf{x} \text{ is normal instance}\right)$

and $\alpha_2 = P\left(\sum_{i=p-r+1}^{p}\frac{y_i^2}{\lambda_i} > c_2 \middle| \mathbf{x} \text{ is normal instance}\right)$.

Under other circumstances, Cauchy-Schwartz inequality and Bonferroni inequality provide a lower bound and an upper bound for the false alarm rate α [15].

$$\alpha_1 + \alpha_2 - \sqrt{\alpha_1\alpha_2} \le \alpha \le \alpha_1 + \alpha_2 \tag{11}$$

The values of α_1 and α_2 are chosen to reflect the relative importance of the types of outliers to detect. In our experiments, $\alpha_1 = \alpha_2$ is used. For example, to achieve 2% false alarm rate, Equation (10) gives $\alpha_1 = \alpha_2 = 0.0101$. Since the normality assumption is likely to be violated, we opt to set the outlier thresholds based on the empirical distributions of $\sum_{i=1}^{q}\frac{y_i^2}{\lambda_i}$ and $\sum_{i=p-r+1}^{p}\frac{y_i^2}{\lambda_i}$ in the training data rather than the chi-square distribution. That is, c_1 and c_2 are the 0.9899 quantile of the empirical distributions of $\sum_{i=1}^{q}\frac{y_i^2}{\lambda_i}$ and $\sum_{i=p-r+1}^{p}\frac{y_i^2}{\lambda_i}$, respectively.

4 Experiments

Several experiments were conducted to evaluate the performance of the PCC method by comparing it with the density-based local outliers (LOF) approach [3], the Canberra metric, and the Euclidean distance. The method based on the Euclidean distance is, in fact, the k-nearest neighbor (KNN) method. We choose k=1, 5, and 15 for the comparative study.

The KDD CUP 1999 data set [16] is used in these experiments, which includes a training data set with 494,021 connection records, and a test data set with 311,029 records. Since the probability distributions of the training data set and the test data set are not the same, we sampled data only from the training data set and used in both the training and testing stages in the experiments. A connection is a sequence of TCP packets containing values of 41 features and labeled as either normal or an attack, with exactly one specific attack type. There are 24 attack types in the training data. These 24 attack types can be categorized into four big attack categories: DOS – denial-of-service, Probe – surveillance and other probing, u2r – unauthorized access to local superuser (root) privileges, and r2l – unauthorized access from a remote machine. However, for the purpose of this study, we treat all the attack types the same as one attack group. These 41 features can be divided into (1) the basic features of individual TCP connections, (2) the content features within a connection suggested by the domain knowledge, and (3) the traffic features computed using a two-second time window. Among the 41 features, 34 are numeric and 7 are symbolic, where the 34 numeric features are used in our experiments.

4.1 Performance Measures

The experiments are conducted under the following framework:

1. All the outlier thresholds are determined from the training data. The false alarm rates are varied from 1% to 10%. For the PCC method, the thresholds are chosen such that $\alpha_1 = \alpha_2$.
2. Both the training and testing data are from KDD CUP 1999 training data set.
3. Each training data set consists of 5,000 normal connections randomly selected by systematic sampling from all normal connections in the KDD CUP 1999 data.
4. To assess the accuracy of the classifiers, we carry out five independent experiments with five different training samples. In each experiment, the classifiers are tested with a test set of 92,279 normal connections and 39,674 attack connections randomly selected from the KDD CUP 1999 data.

The result of classification is typically presented in a confusion matrix [5]. The accuracy of a classifier is measured by its misclassification rate, or alternatively, the percentage of correct classification.

4.2 Results and Discussions

In an attempt to determine the appropriate number of major components to use in the PCC, we conducted a preliminary study by varying the percentage of total variation that is explained by the major components. A classifier of only the major components (r=0) is used. Let c be the outlier threshold corresponding to the desired false alarm rate, if $\sum_{i=1}^{q} \frac{y_i^2}{\lambda_i} > c$ then $\mathbf{x}$ is classified as an attack, and if $\sum_{i=1}^{q} \frac{y_i^2}{\lambda_i} \leq c$, $\mathbf{x}$ is classified as normal.

Table 1. Average detection rates of five PCCs without minor components at different false alarm rates with standard deviation of the detection rate shown in the parenthesis.

False Alarm	PC 30%	PC 40%	PC 50%	PC 60%	PC 70%
1%	67.12 (±0.98)	93.68 (±0.43)	97.25 (±1.06)	94.79 (±0.46)	93.90 (±0.01)
2%	68.97 (±1.52)	94.48 (±1.87)	99.05 (±0.01)	98.76 (±0.60)	96.07 (±0.53)
4%	71.07 (±1.00)	94.83 (±2.46)	99.23 (±0.05)	99.24 (±0.03)	99.24 (±0.04)
6%	71.79 (±0.28)	94.91 (±2.44)	99.33 (±0.00)	99.45 (±0.02)	99.44 (±0.07)
8%	75.23 (±0.28)	98.85 (±0.84)	99.34 (±0.00)	99.49 (±0.03)	99.58 (±0.10)
10%	78.19 (±0.34)	99.26 (±0.24)	99.35 (±0.00)	99.53 (±0.05)	99.65 (±0.08)

Table 1 shows the average detection rates and the standard deviations of five classifiers with different numbers of the major components based on 5 independent experiments. The standard deviation indicates how much the detection rate can vary from one experiment to another. The components account for 30% up to 70% of the total variation. We observe that as the percentage of the variation explained increases, which means more major components are used, the detection rate tends to be higher except for the false alarm rates of 1% to 2%. The PCC based on the major components that can explain 50% of the total variation is the best for a low false alarm rate, and it is adequate for a high false alarm rate as well. The standard deviation of the detection rate is also very small. This suggests the use of q = 5 major components that can account for about 50% of the total variation in the PCC method.

To advocate the selection of r minor components whose eigenvalues are less than 0.20 in the PCC approach, we run some experiments for the value of eigenvalues from 0.05 to 0.25 with major components accounting for 50% of the total variation. The average detection results from 5 independent experiments are very encouraging as shown in Table 2. It can be seen that the detection rates for eigenvalues 0.15 to 0.25 are mostly the same or very close. The differences are not statistically significant. This comes from the fact that all these eigenvalue thresholds produce roughly the same set of minor components. In other words, the results are not overly sensitive to the choice of the eigenvalue threshold as long as the value is not too small to retain a sufficient number of minor components that can capture the correlation structure in the normal group. Table 3 shows that the observed false alarm rates are mostly not significantly different from the specified values, and a few are lower.

Table 2. Average detection rates of PCCs with different criteria for selecting minor components and 50% of the total variation accountable by major componernts with standard deviation of the detection rate shown in the parenthesis.

False Alarm	$\lambda<0.25$	$\lambda<0.20$	$\lambda<0.15$	$\lambda<0.10$	$\lambda<0.05$
1%	98.94 (±0.20)	98.94 (±0.20)	98.94 (±0.20)	97.39 (±1.53)	93.80 (±0.20)
2%	99.15 (±0.01)	99.14 (±0.02)	99.14 (±0.02)	98.31 (±1.10)	97.71 (±1.13)
4%	99.22 (±0.02)	99.22 (±0.02)	99.22 (±0.01)	99.18 (±0.03)	99.17 (±0.05)
6%	99.28 (±0.04)	99.27 (±0.02)	99.27 (±0.02)	99.28 (±0.02)	99.42 (±0.16)
8%	99.44 (±0.06)	99.41 (±0.04)	99.42 (±0.03)	99.47 (±0.07)	99.61 (±0.11)
10%	99.59 (±0.11)	99.54 (±0.06)	99.55 (±0.04)	99.60 (±0.07)	99.70 (±0.09)

Table 3. Observed false alarm rate of PCCs from 92,279 normal connections with different criteria for selecting minor components and 50% of the total variation accountable by major componernts with standard deviation of the detection rate shown in the parenthesis.

False Alarm	$\lambda<0.25$	$\lambda<0.20$	$\lambda<0.15$	$\lambda<0.10$	$\lambda<0.05$
1%	0.91 (±0.04)	0.92 (±0.06)	0.95 (±0.06)	1.02 (±0.09)	1.00 (±0.10)
2%	1.89 (±0.10)	1.92 (±0.10)	1.95 (±0.05)	1.98 (±0.01)	1.95 (±0.11)
4%	3.89 (±0.06)	3.92 (±0.08)	3.92 (±0.08)	4.07 (±0.14)	4.27 (±0.30)
6%	5.77 (±0.23)	5.78 (±0.24)	5.78 (±0.24)	6.10 (±0.22)	6.76 (±0.50)
8%	7.05 (±0.28)	7.06 (±0.29)	7.09 (±0.25)	7.57 (±0.42)	8.56 (±0.29)
10%	8.53 (±0.32)	8.49 (±0.29)	8.52 (±0.24)	9.10 (±0.56)	10.40 (±0.40)

A detailed analysis of the detection results indicates that a large number of attacks can be detected by both major and minor components, some can only be detected by either one of them, and a few are not detectable at all since those attacks are not qualitatively different from the normal instances. An example is some attack types in category Probe. The detection rate in this category is not high, but it does not hurt the overall detection rate due to a very small proportion of this class in the whole data set, 414 out of 39,674 connections. We use the Probe group to illustrate the advantages of incorporating the minor components in our detection scheme. Figure 1 gives the detailed results of how the major components and minor components alone perform as compared to the combination of these two in PCC using the receiver operating characteristic (ROC) curve, which is the plot of the detection rate against the false alarm rate. The nearer the ROC curve of a scheme is to the upper-left corner, the better the performance of the scheme is. If the ROCs of different schemes are superimposed upon one another, then those schemes have the same performance [22]. For this attack category, the minor component function gives a better detection rate than the major component function does. Many more attacks are detected by the minor components but would otherwise be ignored by using the major components alone. Hence, the use of the minor function improves the overall detection rate for this group.

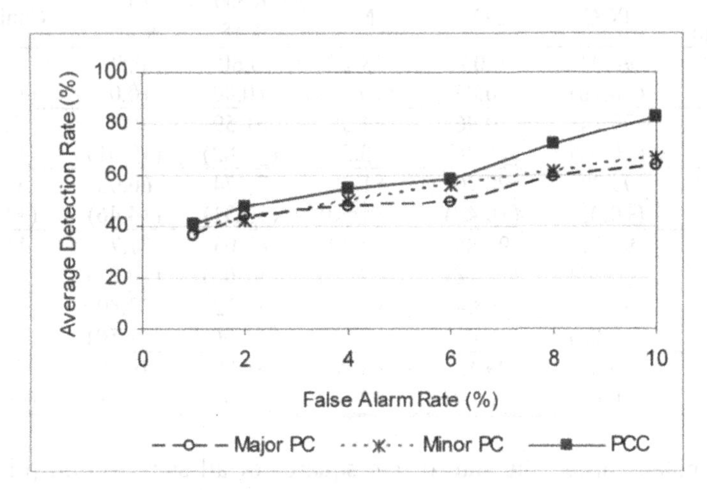

Fig. 1. Average detection rates in Probe attack type by PCC and its major and minor components

Next, the detection rates of PCC with both major and minor components at different false alarm levels were compared to the other six detection methods (shown in Table 4). The results are the average of five independent experiments. As seen from the table, the results of some methods vary wildly, e.g., when the false alarm is 6%, the NN method (k=1) has 9.86% standard deviation, and the detection rate from the 5 experiments ranges from 70.48% to 94.58%. In general, the Canberra metric performs poorly. This result is consistent to the results presented in [6] that it does not perform at an acceptable level. The PCC has a detection rate about 99% with a very small standard deviation at all false alarm levels while the LOF approach does well only when the false alarm rate is not too small. For the k-nearest neighbor method, k=1 has a better performance than a larger k. The PCC outperforms all other methods as clearly seen from this table. It is the only method that works well at low false alarm rates. The LOF approach is better than the nearest neighbor method when the false alarm rate is high, and vice versa. These experimental results show that our anomaly detection scheme based on the principal components works effectively in identifying the attacks. Additionally, the PCC also has the ability to maintain the false alarm at the desired level. The only comparable competitor in our study is the LOF approach, but that is when the false alarm rate is 4% or higher.

Table 4. Average detection rates of six anomaly detection methods with standard deviation of the detection rate shown in the parenthesis.

False Alarm	PCC	LOF	NN	KNN k=5	KNN k = 15	Canberra
1%	98.94 (±0.20)	0.03 (±0.03)	58.25 (±0.19)	0.60 (±0.00)	0.59 (±0.00)	4.12 (±1.30)
2%	99.14 (±0.02)	20.96 (±10.90)	64.05 (±3.58)	61.59 (±4.82)	0.60 (±0.01)	5.17 (±1.21)
4%	99.22 (±0.02)	98.70 (±0.42)	81.30 (±8.60)	73.74 (±3.31)	60.06 (±4.46)	6.13 (±1.14)
6%	99.27 (±0.02)	98.86 (±0.38)	87.70 (±9.86)	83.03 (±3.06)	71.72 (±0.01)	11.67 (±2.67)
8%	99.41 (±0.02)	99.04 (±0.43)	92.78 (±9.55)	87.12 (±1.06)	72.80 (±1.50)	26.20 (±0.59)
10%	99.54 (±0.04)	99.13 (±0.44)	93.96 (±8.87)	88.99 (±2.56)	79.10 (±1.59)	28.11 (±0.04)

As noted earlier, the sum of the squares of all standardized principal components $\sum_{i=1}^{p} \frac{y_i^2}{\lambda_i}$ is basically the Mahalanobis distance. By using some of the principal components, the detection statistic would have less power. However, our experiments show that PCC has sufficient sensitivity to detect the attacks. Also, unlike Mahalanobis, PCC offers more information on the nature of attacks from the use of two different principal component functions. One more benefit of PCC is that during the detection stage, the statistics can be computed in less amount of time. This is because only one third of the principal components are used in PCC, 5 major principal components which explain 50% of the total variation in 34 features, and 6 to 7 minor components that have eigenvalues less than 0.20.

5 Conclusions

In this chapter, we presented a novel anomaly detection scheme called PCC. PCC is capable of detecting an attack or intruder in the computer network system with the benefit that it distinguishes the nature of the anomalies whether they are different from the normal instances in terms of extreme values or different correlation structures. Based on the outlier detection concept, PCC is developed from the major principal components (explaining about 50% of the total variation) and minor components (eigenvalues less than 0.20) of the normal connections. We have conducted several experiments using the KDD CUP 1999 data and the experimental results demonstrate that PCC performs consistently better than the other techniques regardless of the specified false alarm rates.

Acknowledgements

For Mei-Ling Shyu, this research was supported in part by NSF ITR (Medium) IIS-0325260. For Shu-Ching Chen, this research was supported in part by NSF EIA-0220562 and NSF HRD-0317692. For Shu-Ching Chen and Mei-Ling Shyu, this research was supported in part by Naval Research Laboratory (NRL)/ITT: 176815J.

References

1. Aggarwal CC, Yu PS (2001) Outlier detection for high dimensional data. In: The ACM SIGMOD Conference, Santa Barbara, California, pp 37-46
2. Bouzida Y, Gombault S (2003) Intrusion detection using principal component analysis. In: Proc. of the 7th World Multiconference on Systemics, Cybernetics and Informatics, Orlando, Florida.
3. Breunig MM, Kriegel H-P, Ng RT, Sander J (2000) LOF: Identifying density-based local outliers. In: Proc. of the ACM SIGMOD Conference, Dallas, Texas, pp 93-104

4. Denning DE (1987) An intrusion detection model. IEEE Transactions on Software Engineering, SE-13, pp 222-232
5. Dokas P, Ertoz L, Kumar V, Lazarevic A, Srivastava J, Tan P-N (2002) Data mining for network intrusion detection. In: Proc. of the National Science Foundation Workshop on Next Generation Data Mining. pp 21-30
6. Emran SM, Ye N (2001) Robustness of Canberra metric in computer intrusion detection. In: Proc. of the IEEE Workshop on Information Assurance and Security, United States Military Academy, West Point, New York, pp 80-84
7. Gnanadesikan R, Kettenring JR (1972) Robust estimates, residuals and outlier detection with multiresponse data. Biometrics, 28: 81-124
8. Hawkins DM (1974) The detection of errors in multivariate data using principal components. Journal of the American Statistical Association, vol 69, no 346: 340-344
9. Heady R, Luger G, Maccabe A, Servilla M (1990) The architecture of a network level intrusion detection system, Technical Report, Computer Science Department, University of New Mexico
10. Hu W, Liao Y, Vemuri VR (2003) Robust support vector machines for anomaly detection in computer security. In: Proc. of the International Conference on Machine Learning and Applications (ICMLA), Los Angeles, California, pp 168-174
11. Javitz HS, Valdes A (1991) The SRI statistical anomaly detector. In: Proc. of the IEEE Symposium on Research in Security and Privacy, Oakland, California, pp 316-326
12. Jobson JD (1992) Applied multivariate data analysis, Volume II: Categorical and multivariate methods. Springer-Verlag, New York
13. Johnson RA, Wichern DW (1998) Applied multivariate statistical analysis, 4th edn. Prentice-Hall, New Jersey
14. Jolliffe T (2002) Principal component analysis, 2nd edn. Springer-Verlag, New York
15. Kendall MG, Stuart A, Ord JK (1987) Kendall's advanced theory of statistics, V. 1 Distribution theory, 5th edn. Oxford University Press, New York
16. KDD Cup 1999 Data, Available on: http://kdd.ics.uci.edu/databases/kddcup99/kddcup99.html
17. Lazarevic A, Ertoz L, Kumar V, Ozgur A, Srivastava J (2003) A comparative study of anomaly detection schemes in network intrusion detection. In: Proc. of the 3rd SIAM Conference on Data Mining
18. Lee W, Stolfo SJ (2000) A framework for constructing features and models for intrusion detection systems. ACM Transactions on Information and System Security, 3:227-261
19. Lippmann RP, Fried DJ, Graf I, Haines JW, Kendall KP, McClung D, Weber D, Webster SE, Wyschogrod D, Cunningham RK, Zissman MA (2000) Evaluating intrusion detection systems: The 1998 DARPA off-line intrusion detection evaluation. In: Proc. of the DARPA Information Survivability Conference and Exposition (DISCEX) 2000, vol 2, IEEE Computer Society Press, California, pp 12-26

20. Markou M, Singh S (2003) Novelty detection: A review, Part 1: Statistical approaches. Signal Processing, 83, 12:2481-2497
21. Markou M, Singh S (2003) Novelty detection: A review, Part 2: Neural network-based approaches. Signal Processing, 83, 12: 2499-2521
22. Maxion RA, Tan KMC (2000) Benchmarking anomaly-based detection systems. In: Proc. of the 1st International Conference on Dependable Systems & Networks, pp 623-630
23. Shyu M-L, Chen S-C, Sarinnapakorn K, Chang L (2003) A novel anomaly detection scheme based on principal component classifier. In: Proc. of the IEEE Foundations and New Directions of Data Mining Workshop, in conjunction with the Third IEEE International Conference on Data Mining (ICDM'03), Melbourne, Florida, pp 172-179
24. Ye N, Chen Q (2001) An anomaly detection technique based on a chi-square statistic for detecting intrusions into information systems. Quality and Reliability Eng. Int'l, 17:105-112
25. Ye N, Emran SM, Chen Q, Vilbert S (2002) Multivariate statistical analysis of audit trails for host-based intrusion detection. IEEE Transactions on Computers, 51:810-820

Making Better Sense of the Demographic Data Value in the Data Mining Procedure

Katherine M. Shelfer, Xiaohua Hu

College of Information Science and Technology, Drexel University Philadelphia, PA 19104, USA

ABSTRACT

Data mining of personal demographic data is being used as a weapon in the War on Terrorism, but we are forced to acknowledge that it is a weapon loaded with interpretations derived from the use of dirty data in inherently biased systems that mechanize and de-humanize individuals. While the unit of measure is the individual in a local context, the global decision context requires that we understand geolocal reflexive *communal* selves who have psychological and social/societal relationship patterns that can differ markedly and change over time and in response to pivotal events. Local demographic data collection processes fail to take these realities into account at the data collection *design* stage. As a result, existing data values rarely represent an individual's multi-dimensional existence in a form that can be mined. An abductive approach to data mining can be used to improve the data *inputs*. Working from the "decision-in," we can identify and address challenges associated with demographic data collection and suggest ways to improve the quality of the data available for the data mining procedure. It is important to note that exchanging old values for new values is rarely a 1:1 substitution where qualitative data is involved. Different constituent user populations may require different levels of data complexity and they will need to improve their understanding of the data values reported at the local level if they are to effectively relate various local demographic databases in new and different global contexts.

1. INTRODUCTION AND OVERVIEW

In 1948, the UN General Assembly declared personal privacy and the associated freedoms of association, belief, inquiry and movement to be universal (UN 1948). They are guaranteed by the U.S. Constitution. According to Andrew Ford of Usenet, *"Without either the first or second amendment, we would have no liberty."* At the same time, data mining of personal demographic data is increasingly used as a weapon to deny opportunity to individuals, even though it is a weapon loaded with interpretations that are derived from the use of dirty demographic data containered in inherently biased systems that mechanize and de-humanize individuals.

When demographic data on individuals is mined, individuals are judged on the basis of data associations that form certain links and patterns. This means that the use of data mining in law enforcement triggers a response based on an individual's presumptive *guilt* even though a nation may espouse the *de jure* presumption of *innocence.* Since presumption of guilt can victimize the innocent even in the most traditional of legal proceedings (Yant 1991), there was a public outcry over the Total Information Assurance initiative that would have authorized secret law enforcement data mining "fishing expeditions" using combined heterogeneous demographic databases (e.g., credit card companies, medical insurers, and motor vehicle databases). DARPA has asserted its intention to protect constitutional guarantees in relation to data mining activities (DARPA 2002). However, given the role that data mining already plays in risk-based assessments of all types (PCI 2004), it is imperative to provide the most accurate interpretation of demographic data values possible.

Regardless of how we approach demographic data mining, we should remain aware that a number of specific federal and state laws and regulations address the control and/or use of personal data. There are reasons for formalizing such protections that take precedence over technological and resource constraints focused on economies of scope and scale. For example, researchers and applied workers cannot seem to agree on the relative merits of the various statistical methodologies that support important data mining applications (e.g., credit scoring) in use at the present time, yet these systems are sometimes used as the sole decision criteria to adjudicate "guilt" and deny opportunity to individuals and classes of individuals in a range of otherwise unrelated risk-based contexts.

It is axiomatic that individuals exist simultaneously on *at least* three planes: physical, psychological and social. There is a growing recognition of the existence of geolocal relationships; that is, "communities" exist that can be simultaneously local and transnational. These communal associations are comprised of individuals who are mutually reflexive. As a result, we know that psychological and social/societal relationship patterns can differ markedly over time and in response to psychographic stimuli including pivotal events. Since the local data collection effort centers on capturing one-dimensional attributes of an individual, the existing data values rarely represent such multi-dimensional existences in a form that can be mined. In other words, the unit of *measure* is the individual in a local context, yet the unit of *interest* is the mutually reflexive communal self in a global context. This is one reason that we must be on guard against the fallacies associated with the use of "collective" terms. (Reid 1981)

Today, data mining activities can be used to harm individuals who have no right (and no avenue) of appeal (e.g., immigration visa approvals). Moreover, these procedures fail to identify many high-risk individuals, including terrorists who have hidden in plain sight. Unless law enforcement can find ways to improve the data mining of demographic data, associated problems with interpretation and use could undermine the fundamental freedoms that military and law enforcement agencies are tasked to protect. If law enforcement is to retain access to data mining as a weapon and extend its effective range, then the problems associated with turning individuals into collateral damage must be addressed. This damage is of two types: (1) failure to identify a dangerous individual (e.g., a terrorist) in a timely fashion, resulting in harm to others; and (2) misidentification as a dangerous individual (e.g., a terrorist) of an individual who is not dangerous, resulting in a waste of analytical attention at a time when existing information systems and analytical resources are strained. To have any chance of relieving the pressure, we must find better ways to gather, record and report multi-dimensional and relational data associations between individuals at the outset. *Efforts need to be concentrated on trying to resolve problems rather than trying to dissolve them.* (Fergus and Reid 2001)

Data mining that works from the "data-out" derives meaning from existing data values. Unlike traditional inductive and deductive approaches to data mining that focus on data *outputs*, the abductive "decision-in" approach allows us to challenge the interpretation and use of data *inputs*. By questioning the questions used to gather the demographic data available for the data mining procedure, we are better able to recognize and address existing problems with demographic data collection instruments *at the de-*

sign stage. This, in turn, allows us to use available resources to improve the quality of the data available for the data mining procedure. This results in improving the interpretations derived from it. It is important to note that exchanging old values for new values is rarely a 1:1 substitution where qualitative data is involved. We must also recognize that different constituent user populations may require different levels of data complexity, or they may need to relate data values in new and different ways.

The immediate benefits obtained by taking an abductive approach to the problem-solving process can be observed in several knowledge domains. For example,

1. Psychotherapists recognize that individuals associated in family units are mutually reflective and supportive. It has been documented that an individual's demographic and psychographic attributes can change as a result of contact with others. (Fergus and Reid 2001) Data mining that works from the "decision-in" can be used to guide improvements in data collection and management to allow us to recognize land address the inherent "we-ness" of individuals. (Acitelli 1993)
2. Library collections have grown exponentially and changed formats dramatically, first with the advent of printing and again with the introduction of computer-based knowledge transfer technologies. Librarians have changed subject classification systems to improve logical associations between items in their collections.

This chapter discusses the use of demographic data mining to protect personal identity and prevent identity fraud, followed by discussions on the problems associated with the use of demographic data mining using the war on terrorism and credit scoring as a focus. We conclude with some suggestions for future research associated with improving demographic data mining at the data collection *design* stage.

2. PERSONAL IDENTITY AND IDENTITY FRAUD

2.1 Personal Identity

Demographic data is associated with both individuals and households, collected in databases and mined for meaning. However, insufficient theoretical attention has been applied to the practical problems associated with consolidating separately "containered" data collections that are used for data mining. The unit of measure remains the individual. This has led to systems of data collection and data mining that mechanize and de-individualize individuals. The impact of interpersonal and intra-personal contexts, even where they are available at the point of origin of the data, are ignored. The literatures of the social sciences provide extensive insights into what constitutes an "individual." (Acitelli 1993; Mead 1934; Miller 1963) At the data collection stage, there are unrecorded psychosocial, geopolitical, sociological and systemic contextual interpretations attached to recorded demographic data values. Data mining attempts to restructure these relational dynamics. However, the process breaks down when data mining fails to acknowledge that the highly customized relationships that participants create can cause us to miss and/or misrepresent the points at which the proportion of risk become less predictable (Fergus and Reid 2001).

Improving the quality of existing databases requires a substantial change in approach. This is necessary if we are to be sure that we learn what we need to know and can correctly interpret what we learn. At this time, localized law enforcement threat assessment systems are working on the problem from the perspective of a foreigner in an unfamiliar culture, because data values do not actually tell them what they need to know. We have observed that terrorists, for example, are part of a larger organism that takes the form of one or more terrorist cells. As a result, the terrorist may engage in an individual terrorist act that is actually part of a larger operation. As a result, an individual terrorist may act alone but s/he does not act in a vacuum. The unit of understanding for an individual terrorist, then, is actually the terrorist cell.

If we expect the data mining procedure to accurately divine a particular individual's intentions, we must first understand whether there is, in fact, some overarching transnational "self" that has, in fact, become a substitute for an individual's demographics in terms of interpretive value. The culture of terrorist cells and their individual members are mutually reflective and supportive. To improve our ability to identify terrorists, we must find ways to acquire and record a deeper, richer sense of an individual's relational space (Josselson 1994). We must find ways to capture and record data that bridges these personal, psychological and social domains.

The systemist-constructivist approach is useful, because it emphasizes shared frames of reference over the restructuring of relational dynamics. (Shotter 2001). This approach is based on the assumption that change within systems and individuals is mutually reflective. Failure to capture systemic understanding and link it to individuals and groups continues to constrain our understanding. Despite the potential difficulties involved in capturing and recording this elusive sense of "we-ness," it is important to locate the point at which terrorist and terrorist cells and systems converge, merge and diverge geographically and psychographically. It should be understood that the creation of contextual data elements is not simply a premature aggregation of existing data. Rather, it involves the identification of *new* data elements as well as *re*-aggregation of data in ways that recognize the multiple dimensions and relational spaces that are inherently associated with any individual or group.

2.2 Identity Fraud

The bombing of the World Trade Center on 11 September 2001 raised our awareness of the seriousness of this problem, especially as it relates to the activities of terrorists. The use of fraudulent identities and documents provided those terrorists with the prefect cover. Identity (ID) fraud is an enormous and rapidly growing problem, affecting over 750,000 separate individuals per year. ID fraud involves the use of a victim's identity to acquire new accounts and/or take control of existing ones.

In October 1998, Congress passed the Identity Theft and Assumption Deterrence Act of 1998 (Identity Theft Act) to address the problem of identity theft. Specifically, the Act amended 18 U.S.C. § 1028 to make it a federal crime to knowingly transfer or use, without lawful authority, a means of identification of another person with the intent to commit, or to aid or abet, any unlawful activity that constitutes a violation of Federal law, or that constitutes a felony under any applicable State or local law. Identity verification contributes to our ability to identify and track terrorists and those who contribute to their support. There are three forms of verification:

- **Positive verification**. This requires one or more sources of independent validation and verification of an individual's personal information. Data mining cannot contribute useful interpretations where underlying demographic data values cannot be accurately cross validated.
- **Logical verification**. We ask whether a specific transaction makes sense. Data mining on past group behaviors must include ways to identify and extract the "nonsense" of sense-making in order to avoid harming individuals who are going to be "outliers" in a statistical sense, but who should *not,* for various reasons, be denied opportunity.

- **Negative verification.** This process relies on the use of historical or other risk databases, in which past incidents are used to predict future risks. Given that past databases are filled with dirty data on both individuals and outcomes, there must be a way to alert applied workers who perform the data mining procedure, so that the "dirt" in the database does not "rub off" on innocent victims.

Where data mining is used in ID fraud detection, we must be very careful not to victimize the innocent. Here is an example of the types of situations that are already happening. A data mining application is used to support a credit card "fraud detection" system. The system targets a traveler's company-purchased airline ticket as "suspect." The traveler is stranded mid-journey with a revoked ticket when the company is closed on a holiday and the telephone call to authenticate the ticket purchase cannot be completed. The traveler's *rescuer* is victimized when replacement credit card information, vocalized by airline staff, is stolen and used to purchase additional flights. When this second credit card is canceled, the traveler expends funds earmarked for post-holiday credit card payments to complete the business trip. As a result, these payments are late while the travelers awaits additional reimbursements for the trip. The "late" status of the payments enters the credit scoring system, where the traveler continues to be victimized by data mining applications in future—and otherwise unrelated—contexts that rely on credit scoring in the decision process.

3. DATA MINING AND THE WAR ON TERRORISM

Data mining is the automated extraction of hidden predictive information from databases. However, the data must first be available to be mined. The purpose of building models is to make better *informed* decisions, not just faster ones, especially where the data mining procedure cannot be monitored or even understood by the majority of those impacted by it. If predictive models are carelessly constructed, and data mining is undertaken with problematic decision rules based on limited information or insight into the problem context, then the results are likely to cause more harm than having no information at all. Even worse, developing and deploying such systems diverts substantial resources from higher and better uses.

3.1 Risk Management in Global Law Enforcement Contexts

We know that the most important goal when building models is that the model should make predictions that will hold true when applied to unseen data. We also know that a *stable* model is one that behaves similarly when applied to different sets of data. This is unlikely to happen where systems are polluted with data that represents hidden assumptions and ethnocentric biases. There must be time to think about what the data means—and does not mean—in its various contexts of use.

At the present time, the key concern that drives law enforcement to mine demographic data is the need to identify individuals and groups of individuals who pose a significant *transnational risk.* By this, we mean those individuals who ignore the invisible lines used to describe national boundaries, who aggregate to form a geographically-disbursed communal "self" that fails to register on the radar screens of "nation"-centered data collection processes and the associated analytical frameworks.

In the law enforcement context, data mining supports three objectives:

1. To Identify and prioritize the use of limited resources;
2. To effectively apply these limited resources to improve national security by improving law enforcement's ability to identify dangerous individual and deny them the opportunity to harm others; and
3. To accurately identify and predict the potential risks associated with a single *specific* individual from a population of *unknown* individuals.

This means that data mining is not simply an interesting exercise for "numbers junkies." That is, law enforcement analysts cannot legitimately defend efforts to quantify and objectify decisions when they are based on subjective ethnocentric decision rules governing the use of demographic data that have no meaning or relevance in the new game. For this reason, successful data mining depends on access to the *right* data, even though this may require changes in laws and regulations related to demographic data collection processes. We know that it will never be possible to correctly identify each individual terrorist, as other individuals with apparently "clean" records will rise up to take their places. It will never be possible to eliminate every threat. However, the biases inherent in current systems that constrain our understanding must be reduced.

Mining data in systems that fail to retrospectively re-interpret past insights in light of new information can be more dangerous than helpful. If a sufficient number of individuals are victims of "friendly fire," laws will be passed to eliminate or at least reduce the use of data mining to support decision processes. This is already happening as consumers learn about the role of credit scoring in insurance, for example. When systems become suspect, the general public is likely to resist their use long after the data is cleaned and the embedded data mining procedures and interpretations have been improved. The price of continued dependence on dirty systems is also high because reliance on data that represents misinterpretations or patterns of discrimination can disguise important changes in pattern and process in different settings and contexts over time. At this point, it should again be emphasized that most systems are not statistically credible or useful when they use demographic data that is not retrospectively re-interpreted and cleaned, because this results in interpretations that support self-fulfilling prophecies.

3.2 Risk Management Models and Myopia

The primary steps of the Risk Assessment & Risk Mitigation Process Model are: (1) Gather the data; (2) Model and Score the risks; (3) Predict the outcomes; and (4) Embed risk mitigation into decision processes. (www.ncisse.org/Courseware/NSAcourses/lesson4/lesson4.PPT). This model actually constrains our thinking, because it fails to emphasize that *planning and direction* in the problem context should always precede data collection. Planning and direction looks for ways to maximize the return on investment derived from all subsequent stages of the risk management process. We can apply this logic to data mining processes as well. By taking an *abductive* approach (focusing on *inputs* rather than the outputs), we can improve demographic data available for the data mining procedure.

We manage risk through the identification, measurement, control and minimization of security risks in systems to a level commensurate with the potential impact of adverse effects. Based on the existing evidence, law enforcement data mining, models and scores have not learned as much as they could from the existing nation-based intake systems that center on individuals rather than loosely coupled transnational groups. For our purposes, the models have not been designed to support the right problem. Radical restructuring is required. Such efforts call for an unprecedented level of collaboration between experts across knowledge domains as well as between agencies of nations if global terrorism threat assessments are to be optimized *at all times in all locations for all nations.*

Where data quality and data mining that results in models and interpretive scores is used to provide/deny opportunity to *individuals* as well as groups, it will not be possible to make the correct determination that any specific individual is (or is *not*) a risk, e.g., criminal; terrorist) each and every time. Expert analysts will continue to be limited by their knowledge base, local experiences and lack of access to transnational expertise. They will continue to have limited access to systems that are able to combine and process large volumes of reliable and useful transnational data. For this reason, we must learn to manage customer expectations in situations where threshold decisions about individuals are based on levels of "acceptable" risk that are constrained by available resources. We must also provide a means for individuals to appeal decisions that are based on statistical generalizations rather than evidence.

3.3 Four Errors of Interpretation that Affect Data Mining.

Statistical abuses occur when interpretations of sample data are used as valid predictors for the wrong populations. For example, an individual terrorist may pose a significant risk to the USA, yet pose no threat to another nation, e.g., Switzerland; China. If we are to understand and effectively cross-validate demographic data in a global context, we need to address the four potentially harmful errors of interpretation that occur when local (heterogeneous) databases are combined and mined in a global context. The lesson to be learned is that knowledge of the original environmental context is required if we are to develop appropriate data associations and protect original interpretations in new contexts of use.

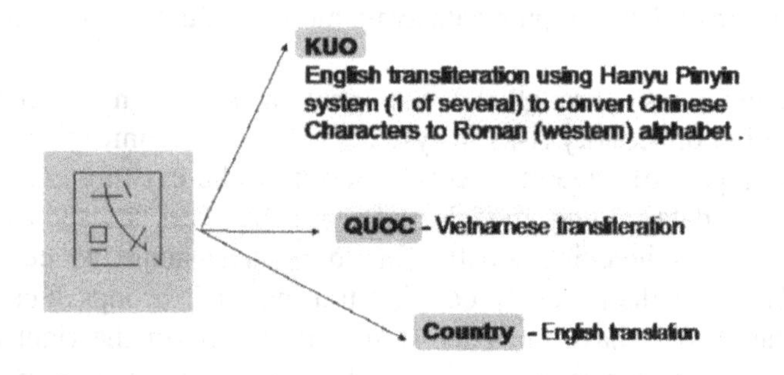
KUO
English transliteration using Hanyu Pinyin system (1 of several) to convert Chinese Characters to Roman (western) alphabet.
QUOC - Vietnamese transliteration
Country - English translation

1st of a 2-character personal "first" name of person born in Hanoi to (Han) Chinese parents; uses Chinese name in social and cultural settings; uses Vietnamese transliteration for legal documents; also uses self-selected "American" first name (e.g., Joe) interchangeably.

Fig. 1. Data Validation

1. **Data that is an alpha-numeric match can be an interpretive mis-match**. For example, an individual who resides in Cairo may live in a very small town in Georgia (USA) or a very large city in Egypt—or even both, if the individual is a student from one city residing in the other. To scale a local database to support global data mining contexts, additional associative values must be used and data values extended *asymmetrically* to identify the specific Cairo of interest. For Cairo, Egypt, only a 2-digit ISO country data value may be required. For Cairo, Georgia, USA, however, both a state and a country code may be required, since several other states in the USA also have towns with the same data values in the name field.
2. **Data that is an alpha-numeric mis-match can be an interpretive match**. For example, in a global immigration/visa application processing system, an accurate global cross-validation that recognizes the language/transliteration/translation systems associated with the data values. For example, the "name" field depends on a good understanding of linguistics and naming conventions as well as a good understanding of how the applicant interpreted a specific question in the original context and data collection process. An applicant is likely to use different data values in different contexts of use. For example, *Mr. Wei* = *Mr. Nguy* = *Mr. Nation*, depending on the original context and the original data collection process, including (see Fig. 1)
3. **Data that is an alpha-numeric and interpretive match may not be associated with (or assessed) in the appropriate context**. For example, two girls, both named "Kathy Mitchell", live in the same town, know each other and have the same friends, but one of them is also a star basketball player. Any deeper distinction may be unnecessary in the context of "number of females in sports." However, failure to identify the *correct* one where additional insight about "immigration status" is required could result in the delivery of a notice of "intent to deport" from the immigration service to the *wrong* one, leaving one girl upset and one girl unwarned of potentially negative (and permanent) consequences. One of these two girls later marries Larry Conan Doyle and shares her name with a *different* Kathy, this time Kathy Doyle, who is Tom Doyle's wife. Both of them may be teachers in the same town and their husbands may work for the same company. In yet another context, such as a third party search for a birth parent, the available data associations for *any* of these three individuals named "Kathy" may not be useful, but delivery of their demographic data as "possible matches" to the information seeker could damage or even destroy entire families,

even though the knowledge of interest to the third party remains undiscovered. The fundamental issue, then, is one of trust. We must be able to trust the data, just as we be able to trust those who mine it and those who use it. In the first instance, correct identification depends on knowledge of age or date of birth and a specific immigration status. In the second instance, knowledge of spouse is required to correctly identify which Mrs. Doyle was originally Miss Mitchell and which one was originally a well-known local basketball player. In the third instance, we must be certain or remain silent. In each case, the intended context of use determines the relative importance of data associations and data sharing needed for the data mining procedure. .

4. **Data that is an alpha-numeric and interpretive mis-match may not be associated with (or assessed) in the appropriate context**. For example, Ms. Susan Mitchell = Mrs. John Nguyen = Mrs. Lawrence Conan Doyle, with all of the data values representative of a single individual, but two of the data values can be correct and simultaneous and accurately reported in the same database and carry the same semantic meaning of the original context, and at times represent two different individuals who are entitled to use the same name at the same time--Mrs. John Nguyen. Additional insight is needed to determine which individual matches the other criteria of our investigation and whether, in fact, these conflicts are normally occurring or indicators of a potential threat. The correct interpretation in this case is context-dependent, time-sensitive and requires awareness of social etiquette as well as the laws governing name changes, bigamy and divorce in the USA.

4. Key Questions of Demographic Data Mining

Data Mining uses existing data to answer context-related questions[1], each of which has its own set of information requirements. A law enforcement example is discussed below:

Who	How do we know who they are?
What	What (if any) connections are there to known terrorists or criminal organizations?
When	When do pivotal events occur? How long do they last?

[1] The data requirements that are used to answer the questions of who, when, where, how and why are based on discussions with Dr. Gary Gordon, Executive Director, Economic Crime Investigation Institute of Utica College

Where and	Where will they go?
With Whom	Who, if anyone, travels with them?
Why	What is the real/purported purpose of this activity?
How	How will they be funded?

Information Requirements – The *WHO*

- Good authentication data sets that allow for cross-checking of multiple identity factors (refer to Fig. 1), such as databases and "Watch lists" of dangerous individuals/groups
- Foreign data sets that enable us to recognize and address our own ethnocentric biases inherent in our current models and scores

Information Requirements – The *WHAT*

- Databases that combine work products of analysts who perform similar tasks in a range of settings associated with a variety of social/societal contexts
- Global data sets useful for testing the value of pattern and link analysis tools

Information Requirements – the *WHEN*

- Access to information on individuals traveling to destinations at the same time
- Databases useful for determining temporal "norms", e.g., time periods associated with specific events (academic semesters, segments of flight-school training, etc.)

Information Requirements – the *WHERE*

- Access to local databases useful for verifying data values such as personal names; types of locations, specific street addresses, etc.
- An understanding of the semantic meanings associated with the reported data values in the original context that can affect the usefulness in global contexts.

Information Requirements – the *WHY*

- Information that connects individuals with similar purposes
- Ability to authenticate information and verify the "story"

Information Requirements – the *HOW*

- Access to information to verify sources, amounts and uses of available financial support (credit cards, debit cards, bank accounts)

- Access to databases used to assess money laundering activity

When we begin to think globally, we recognize the need for a paradigm shift, because not all systems are not equally useful in all settings. New ways to develop better intelligence requires that we identify new approaches to demographic data acquisition and use. This requires a team approach.

4.1 Collateral Damage Control

Much of today's commercial activity depends on risk-based models and scores. Metrics have been developed and tested in Asian settings as well as in Euro-centric business environments. In the USA, the purpose of the data mining procedure is to assign financial risk assessment scores in a consistent and *nondiscriminatory* manner. That means one that is based on *process* and *pattern*, not on demographic attributes such as *person*, *race*, *color*, or *creed*, even if these types of variables are included in the model itself. This is because demographic characteristics describe individuals, but not *intentions*. Use of demographic data requires caution. For example:

1. Discrimination based on demographic characteristics is illegal in the USA, but not necessarily illegal in countries that develop and report on data mining research. For this reason, results may not be statistically generalizable or useful in other contexts.
2. Some demographic data is not reported in a useful way (a "hotspot" geolocator may encompass parts of several countries, the individual may only be able to claim citizenship for one of them, yet travel in all of them and owe allegiance to none of them.
3. Individuals are mutually reflexive, aggregating based on shared motivations. The unit of interest includes this association, but the unit of measure is the individual.

For these reasons, merely knowing something about the personal demographic characteristics of specific individuals is not enough. We need to know what the data really means, we need to let people tell us what we really want to know at the data collection stage, and we need to be able to predict individual intentions if we are to improve our insight into reflective behaviors that identify terrorists and their *relationally co-located* communal selves who, *if given sufficient motivation, resources, support and context*, will choose to harm others. To accomplish this goal, we must begin building into our data mining procedures an awareness of data *contexts*. We must incorporate ethical and legal and intuitive safeguards that protect initials from becoming victims of guilt by association, even though external events might be used to pressure us to do otherwise.

5 Data Mining Approaches – "Data out" and "Decision In"

A chain is only as strong as its weakest link. To have any chance of success, global law enforcement initiatives, such as the war on terrorism, must focus on protecting nations, critical infrastructures and economies from transnational, asymmetric threats. Links between commerce and national security have encouraged the use of credit scoring applications in law enforcement analysis and decision making. Successful outcomes will depend on sharing valid interpretations of useful information among and between agencies and allies. In the area of demographic intelligence, we know that knowledge management and commercial/military/government intelligence analysis now converge in the data mining procedure. There are two possible directions from which to approach data mining: *data-out* and *decision-in*. The choice of direction must be driven by the goal:

1. Will we focus on leveraging and re-use existing knowledge to generate new insights about old data? Will we try to force new data values into old containers?
2. Will we try to develop new containers that can adapt to new data?
3. Will we try to determine if the knowledge base that informs our decisions has been outgrown and it is now inadequate or even dangerous to use?

5.1 The "Data Out" Approach

When data mining approaches a problem from the *data out*, it mines the existing data values (the "what") contained in the databases being used for decision support. Data mining queries are of the following types:

1. What do we know?
2. Where is it?
3. How fast can we maximize our use of it?
4. As we learn, how will we share our new insights with appropriate others?

This approach assumes that data values *do not change in meaning or interpretation over time*. This approach treats interpretations as quantitative, not qualitative, viewing them as unaffected by changes in context, location or timing of the query. Data mining algorithms, while often complex, are relatively free of interpretive bias where quantitative data values have quantitative interpretations associated with them. However, data values associated with individuals can--and certainly *do*--change, as do the interpretations associated with them. For example

- "Personal weight" may change from time to time, as a condition of aging and change in lifestyle (which is categorical and generally predictable) but also as a result of an undiagnosed medical conditions that are not predicted with human genetics.
- "Income" data values do not mean the same for all individuals, because the impact of the data value depends on unique combinations of personal values, sociocultural norms, available support structures, open options, individual judgment, forward planning and external impact factors that can include catastrophic events.

When data mining processes are used to assess risk in contexts that have no right (or avenue) of appeal, the general public has every right to be wary of these systems and every need to become informed about how their own data is represented in them. [http://www.fico.com; www.riskwise.com] As we have discussed above, law enforcement agencies use cross-validation of databases for purposes of identity verification, and authentication. The resulting insights can certainly have predictive value, regardless of whether they meet the requirements for statistical generalizability. We must keep three points in mind, however:

1. **Cross-matches mix data collected for one purpose with data collected for another**. This is helpful, because cross-validation is more secure than reliance on single databases. Data errors as well as attempted frauds are also easier to catch.
2. **The results on an individual are based on comparison with predefined profiles of "good" and "bad" risks statistically derived from the past behavior of *others*.** Data values incorporate flawed decision processes that result in self-fulfilling patterns of illegal discrimination and socioeconomic and geopolitical ignorance. Results of the data mining procedure predict, but *do not report* the future capability, capacity or intentions associated with any specific individual being assessed.
3. **Data values, including basic law enforcement crime statistics are sometimes deliberately mis-reported on a large scale over a long period of time, sometimes for political reasons.** This happened in Philadelphia, Pennsylvania, The need for corrections to the crime statistics databases were widely and publicly acknowledged, yet the correc-

tions were mainly anecdotal, limited to newspaper accounts of test cases reviewed, and few corrections were made to the database or carried forward to revise models and scores. It is important to be careful about making claims about the accuracy of interpretations derived from mining demographic data unless the databases themselves are clean and the data mining procedures have access to both data interpretations and data values for use in the data mining procedure.

5.2 The "Decision In" Approach.

Using the revised risk management process model suggested above, we can identity a second, less widely understood method for acquiring and examining data that works from the "decision in." When this approach is used, data mining begins with the decision to be made. It next examines the existing data (the inputs) in order to assess the interpretive weaknesses and opportunities associated with mining it. Queries are of the following types:

1. Do we have valid data to mine? How do we know?
2. Do we have all the relevant data we need to mine? *If not, where is it and how can we obtain it? And how can we integrate that insight into our system?*
3. Do we know what our data actually means? What are the interpretive loads on data values? Do these interpretive loads change? If so, to what? Why? How? Where? When? And to what purpose? And does this happen because of (or before) some other factor or pivotal event?
4. If our data is false, do we know if it is purposefully false, accidentally false or forced by system constraints to be false because there is no way for the data to be true?

If we know that the data is false, we should clean it before mining it rather than relying on statistical processes to offset the problem. For example, in the case of a deliberate lie, the application for an immigration visa asks if an individual has ever committed a crime. The only possible response is "no," which is a system-enforced choice, not necessarily a correct response. Mining on that value tells us nothing useful. In another instance, on that same application, we may assume a single Arab male is "potential terrorist" because he reports that he is single and intends to attend a flight school in the USA. The reality may be that he is working as a flight mechanic for an international airline and he wants to learn to fly the plane, but he knows that flight school training is only for the military and the elite in his home country and flight school training in the USA is much more accessible. We must be able to distinguish innocent individuals from terrorists hiding in plain sight.

5.3 System Constraints on Interpreting Demographic Data Values

System constraints can actually force a response to be "false" because it limits a recorded response to a single data value. That data value may only be true a part of the time. For example, "primary residence" information for a bi-national who could justifiably report multiple "primary residences" depends on the local context of the query, yet system mis-matches receive higher risk assessment scores, even where multiple simultaneous data values may be correct.

- **How do the answers to these queries constrain and direct our data mining activities and how will it affect the interpretation of our results?**

It should be noted that the existence and value of databases is context-dependent. We cannot assume that the same records are created in the same manner for the same purpose with the same general quality in all local contexts. For example, we can successfully cross-match individuals using property records databases in the USA, but these records do not exist in many countries. For example, in Greece, individuals may own specific chestnut trees in a village, and extended families informally share a cottage in an ancestral village. This country is almost 50 percent agrarian, does not have a property tax, and the majority of the citizens' meager incomes and limited cash savings could not support it.

Content creators are often unwilling to commit the resources required to verify and correct specific mistakes, even when they are identified at the point of need. As a result, individual data errors and interpretive mistakes go uncorrected. According to private conversations with a senior manager of a large credit scoring company, the current approach is to embed a score that reflects what is known about the "dirty-ness" of the specific database used in the data mining procedure and the consequences to specific individuals are ignored. The degree of significance and the amount of human collateral damage may well vary, but a relative proportion of "highly significant" errors will continue to go unrecognized and unchallenged, given that those who develop the "data quality assessment factors" lack local insight into the significance of the specific situations involved and lack any incentive to change.

- **At what point are guarantees of civil liberties (freedoms of association, *de jure* presumptions of innocence, etc.) and due process (the right to face one's accuser, for example) *de facto* vacated because of decisions that rely on data mining?**

Errors can appear to have the same degree of significance, yet differ in orders of magnitude in the context of other data associations. In a country that considers the dignity of the individual to be a useful unit of measure, cleaning dangerously dirty data offers potentially more value than simply assigning a weight that tells us that a database contains data that is "mostly right most of the time." For example, in a war gaming exercise, the data value "age of the street map" will not matter. When the results of that war game are used to mistakenly bomb another country's embassy, however, a single data value can bring nuclear superpowers to the brink of war. There are some kinds of trouble that a simple apology will not offset.

It should also be pointed out that the purpose of business systems used to score individuals are intended to serve a purpose that is *unrelated* to the need to identify and thwart specific unknown individuals. To a degree, the same logical outcomes are involved—providing opportunity to "good" risk individuals and denying it to "bad" risk individuals. However, only the *what* values are effectively measured. Commercial risk assessment is about revenue prediction and loss control, so specific individuals do not matter. Law enforcement, however, is about understanding the socialization of otherwise unrelated individuals who simply agree on the *why*..

- **How do systems account for external factors outside an individual's control?**

Business risk assessment systems are designed to interpret past outcomes as predictors of future *intentions*. There is a general disregard for the fact that past outcomes can be (and often are) due to external factors that can overset the best of individual intentions. For example, there may be a sudden loss of income or a catastrophic illness that drains resources for an unpredictable period of time. This kind of credit scoring is sufficient for firms that must simply improve the prediction of revenue streams. The safety bias inherent in these systems is that people who are denied credit are unlikely to post a serious threat to society, so equal treatment under the law is preserved, yet individuals can be inequitably regarded as a risk for reasons and results that are simply beyond their control.

Hidden in these business systems is the assumption that negative outcomes are always under an individual's direct control. In other words, if an individual fails to pay, this is treated as a reflection of *intent*. Such an approach is a dangerous trivialization of the *why* value which is critical to successful identification of terrorists and their supporters. In the war on terrorism, we can never disassociate the *why* from the *what*. For this reason, law enforcement use of credit scoring systems must find ways to offset the lack of "motivation" data. For example, an individual may overdraft a bank account often enough to catch the attention of the law enforcement community. This may signal an intend to defraud. However, this may also be due to extenuating circumstances such as identify theft that depletes an account for a U.S. serviceman stationed abroad who is in no position to monitor bank accounts on a regular basis. Regardless of the combination of factors involved in creating the recorded transaction history, what is captured in the data mining procedure is the data value that represents the *result*, not the *intent*.

When we begin the risk management process at the planning and direction stage, we discover a need for data values that reflect the *what if* proposition. That is, we discover a need for data values that re-value risks in tandem with changes in availability of sufficient motivation and/or resources and/or opportunity. By starting at the planning and direction stage, we quickly recognize that there are social/societal differences, such as situational ethics or gender norms, that can limit our ability to separate "apples from tomatoes."

The *decision in* approach is superior to the data out approach in that it treats data as capable of carrying multiple, simultaneous, and often quite different interpretive loads with values that would each be equally "true," but only for a given moment in time. This approach calls for the input of experts who know when data value "X" will *only and always mean "X"* and when it is only "X" *during a certain time period, or in certain contexts*. This approach also recognizes when data that is housed in container "X" should be scored *as if* it were housed in container "Y" or when the value should actually reflect a combination of containers that might—or might not—include either "X" or "Y."

The *decision-in* approach is used to identify and study the strengths and limitations of underlying databases in contributing information systems so that we can identify and capture in the data mining algorithm itself the various interpretive possibilities embedded in-- or associated with—specific data values. This approach acknowledges that the data value is the equivalent of a 2-dimensional photograph in a multidimensional world that does not always share the same belief systems or attach the same significance to specific kinds of data. Since the purpose of data mining in law enforcement is to predict individual intent (or lack of intent) to harm, the decision-in approach enables us to develop algorithms that are globally scalable, or at least to more readily recognize when they are not, before these systems are used to undermine the very societies they purport to protect.

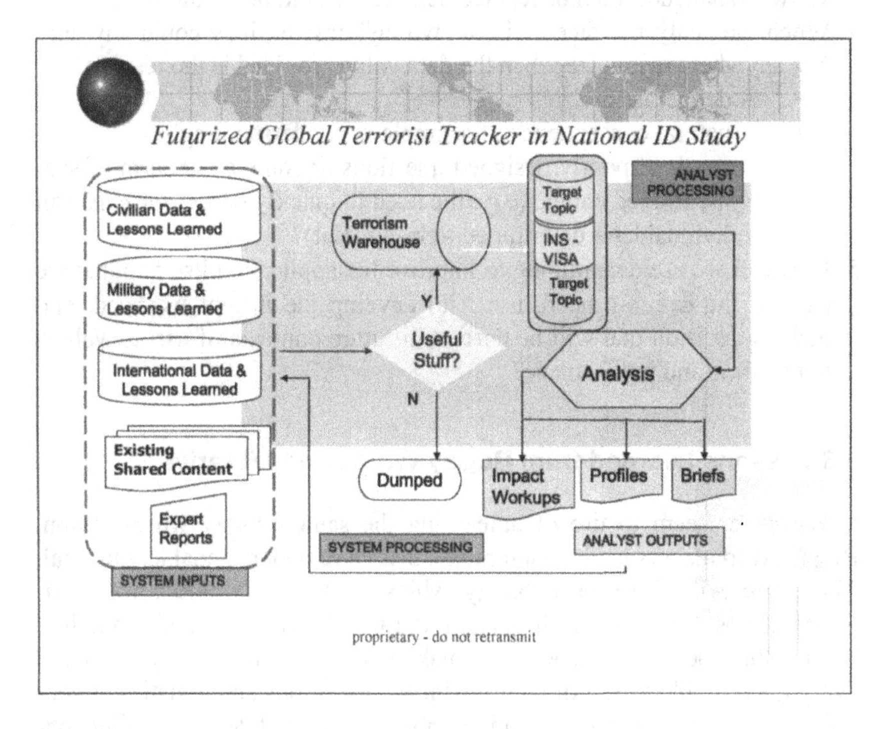

Fig. 2. Futurized Global Terrorist Tracker. Source: Katherine M. Shelfer and J. Drew Procaccino. National ID Cards: A Solution Study. © 2003. Used by permission.

For example, in the case of visa applications, the *decision-in* approach would be operationalized by working from the immigration visa applicant pool (at the point an application is submitted to the visa analyst) back through the decision process to the data values entered into the database. The "correctness" of the decision itself is treated as the focus of the data mining initiative. In the case of visa applications, for example, queries might be of the following types:

1. Is the Applicant the "real" applicant? How can analysts located elsewhere and looking at the data in other contexts actually know?
2. Did the application ask all that the analyst needed to know to make a "correct" decision? Would any analyst anywhere come to the same conclusion? Could other analysts accurately interpret the data provided? If not, what additional or replacement data or other forms of decision support would be required?
3. Did the applicant provide a truthful response? How could all analysts know? What additional or replacement data would be required?
4. When the analyst is faced with a "wrong" answer, how could any analyst anywhere recognize when the data value provided is the result of ...
   - A deliberate lie
   - A natural consequence of a simple misunderstanding
   - A result of poorly designed questions or some other form of system-enforced choice (e.g., the need to quickly deport *for cause* an individual later determined to be a threat)?
5. How can we use data mining to improve the application itself so that we capture the data we need, use it to develop the insight we need, and make a decision that will be correct in future contexts of use as well as at this time and in this place?

5.3 Lessons Learned from Buggy Whips and Libraries

We never seem to tire of unlearning the same "buggy whip" lesson. Buggy whip makers made assumptions that were understandable, but fatal. There had ALWAYS been buggy whips, and there would always BE buggy whips. So buggy whip makers did not focus on strategic thinking, but on the operational aspects of making buggy whips. They mastered every quantifiable aspect of their business--operations and logistics, inventory management, production and distribution, product design, quality control, marketing and advertising. Since there was always a finite right answer to operational questions, the strategic limits of the questions themselves were never explored. As a result, buggy whip makers missed the implication of automobiles. There was nothing wrong with what they did, but there were long-term implications for the quality of their lives that they were not mentally prepared to see. They missed the significance of thinking about amorphous stuff. They were too busy with the daily grind of making buggy whips. As a result, they failed.

In the world of intelligence, a problem solved is a problem that does not resurface. The next problem will not have the same flavor--or the same answer--as a previous one. A formulaic approach absolutely will not work. For this reason, it is important to step away from a process to look at the goals from a different perspective—the customer's need to decide and then defend that decision. In other words, it is important to recognize when the "right" answer for one data value and resulting interpretation is the very *wrong* one for another situation or context, because the relational variables that feed that equation are often quite different. In the case of terrorism, there is no safe zone. So we must think about every aspect of the problem. We can no longer afford the luxury of unwarranted assumptions simply because dealing with them is uncomfortable.

By focusing from the decision-in, we learn to stretch how we think about a problem. We learn to identify assumptions we did not know we had made. We learn to find the gaps in our insights and data collection and data mining strategies. The goal is *process*, not *product*, but it requires that each of the players willingly contribute useful data. Frankly, this technique works when everyone participates--and it doesn't when they don't. Without anything to compare, there is no insight to be gained into how others approach the same problem, given the same limitations and unknowns.

As you can imagine, the military and law enforcement intelligence agencies have the best tools in the world, but they still need to think about whether the data is sufficient or of the right type, or biased, or what *else* may be needed, or how any of the "peripheral players" might respond or how such a response might change the actions of others. And when management consultants ask a company to answer a question like "what business are you in?" they are going after the same sort of thing--they are pushing the boundaries.

5.4 The Principle of Co-Location and the Policy of Superimposition

To promote the usefulness of their collections, librarians adopted a digital library catalog because it allowed them to add new access points to individual records without disturbing the existing ones. Based on that field's Principle of Collocation, the old and the new data values have been electronically brought together (co-located), yet they continue to co-exist, with each data value having its own separate and unchanged existence that is relevant in its original context of use. Based on that field's policy of superimposition (discussed in more detail later in this paper), the new items are assigned more useful interpretative meta data values that electronically co-exist and are cross-referenced with the old data and meta data values. As the new items gradually outnumber the old items, the interpretive quality of the entire database is improved over time. Through the use of global ontologies and meta data registries, local library databases can be successfully combined and usefully mined for meaning in global contexts because all of the available interpretive meanings can be incorporated into the procedure.

The successful survival of an overstressed system with incorrect, inadequate and/or missing data values depends on (1) replacing existing data with new data; and (2) establishing and acquiring new data values that can be more effectively mined and interpreted. As indicated in earlier discussions, library and information science offers some encouragement at learning to survive the potentially staggering impact of radical change. The *Principle of Superimposition* acknowledges that

- The old data values are no longer useful and should be replaced.
- The new data values are better than the old data values, so they should be used.
- There are insufficient resources to convert the old data values to the data values.
- Adding a new field to a data base and assigning new values effectively must realistically begin at the "zero" point.
- Over time, the proportion of the database with the new data values will change, moving from 0 new/100 old toward 100 new/0 old.

One example of surviving such an enormous transition is found in the way that large libraries successfully converted from the Dewey Decimal system to the Library of Congress Classification System. To implement a change of this magnitude in collections with millions of volumes was no easy undertaking. Libraries lacked the resources for complete retrospective conversion to the new system. While adherence to either system alone would be a mistake and physically housing a single item in two locations was impossible, it was *electronically* possible to house the same item in a variety of containers. For this reason, librarians—as a group—elected to look to the future instead of the past. They developed a shared understanding that

1. The two systems necessarily represent different proportions of the whole;
2. The relative proportions will eventually favor the new system;

3. Retrospective conversions of existing data are less important than improving our insights about newly acquired data; and
4. Retrospective conversions will be prioritized and processed over time as needs dictate and resources permit.

It should be noted that library catalogs are also an early example of successful distributed database creation and use. The OCLC system credits/debits creation and use of its content. The database provides for examining and crediting alternative interpretations of the same data element(s) associated with any given item based on local contexts of use. Lessons learned from libraries are that

1. Even static content can, over time, outgrow the interpretive container in which it is housed;
2. Data that looks 100% alike may be different. Data that looks 100% different may be alike. Data that is cross-validated may combine elements of both. How will we know which is which?
3. New data may surface for which new containers must be found. Old data may need to be placed into a different container on a priority basis. The same data may need to be simultaneously accessible through more than one container at the same time.
4. Data may need to be shifted between containers as a result of shifts in context, location, purpose and source of the query at any given point in time.

Libraries teach additional lessons about data values, data cleaning, authority control and change management:

1. Retrospective conversions of old values to new values is needs-based and can only be accommodated as resources become available.
2. When records are cleaned, the cleaner's identity is captured, the cleaning is based on evidence related to that specific item and the data value is recorded using some overarching standard for authority control. However, accommodation in the item record itself is also made for insight at the local level that may be different and/or more informative than is reflected in the standard.
3. When new access control points are added, they are placed into their own data containers. They are not superimposed over old values housed in pre-existing containers. This prevents confusion about which value (old or new) is represented in which container. This provides built-in accommodation for the future creation and use of even more useful values.

4. Digital item records will always capture and retain each and every access point.

6. DATA MINING and Credit SCORING Models and Scores

When credit scoring models and scores are the anticipated outcomes of the data mining procedure, there are usually four types of models that are used to test for differential/discriminatory treatment with adverse impacts, each of which is discussed below. *All of these approaches share significant statistical problems, and recent efforts ...provide* no *solutions,* which partially explains resistance to using data mining processes as the sole criteria for risk-based pricing of products and services. (http://www.fanniemaefoundation.org/programs/jhr/pdf/jhr_0402_carr.pdf. To offset the limitations of these models, improvements in demographic data collection are urgently needed. For example, combining and mining demographic data should allow interest parties to reverse the process to discover whether the score is based on personal intentions or external events or even corporate culpability due to variations in customer relationship management practices and/or debt collection practices.

6.1 Opportunity Rationing – a/k/a Redlining

Disadvantage: Data models of this type assume that discriminatory practices will manifest themselves when neighborhood data is disaggregated and compared against the aggregate volume of activity. However, this fails to consider that there features of the lending process itself, such as the use of prohibited demographic data values (race, gender, age) that cannot be cleaned out of the aggregate data set. The use of these factors impacts other variables that would otherwise appear to be legitimate measures of risk, such as credit history. In other words, where past practices have been outright rejection rather than risk-based pricing, entire classes of individuals have been procedurally denied opportunity in ways that have become invisible and self-perpetuating. http://www1.fee.uva.nl/fo/mvp/incapcesifo290103.pdf.

Example: Data models of this class assume that the mortgage flow data represents a legitimate response to risk that is based on unbiased data, when in fact individuals who are unable to obtain credit at the same rate of the aggregate data set, if they can obtain it at all, are victims of biases that infect all future data in the decision process and cannot be extracted from the data used in the data mining process because we cannot know the outcomes where credit that is not extended (unfair denials).

6.2 Crowding Out

Disadvantage: Data models of this type assume that the data represents equally empowered individuals who have been treated alike, even with evidence to the contrary. This model would not be a good choice for data mining initiatives that result in denial of opportunity to individuals based on external factors such as events and beliefs that are subject to change. Such a model is unlikely to recognize the emergence of a new threat in a timely fashion because the model is not structured to recognize unevenly applied changes in semantic meaning over time. Demographic designations may only represent system- or socially-enforced choice. For example, use of attributes such as age, weight or race or may not scale equally well, because data values do not carry the same semantic meaning in all contexts of use.

Example: This model creates the Prisoner's dilemma, because it uses suspect data with unrecognized subjective loads. For example, data values for race have long been used to deny opportunity, yet data values for race are unevenly and erroneously assigned and often based on racialism that pre-dates the 19th century. Individuals classified as a poor risk based on demographic data values such as race are forced to pay more for less, which reduces their capacity and pitches them into an uphill battle to reverse the subjective judgment that caused the problem. This kind of reasoning can have serious repercussions, e.g., barring safe havens to Jews fleeing Germany in World War II; deprivation and concentration for Americans of Japanese Ancestry during that same time period and for single Arab males today).

6.3 Default

Disadvantage: Data models of this type are used to test whether outcomes are related to the insertion/omission of prohibited variables at the decision stage. However, we cannot offset underlying (and fundamental) differences in standards/criteria applied by individual analysts, nor can we confirm the correctness of a decision when outcomes cannot be measured. Without insight into what *would* have happened, we are unable to verify that the data associations that resulted in that denial were appropriate, relevant or correctly interpreted. Without such insights, we may very well continue to make the same mistakes without any way to identify what these mistakes actually are.

Example: Where humans with negative sentiments are concerned, such loss of opportunity may push them from belief to action, which makes a bad situation even worse. For example, when denied legal opportunity to enter the USA, many enter illegally and their actions cannot be monitored. Nor can we know that point (pre/post decision) at which expressed desire to benefit became intent to harm.

6.4 Rejection Probability Models

Disadvantage: Data models of this type are used to compare detailed individual profiles to group profiles so that we can determine whether improper denials or too lenient approvals are have taken place. First, we do not have the long-term data for the necessary panel studies. Most of what we are trying to do is entirely too new. Second, we can capture and analyze mistaken approvals, but we cannot capture and analyze mistaken *denials*. We cannot observe what *would* have happened if the decision had been reversed.

Example: To remove the "guilt by association" that is built into the earliest "bad risk" decision profile, we would have to deliberately provide opportunity to "bad risks" to validate our model. Where there is even the slightest suspicion of guilt, however, we can't begin to justify putting a nation at risk by letting a potential terrorist into the country just to see what might happen. One positive outcome of such an experiment (even if it were to be conducted and the results could be monitored and fed back into the system in a timely and useful fashion) would not be statistically indicative that the initial insight was flawed. Since we are unable to safely obtain any evidence to the contrary, an individual's "guilt by association" remains. The fundamental right to be presumed innocence and afforded due process of law vanishes into the ether.

7. THE ROLE OF ANALYSTS IN THE DATA MINING PROCEDURE

It should be specifically pointed out that analysts do not develop the decision frameworks that guide their decisions, even though they usually abide by them. This is partly because of time constraints associated with the number of decisions and the available time to decide each case. However, it is also partly because they are not asked at the data mining stage. The local expertise of analysts must be captured where it is the most useful—in the data mining procedure itself—if the use of data models and scores will <u>*ever*</u> provide effective decision support. In no way the decision support system substitute for the expertise of front-line local analysts who deal daily with complex decisions. However, if decision support is to be useful in a global environment, we are going to need to develop some fundamental decision rules, which is where data mining is most useful.

Data models and scoring systems are intended to improve our ability to capture and share analyst expertise very quickly. Basically, the purpose of such systems is to free humans--the analysts--to develop *more* expertise. In turn, that new insight needs to be captured, fed back into the system, and leveraged and re-used to improve system support for other analysts. Current and future situations are going to call for new interpretations--or more refined understanding--about past decisions. Analysts are important first points of awareness and sharing, so they need to have immediate access to new insights about the outcomes of past decisions.

The only way to develop a useful and usable decision support system—one that will actually support analysts in ways that surface as improved decision quality--is to establish a shared understanding of how the complete process actually works. Analysts have a multi-sensory in-take system that accepts environmental inputs as well as data. We need to understand what an analyst sees when s/he spots an "easy" decision (either approval or denial). And we need to understand how different analysts work through the process--what they actually think and do--to arrive at a decision. It would be helpful if we could capture how analysts with deep insight and substantial expertise share this knowledge with others. One challenge in building useful predictive models is that it is hard to collect enough pre-classified data to develop and test a decision support model.

To offset this limitation, we need to learn exactly how analysts make so many tough decisions every day, yet get so many rights. In short, working form the "decision in" helps us to quickly address three major challenges:

1. To learn what information analysts need, what information they use, and whether this information is easy or hard to acquire in a useful format and a timely manner.
2. To identify relevant NEW information that analysts do not have, so that we can locate it and embed it into the process.

3. To identify the warning signals that cause analysts to seek out new information or second opinions or insights that trigger an alarm.

Data mining helps us discover new indicator variables. Cross-validation helps us test the validity of our discoveries. However, data integration about transnational risks takes transnational skill and insight. We must find ways to know the point at which (A+B) =C. For example, as shown in Fig. 3, below, accurate interpretation of cross-validated information assumes that the information input from one database will locate associated matching values bi-/multi-directionally. In other words, if we run a phone number through a database, we should get a name and address in the new database that matches what we have in the database from which we took the phone number. If we run the name and address back through the first database, we would expect to retrieve the phone number. There are many hidden assumptions, but the most dangerous are these:

1. That the data exists in the necessary and accessible format to create a match.
2. That the data in each database is equally valid at the same time.
3. That the data values, whether alpha-numeric matches or mis-matches, are correctly interpreted in the cross-validation process.
4. That data equivalents, regardless of data structures, are appropriately included in the cross-validation process.

Such assumptions can only hold true when data used in the cross-validation process is error-free, represents the same period of time, is correctly interpreted, and is appropriately included/excluded in the cross-validation process. Otherwise, we can never be sure that we know what we think we do, or that our interpretations and our decisions are correct.

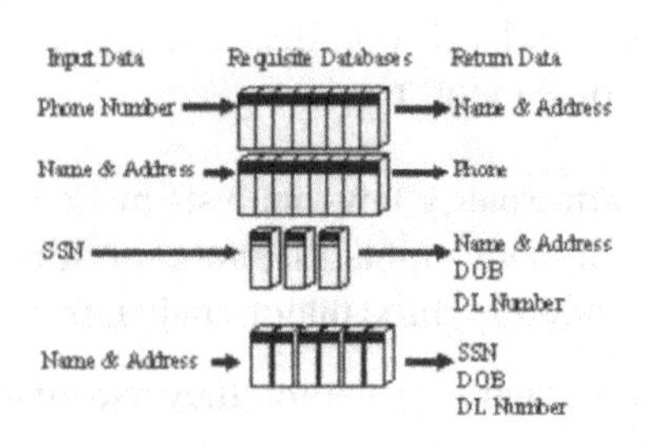

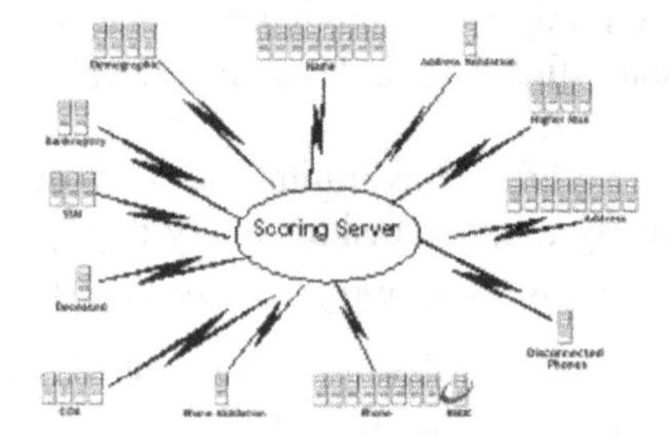

Fig. 3. Cross-validating demographic data in databases

8 CONCLUSION AND IMPLICATIONS FOR FUTURE RESEARCH

Where we know that data mining interpretations rest on data that can be partial, invalid, wrong and biased, it would be a mistake to treat demographic data on individuals as if it was whole, valid, correct and unbiased until it actually meets those criteria. Because there are issues surrounding database business models and available resources of local content providers, we must work from the decision-in, because it is the best direction for identifying and prioritizing data for cleaning. This effort becomes a moral mandate as we seek to scale local databases to support non-local and even global contexts of use that were never envisioned when the databases were originally created. It would be a mistake to deny legal protections to individuals in order to protect economies of scope and scale. Where other individuals—and other institutions--will pay the price for a bad decision, it is important to improve the quality of data before going data mining for meaning in demographic databases.

9. REFERENCES

Acitelli LK (1993) You, Me and Us: Perspectives on Relationship Awareness. In Duck, S. *Individuals in Relationships*. London: Sage.

DARPA Defense Advanced projects Agency (2002) Located at URL: http://www.darpa.mil/body/tia/tia_report_page.htm. For a mirror site on the Total Information Assurance initiative that has since been removed from the DARPA website, see http://www.computerbytesman.com/tia/index.htm

Fergus KD, Reid DW (2001) The Couple's Mutual Identity and Reflexivity: A Systemic-Constructivist Approach to the Integration of Persons and Systems. *Journal of Psychotherapy Integration* 11.3 (September): 385-410.

Freedman MJ. http://www.mjfreedman.org/articles/openingacatalog.pdf)

Josselson R (1994) Identity and Relatedness in the Life Cycle. In Bosma HA, Graafsma TL, Gropterant HD, de Levita DJ (eds) *Identity and Development: AnInterdisciplinaryh Approach*. (p.81-102). London: Sage.

Mead GH (1934). Mind, Self & Society. Chicago: University of Chicago Press.

Miller DR (1963) The Study of Social Relationships: Situation, Identity and Social Interaction. In Koch S (ed) *Psychology: A Study of a Science*. (vol. 5, p.639-737). New York: McGraw Hill.

Propert and Casualty Insurance Association (2004). White paper on credit scoring with olniks to relevant studies, located at http://www.naii.org/

Reid LW (1981) The Seven Fallacies of Economics. *The Freeman* 31.4 (April) http://www.libertyhaven.com/theoreticalorphilosophicalissues/ austrianeconomics/fallacieseco.html

Shotter J (2001) The Social Construction of an "us": Problems of Accountability and Narratology. In Burnett R, McGhee P., Clarke DD (eds). *Accounting for Relationships (p. 225-247).*

Skitka LJ, Houston DA (2002) When Due Process is of No consequence: Moral Mandates and Presumed Defendant Guilt or Innocence. *Social Justice Ressearch* 14.3 (September): 305-326.

United Nations General Assembly (1948). Universal Declaration of Human Rights.

Yant M (1991) Presumed Guilty: When Innocent People are Wrongly Convicted. New York: Prometheus Books.

An Effective Approach for Mining Time-Series Gene Expression Profile

Vincent S. M. Tseng, Yen-Lo Chen

Department of Computer Science and Information Engineering, National Cheng Kung University, *Tainan, Taiwan, R.O.C.*

(Email: tsengsm@mail.ncku.edu.tw)

Abstract. Time-series data analysis is an important problem in data mining fields due to the wide applications. Although some time-series analysis methods have been developed in recent years, they can not effectively resolve the fundamental problems in time-series gene expression mining in terms of scale transformation, offset transformation, time delay and noises. In this paper, we propose an effective approach for mining time-series data and apply it on time-series gene expression profile analysis. The proposed method utilizes dynamic programming technique and correlation coefficient measure to find the best alignment between the time-series expressions under the allowed number of noises. Through experimental evaluation, our method was shown to effectively resolve the four problems described above simultaneously. Hence, it can find the correct similarity and imply biological relationships between gene expressions.

1 Introduction

Time-series data analysis is an important problem in data mining with wide applications like stock market analysis and biomedical data analysis. One important and emerging field in recent years is mining of time-series gene expression data. In general, gene expression mining aims at analysis and interpretation of gene expressions so as to understand the real functions of genes and thus uncover the causes of various diseases [3, 8, 9, 20, 21, 29, 30, 31]. Since the gene expression data is in large scale, there is a great need to develop effective analytical methods for analyzing and exploiting the information contained in gene expression data. A number of relevant studies have shown that cluster analysis is of significant value for the exploration of gene expression data [3, 4, 11-13, 20, 21].

Although a number of clustering techniques have been proposed in recent years [9, 10, 13, 20, 21-23], they were mostly used for analyzing multi-conditions microarray data where the gene expression value in each experimental condition is captured only at a time point. In fact, biological processes have the property that multiple instances of a single process may unfold at different and possibly non-uniform rates in different organisms or conditions. Therefore, it is important to study the biological processes that develop over time by collecting RNA expression data at selected time points and analyzing them to identify distinct cycles or waves of expression [3, 4, 13, 25, 30]. In spite that some general time-series analysis methods were developed in the past decades [2, 4, 14-18], they were not suited for analyzing gene expressions since the biological properties were not considered.

In the time-series gene expression data, the expression of each gene can be viewed as a curve under a sequence of time points. The main research issue in clustering time-series gene expression data is to find the similarity between the time-series profiles of genes correctly. The following fundamental problems exist in finding the similarity between time-series gene expressions:

Scaled and offset transformations: For two given time-series gene expressions, there may exist the relations of scaled transformation (as shown in Figure 1a) or offset transformation (as shown in Figure 1b). In many biological applications based on gene expression analysis, genes whose time-series expressions are of scaled or offset transformation should be given high similarity since they may have highly-related biological functions. Obviously, frequently used measures in clustering like "distance" can not work well for these transformation problems.

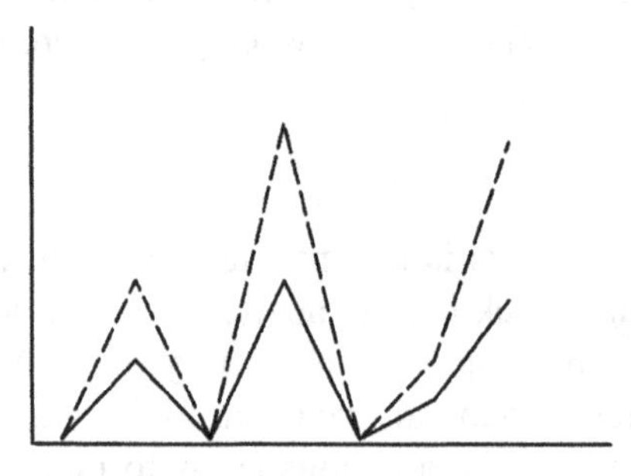

Fig. 1.a. Scaled transformation

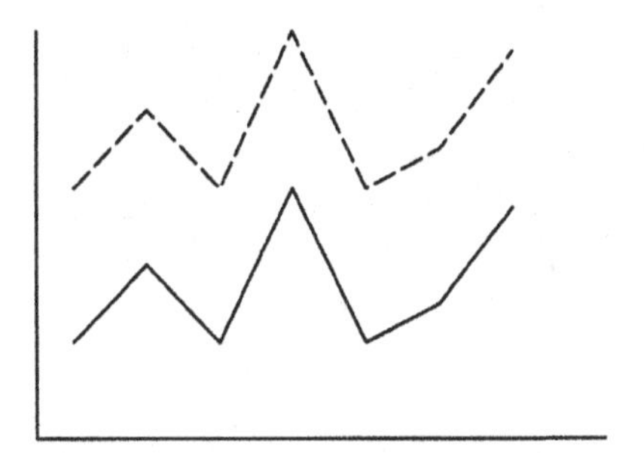

Fig. 1.b. Offset transformation

Time delay: For the time-series expression profile, two genes may have similar shapes but one's expression is delayed by some time points compared to the other's. This phenomenon may be caused by the biological activation function between genes. When this phenomenon exists, it is difficult to cluster these kinds of genes correctly by directly using the existing similarity measures. For example, Figure 2a shows the expressions of genes YLR256W and YPL028W in the α-test microarray experiments by Spellman *et al.* [4], where 18 time points were sampled for each experiment. It was known that YLR256W has activating effect on YPL028W in the transcriptional regulation. However, if Pearson's correlation coefficient is used as the similarity measure, a low similarity as -0.50936 will be resulted for these two genes. In fact, if we make a left-shift on YPL028W's time-series data for one time point and ignore the data points circled in Figure 2a, these two genes exhibit very similar expression profile as shown in Figure 2b (the similarity becomes as high as 0.62328) and this result matches their biological relationship. The above observation shows that the time delay property must be taken into account in dealing with time-series gene expression data.

Noises: It is very likely that there exist noisy data (or outliers) in the time-series gene expressions. The noises or outliers may be caused by wrong measurements or equipment failures. Since global similarity is calculated in measuring the similarity between genes normally, these noisy data will produce biased comparison results, which will be more serious in time-series data than non-time series ones. Therefore, new methods are needed for handling the noisy data in gene expressions.

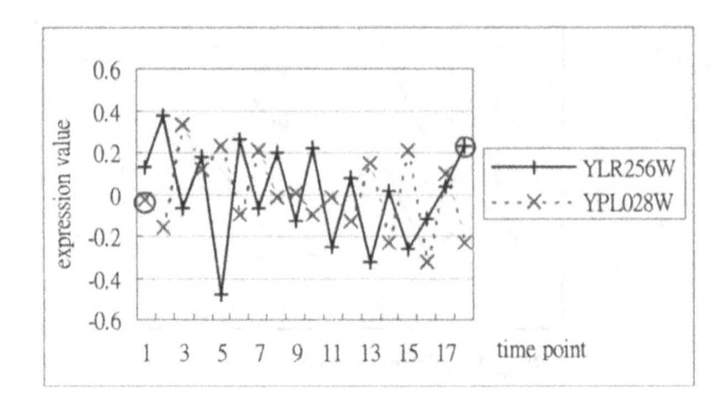

Fig. 2.a. Plotted expression curves of genes YLR256W and YPL028W

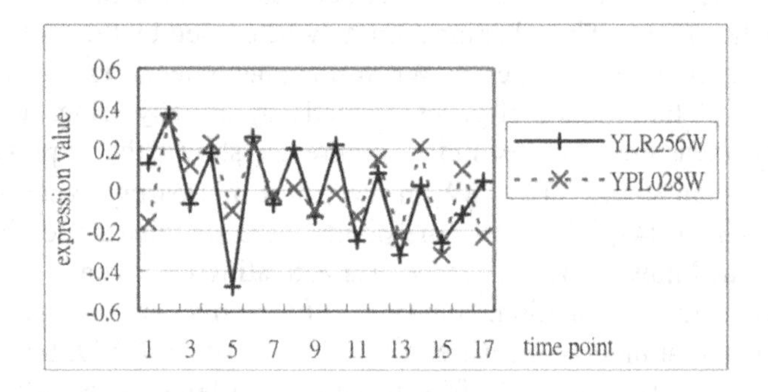

Fig. 2.b. Effect of left-shifting YPL028W's expression by one time point

Although some studies were made on analyzing time-series gene expression data [5, 6, 7, 25-27], they can not resolve the above fundamental problems effectively at the same time (more descriptions of the related work were given in Section 4). In this paper, we propose an effective method, namely *Correlation-based Dynamic Alignment with Mismatch (CDAM)*, for resolving the fundamental problems mentioned above in mining time-series gene expression data. The proposed method uses the concept of correlation similarity as the base and utilizes dynamic programming technique to find the best alignment between the time-series expressions under some constrained number of mismatches. Hence, CDAM can find the correct similarity and implied biological relationships between gene expressions. Through experimental evaluations on real yeast microarray data, it was shown that CDAM deliver excellent performance in discovering correct similarity and biological activation relations between genes.

The rest of the paper is organized as follows: In section 2, the proposed method is introduced; Experimental results for evaluating performance of the proposed method are described in Section 3; Some related studies are described in Section 4, and the concluding remarks are made in Section 5.

2 Proposed Method

As described in Section 1, the key problem in time-series gene expression clustering is to calculate the similarity between time-series gene expressions correctly. Once this is done, the right clustering of the gene expressions can be materialized by applying some clustering algorithms like CAST [10] with the similarity values deployed. In the following, we describe in details our method, *Correlation-based Dynamic Alignment with Mismatch (CDAM)*, for computing the similarity between two time series.

The key idea of CDAM is to utilize the techniques of dynamic programming and the concept of fault tolerance to discover the correct similarity between time-series gene expressions effectively. The input to CDMA is two time series S = $s_1, s_2, \ldots, s_N$ and T = $t_1, t_2, \ldots, t_N$, which represent the expression profiles of two genes under N time points. In addition, a parameter named *mismatch* is also input for specifying the maximal number of data items allowed to be eliminated from each time series in considering the possible noises. The output of our algorithm is the similarity between the given time series S and T, indicating the similarity between the corresponding genes. A higher value of the output similarity indicates a stronger biological relation between the genes.

Our method aims at resolving the problems of scaled transformation, offset transformation, time delay and noises at the same time in calculating the similarity. First, in order to resolve the problem of scaled and offset transformations, we use *Pearson's correlation coefficient* as the base for calculating the similarity between time-series gene expressions. For two time series S = $s_1, s_2, \ldots, s_n$ and T = $t_1, t_2, \ldots, t_n$ with n time points, their *correlation coefficient r* is calculated as follows:

$$r = \frac{1}{n}\sum_{k=1}^{n}\left(\frac{S_k - \overline{S}}{\sigma_X}\right)\left(\frac{T_k - \overline{T}}{\sigma_Y}\right)$$

It has been shown that correlation coefficient may effectively reveal the similarity between two time-series gene expressions in terms of their shape instead of the absolute values [28]. Hence, the problem of scaled and offset transformations can be resolved by using correlation coefficient as the base similarities between genes. To resolve the problems of time delay and noises, we adopt the dynamic programming technique to find the most similar subsequences of S and T. In fact, this problem is equal to finding the best alignment between S and T. Once this is obtained, it is straightforward to get the similarity between S and T by calculating the correlation coefficient on the aligned subsequences.

The main idea of our method is based on the concept of "dynamic time warping" [5, 6] for finding the best alignment of two time sequences. Initially we build up a *NxN* matrix, in which element (i, j) records the distance (or similarity) between s_i and t_j, indicating the best way to align sequences (s_1, s_2, …, s_i) and (t_1, t_2, …, t_j). Based on this approach, the best alignment can be obtained by tracing the warping path from element (1, 1) to element (N, N) in the matrix. One point to note here is that we use *Spearman Rank-Order correlation coefficient* [19] instead of *Euclidean Distance* for calculating the value of element (i, j) in the matrix. The main reason is for resolving the problems of scaled and offset transformations. More details on this issue will be given in later discussions.

The algorithm of CDAM is as shown in Figure 3. The first step of the method is to transform the sequences S and T into the sequences of rank orders Q and R, respectively. That is, the value of each element in the original sequence is transformed into its order in the sequence. For instance, given sequence S as {20, 25, 15, 40, -5}, the corresponding rank-ordered sequence Q is {3, 2, 4, 1, 5}. The next step of CDAM is to calculate the *Spearman Rank-Order correlation coefficient*, r, between Q and R, which is defined as follows:

$$r = 1 - \frac{6D}{N^3 - N} \tag{1}$$

, where N is the length of Q and R, and D is further defined as

$$D = \sum_{i=1}^{N} (Q_i - R_i)^2 \tag{2}$$

, where Q_i and R_i are the *ith* element in sequences Q and R, respectively.

When N is fixed, it is obvious that D is non-negative (from equation (1)) and the larger value of D indicates the lower similarity between S and T (from equation (2)). On the contrary, the smaller value of D indicates the higher similarity between S and T. Hence, our problem is reduced to finding the alignment with minimal D such that S and T has the highest similarity. To achieve this task, the following recursive equation is developed for finding the minimal D through dynamic programming:

$$r(i,j)=\min\begin{Bmatrix} r(i-1,j-1)+(Q_i-R_j)^2 \text{..........}[1] \\ r(i-1,j)+a \text{............................}[2] \\ r(i,j-1)+b \text{............................}[3] \end{Bmatrix} \qquad (3)$$

$1 \leqq i \leqq N,\ 1 \leqq j \leqq N$

In equation (3), r(i, j) represents the alignment with minimal value D in aligning sequences (s_1, s_2, …, s_i) and (t_1, t_2, …, t_j). In the alignment process, three possible cases will be examined:

Case 1: s_i is aligned with t_j. No warp happens in this case.
Case 2: s_i is not aligned with any items in (t_1, t_2, …, t_j). One warp happens in this case.
Case 3: tj is not aligned with any items in (s1, s2, …, si). One warp happens in this case.

```
Input : Two gene expression time series S, T and mismatch M.
Output : The similarity between time series S and T.
Method : CDAM(S, T, M).

Procedure CDAM(S, T, M){
  transform the sequences S and T into rank order sequences Q and R;
  for m = 0 to M{
    calculate r(i, j) for all i, j <= N to find the minimal D of (Q, R);
    alignment (Q', R') with mismatch m ← trace the warping path with minimal D;
  }
  best alignment (S', T') ← the alignments (Q', R') with highest similarity;
  return the similarity of (S, T);
}
```

Fig. 3. Algorithm of CDAM method

Table 1. Two time series S and T

Time point	1	2	3	4	5	6	7	8	9	10	11
S	2	4	5	1	3	9	7	10	8	5.5	0
T	-1	1	3	9	4	0	2.5	8	6	7	4.5

Since both S and T are of length N, the maximal number of total warps during the alignment is 2N. However, if the value of *mismatch* is set as M, the number of total warps must be constrained within 2M and the warps for S and T must be less than M, respectively. Two parameters a and b are thus added in equation (3) for limiting the number of warps within 2M, and they are controlled as follows:

$$a = \begin{cases} 0, \text{ if } warp(i-1, j) < 2M \\ \infty, \text{ otherwise} \end{cases} \tag{4}$$

$$b = \begin{cases} 0, \text{ if } warp(i, j-1) < 2M \\ \infty, \text{ otherwise} \end{cases} \tag{5}$$

Based on the above methodology, CDAM tries to discover the best alignment between Q and R by varying *mismatch* from 0 to M and conduct the calculations describe above. Finally, the alignment with highest similarity will be returned as the result. As an example, consider again the time series S and T in Table 1. If *mismatch* is set as 0, i.e., no mismatch is allowed, both a and b will stay as ∞ during the whole aligning process. Consequently, no warp happens and the resulted similarity is equal to the correlation coefficient between S and T. However, if *mismatch* is set as 2, the $8th$ and $11th$ items in S and the first and $4th$ items in T will be chosen as the mismatches.

3 Experimental Evaluation

3.1 Tested Dataset

To evaluate the effectiveness of the proposed method, we use the time-series microarray data by Spellman *et al.* [4] as the testing dataset. This dataset contains the time-series expressions of 6178 yeast genes under different experimental conditions. In [2], it was found that 343 pairs of genes exhibit activation relationship in the transcriptional regulations by examining the microarray data of alpha test in [4]. Our investigation shows that 255 unique genes are involved in the 343 activated gene pairs, thus the time-series expressions of the 255 genes in alpha test over 18 time points [4] are used as the main tested dataset.

3.2 Experimental Results

The proposed method was implemented using Java, and the tested dataset was analyzed under Windows XP with Intel Pentium III 666 MHz CPU/256MB. We use the proposed method to calculate the similarities of the 255 genes with parameter *mismatch* varied from 0 to 3.

Table 2. Distribution of calculated similarities of 343 pairs of activated genes.

similarity	mismatch=0	mismatch=1	mismatch=2	mismatch=3
> 0.25	97	197	214	227
> 0.33	72	165	181	200
> 0.5	36	88	100	119
> 0.75	9	16	21	26

Table 2 shows the distribution of calculated similarities for the 343 activated gene pairs under different settings of *mismatch*. When *mismatch* is set as 0, only 36 pairs of genes are found to have similarities higher than 0.5. That is, only about 10% of the 343 gene pairs are found to be of high similarity. This shows that using the correlation coefficient directly as the similarity measure can not reveal the correct similarity between genes. In contrast, the number of gene pairs with similarity higher than 0.5 increases to 88 when *mismatch* is set as 1. Moreover, the number of high-similarity gene pairs keeps increased when *mismatch* becomes larger. These observations indicate that our method can effectively find the correct similarities between genes.

Table 3. Average similarity of the 343 gene pairs under different settings of *mismatch*.

	mismatch=0	mismatch=1	mismatch=2	mismatch=3
Average similarity	0.03844	0.299939	0.331175	0.360907

Table 3 depicts the average similarity of the 343 gene pairs under different settings of *mismatch*. The above experimental results show that our method can reveal more accurate biological relationships for the tested genes. Moreover, it is also observed that the average similarity rises with the value of *mismatch* increased. However, the degree of improvement decreases for a larger value of *mismatch*. This indicates that it suffixes to set *mismatch* as large as 3 in our method for this dataset.

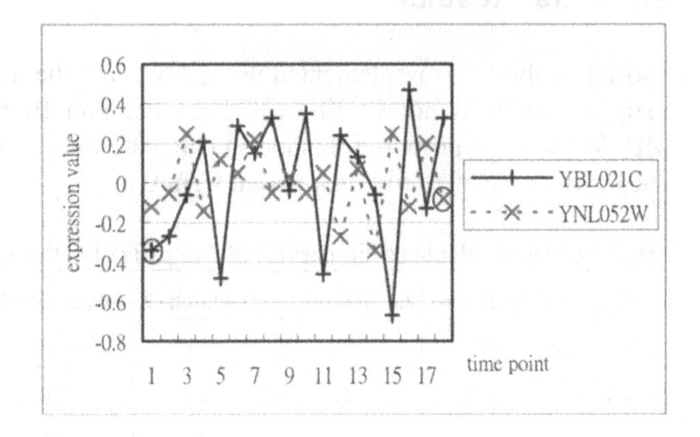

Fig. 4. Time-series expressions of genes YBL021C and YNL052W

3.3 Illustrative Example

Figure 4 shows the plotted curves for the time-series expressions of genes YBL021C and YNL052W over 18 time points. The similarity between these two genes is -0.38456 if correlation coefficient is used directly as the measure for computation. This indicates that these two genes have low similarity, and their activation relationship is not disclosed. In contrast, when our method is applied with *mismatch* set as 1, the items circled in Figure 4 will be selected for elimination so as to obtain the best alignment between the two genes. Consequently, the new similarity turns as high as 0.76106. This uncovers the biological activation relations between these two genes reported in past study [2].

4 Related Work

A number of clustering methods have been proposed, like k-means [13], hierarchical clustering [21], CAST [10], etc. When applied on gene expression analysis [9, 10, 13, 20, 21-23], these methods were mostly used for analyzing multi-conditions gene expressions with no time series.

Comparatively, there exist limited number of studies on time-series gene expression analysis due to the complicatedness. Some general methods were proposed for time-series comparisons and indexing based on the concept of longest common subsequences (LCS) [1, 7, 18]. Agrawal *et al.* [1] extended the concept of LCS by adding the technique of window stitching and atomic matching for handling the noise data in the time series. Hence, the method proposed by Agrawal *et al.* can efficiently extract the highly similar segments among time sequences. Although this kind of method is useful in analyzing large scale time-series data like stock market data, it is not well suited for mining time-series gene expression data since the biological properties were not considered.

For the literatures relevant to time-series gene expression analysis, Filkov *et al.* [2] proposed methods for detecting cycles and phase shift in the time-series microarray data and applied them for gene regulation prediction. Aach *et al.* [25] used Dynamic Time Warping (DTW) [5, 6] technique to find the right alignment between the time-series gene expressions. Some variations of DTW like windowing, slope weighting and step constraints [15, 16, 17] existed in the past literatures. However, the method by Aach *et al.* can not deal with the problems of scale and offset transformations since distance-based measure is used for similarity computation. Moreover, the noise problems were not taken into consideration, either. In comparisons, the method we proposed can efficiently resolve the problems of scale transformation, offset transformation, time delay and noises simultaneously.

5 Concluding Remarks

Analysis of time-series gene expression data is an important task in bioinformatics since it can expedite the study of biological processes that develop over time such that novel biological cycles and gene relations can be identified. Although some time-series analysis methods have been developed in recent years, they can not effectively resolve the fundamental problems of scale transformation, offset transformation, time delay and noises in time-series gene expression analysis.

In this paper, we proposed an effective approach, namely *Correlation-based Dynamic Alignment with Mismatch (CDAM)*, for mining time-series gene expression data. The proposed method uses the concept of correlation similairty as the base and utilizes dynamic programming technique to find the best alignment between the time-series expressions under the constrained number of mismatches. Hence, CDAM can effectively resolve the problems of scale transformation, offset transformation, time delay and noises simultaneously so as to find the correct similarity and implied biological relationships between gene expressions. Through experimental evaluations, it was shown that CDAM can effectively discover correct similarity and biological activation relations between genes. In the future, we will conduct more extensive experiments by applying CDAM on various kinds of time-series gene expression dataset and integrate CDAM with other clustering methods to build up an effective system.

Acknowledgement

This research was partially supported by National Science Council, Taiwan, R.O.C., under grant no. NSC91-2213-E-006-074.

References

[1] R. Agrawal, Lin, K. I., Sawhney, H. S., and Shim, K., "Fast Similarity Search in the Presence of Noise, Scaling, and Translation in Time-Series Databases." *In Proc. the 21st Int'l Conf. on Very Large Data Bases, Zurich, Switzerland,* pp. 490-501, Sept. 1995.

[2] V. Filkov, S Skiena, J Zhi, "Analysis techniques for microarray time-series data", *in* RECOMB 2001: Proceedings of the Fifth Annual International Conference on Computational Biology, *Montreal, Canada,* pp. 124-131, 2001.

[3] R.J. Cho, Campbell M.J., Winzeler E.A., Steinmetz L., Conway A., Wodicka L, Wolfsberg T.G., Gabrielian A.E., Landsman D., Lockhart D., and Davis R.W. "A Genome-Wide Transcriptional Analysis of the Mitotic Cell Cycle." *Molecular Cell, Vol.2, 65-73*, July 1998.

[4] P.T. Spellman, Sherlock, G, Zhang, MQ, Iyer, VR, Anders, K, Eisen, MB, Brown, PO, Botstein, D, and Futcher, B. "Comprehensive identification of cell cycle-regulated genes of the yeast saccharomyces cerevisiae by microarray hybridization." *Mol Biol Cell. 9:3273-3297,* 1998.

[5] D. J. Berndt and J. Clifford, "Using Dynamic Time Warping to Find Patterns in Time Series." *In KDD-94: AAAI Workshop on Knowledge Discovery in Databases.* Pages 359-370, Seattle, Washington, July 1994.

[6] Eamonn J. Keogh and Michael J. Pazzani, "Derivative Dynamic Time Warping." *In First SIAM International Conference on Data Mining*, Chicago, IL, USA, April, 2001.

[7] B. Bollobas, Gautam Das, Dimitrios Gunopulos, and H. Mannila., "Time-Series Similarity Problems and Well-Separated Geometric Sets." *In Proceed-*

ings of the Association for Computing Machinery Thirteenth Annual Symposium on Computational Geometry, pages 454--476, 1997.

[8] B. Ewing, and P. Green, "Analysis of expressed sequence tags indicates 35,000 human genes". *Nature Genetics 25, 232-234, 2000*

[9] A. Brazma, and Vilo, J., "Gene expression data analysis." *FEBS Letters, 480, 17-24. BIOKDD01: Workshop on Data Mining in Bioinformatics (with SIGKDD01, Conference)* p. 29, 2000.

[10]A. Ben-Dor, and Z. Yakhini, "Clustering gene expression patterns." *In RECOMB99: Proceedings of the Third Annual International Conference on Computational Molecular Biology*, Lyon, France, pages. 33-42, 1999.

[11]P. Tamayo, D. Slonim, J. Mesirou, Q. Zhu, S. Kitareewan, E. Dmitrovsky, ES. Lander, TR. Golub "Interpreting patterns of gene expression with self-organizing maps: Methods and application to hematopoietic differentiation." *Proc Natl Acad Sci* USA 96:2907, 1999.

[12]S. M. Tseng and Ching-Pin Kao, "Efficiently Mining Gene Expression Data via Integrated Clustering and Validation Techniques." *Proceedings of the Sixth Pacific-Asia Conference on Knowledge Discovery and Data Mining, PAKDD 2002*, pages 432-437, Taipei, Taiwan, May, 2002.

[13]M. Eisen, P. T. Spellman, Botstein, D., and Brown, P. O., "Cluster analysis and display of genome-wide expression patterns." *Proceedings of National Academy of Science* USA 95:14863—14867, 1998.

[14]D. Goldin and P. Kanellakis, "On similarity queries for time-series data: constraint specification and implementation." *In proceedings of the 1st International Conference on the Principles and Practice of Constraint Programming.* Cassis, France, Sept 19-22. pp 137-153, 1995.

[15]Donald J. Berndt and James. Clifford, "Using Dynamic Time Warping to Find Patterns in Time Series." *In Proceedings of the AAAI-94 Workshop on Knowledge Discovery in Databases.* Pages 359-370, Seattle, Washington, July 1994.

[16]J. B. Kruskall and M. Liberman, "The symmetric time warping algorithm: From continuous to discrete." *In Time warps, String Edits and Macromolecules: The Theory and Practice of String Comparison.* Addison-Wesley, 1983.

[17]C. Myers, L. Rabiner and A. Roseneberg, "performance tradeoffs in dynamic time warping algorithms for isolated word recognition." *IEEE Trans. Acoustics, Speech, and Signal Proc.,* Vol. ASSP-28, 623-635, 1980.

[18]Tolga Bozkaya, Nasser Yazdani, and Meral Ozsoyoglu. "Matching and Indexing Sequences of Different Lengths." *In Proceedings of the Association for Computing Machinery Sixth International Conference on Information and Knowledge Management*, pages 128--135, Las Vegas, NV, USA, November 1997.

[19]E. L. Lehmann, "Nonparametrics: Statistical Methods Based on Ranks." Holden and Day, San Francisco, 1975.

[20]S. Raychaudhuri, P D Sutphin, J T Chang, R B Altman, "Basic microarray analysis: Grouping and feature reduction", *Trends in Biotechnology,* 19(5):189-193, 2001.

[21]S. Tavazoie, Hughes, J. D., Campbell, M. J., Cho, R. J., and Church, G. M., “Systematic determination of genetic network architecture.” *Nature Genetics,* 22(3):281—285, 1999.

[22]E. M. Voorhees, “Implementing agglomerative hierarchical clustering algorithms for use in document retrieval.” *Information Processing & Management,* 22:465-476, 1986.

[23]J.B. McQueen, “Some Methods of Classification and Analysis of Multivariate Observations.” *Proceedings of the fifth Berkeley Symposium on Mathematical Statistics and Probability,* pages 281-297, 1976.

[24]L. Kaufman and P.J. Rousseeuw, “Finding groups in data: an Introduction to cluster analysis.” John Wiley & Sons, 1990.

[25]J. Aach and G.. Church, “Aligning gene expression time series with time warping algorithms.” *Bioinformatics*. Volume 17, pp 495-508, 2001.

[26]Alexander V. Lukashin and Rainer. Fuchs, "Analysis of temporal gene expression profiles: clustering by simulated annealing and determining the optimal number of clusters" *Bioinformatics* 17: 405-414., 2001.

[27]Z. Bar-Joseph, G. Gerber, D. Gifford, and T. Jaakkola. “A new approach to analyzing gene expression time series data.” *In the Sixth Annual International Conference on Research in Computational Molecular Biology,* 2002

[28]Mark S. Aldenderfer and Roger K. Blashfield, “Cluster Analysis.” Sage Publications, Inc., 1984.

[29]M. Schena, D. Shalon, R. W. Davis and P. O. Brown, “Quantitative monitoring of gene expression patterns with a complementary DNA microarray.” *Science* 270:467-470, 1995.

[30]J. L. DeRisi, Penland L, Brown PO, Bittner ML, Meltzer PS, Ray M, Chen Y, Su YA, Trent JM., “Use of a cDNA microarray to analyze gene expression patterns in human cancer.” *Nature Genetics* 14(4):457-60, 1996.

[31]J.L. DeRisi, V. Iyer and P.O. Brown, “Exploring the metabolic and genetic control of gene expression on a genomic scale.” *Science*, 278: 680-686, 1997.